# Progress in Electrorheology

## Science and Technology of Electrorheological Materials

# Progress in Electrorheology

## Science and Technology of Electrorheological Materials

Edited by

Kathleen O. Havelka
*The Lubrizol Corporation*
*Wickliffe, Ohio*

and

Frank E. Filisko
*The University of Michigan*
*Ann Arbor, Michigan*

Springer Science+Business Media, LLC

Library of Congress Cataloging-in-Publication Data

American Chemical Society Symposium on Electrorheological (ER)
Materials and Fluids (1994 Washington, D.C.)
Progress in electrorheology science and technology of
electrorheological materials / edited by Kathleen O. Havelka and
Frank E. Filisko.
p. cm.
"Proceedings of an American Chemical Society Symposium on
Electrorheological (ER) Materials and Fluids, held August 21-22,
1994, in Washington, D.C."--T.p. verso.
Includes bibliographical references and index.
1. Electrorheological fluids--Congresses. 2. Materials--Electric
properties--Congresses. 3. Materials--Magnetic properties-
-Congresses. 4. Electromagnetic fields--Congresses. I. Havelka,
Kathleen O. II. Filisko, Frank E. III. Title.
QC189.5.A1A48 1994
620.1'1--dc20 95-15749
CIP

Proceedings of an American Chemical Society Symposium on Electrorheological (ER) Materials and Fluids, held August 21–22, 1994, in Washington, D. C.

ISBN 978-1-4899-1038-7 ISBN 978-1-4899-1036-3 (eBook)
DOI 10.1007/978-1-4899-1036-3


Originally published by Plenum Press, New York in 1995
Softcover reprint of the hardcover 1st edition 1995

10 9 8 7 6 5 4 3 2 1

## PREFACE

This treatise is a compendium of refereed papers based on invited talks presented at the American Chemical Society Symposium on Electrorheological (ER) Materials and Fluids. ER fluids were first investigated 50+ years ago. These fluids, which change rheology when placed in an electric field, were recognized, from the beginning, for allowing an extremely efficient interface between electrical control and mechanical devices. Critical problems, however, existed with the initial fluids, which prevented them from serious consideration for large-scale applications. While over time some of the critical problems have been solved and activity in ER technology has increased, commercial success has remained elusive.

A recent Department of Energy report concluded that a primary reason for the failure to commercialize this promising technology is due to a lack in understanding the physics and chemistry of how the materials work. The goal of the symposium was to address the issue of understanding how ER materials work and how they can be used. One of the outcomes of the symposium, which we hope is conveyed in this book, is a feeling that if the mechanism of ER is to be fully understood and improved, expertise from diverse fields must be applied to the problem.

The recent explosion of interest in this area stems from the prospect of utilizing the ER effect for vibration control and transfer of energy. Electrorheology has many distinct merits over competitive technologies, including: the ability to achieve a continuously variable wide dynamic range that is energy efficient, fast, reliable, and compact at a reasonable cost. These properties cannot be achieved by conventional electromechanical interfaces.

Considering the recent surge in research activity and its multidisciplinary nature, there is a need for a comprehensive treatise that covers important current developments in this field. This book is intended to fill such a void by providing a broad coverage of recent work by leading international scientists in their respective areas. The book is very comprehensive in scope, covering the status and providing the state-of-the-art in topics ranging from theoretical modeling of the ER effect, to materials development, to applications.

We intend that this book will serve as a useful monograph for scientists and engineers entering the area as well as a valuable reference for those already in this emerging field.

We would like to acknowledge the overwhelmingly positive response from the ER community. The breadth of participation in the symposium is due in part to the support form a number of organizations. In particular, we would like to thank the National Science Foundation, Fluid, Particulate, and Hydraulic Systems Program; ACS Polymer Chemistry Division; and Donors of the Petroleum Research Fund, administered by the American Chemical Society, for their financial support.

We are truly indebted to the numerous authors for their timely effort and to the referees for their critical evaluation of the manuscripts.

Kathleen O'Leary Havelka
The Lubrizol Corporation
Wickliffe, Ohio

Frank E. Filisko
University of Michigan
Ann Arbor, Michigan

## CONTENTS

## MECHANISMS



## MECHANICS

## MODELS

# THE EVOLUTION OF ER FROM DISCOVERY TO APPLICATION

Willis M. Winslow

8025 W. 44th Ave.
Wheat Ridge, Colorado 80033

The first tests of a new gadget becomes evolutionary. In the case of ER fluids, for example, it was found that a brass disc suspended by a thin wire in a rotating cup filled with cornstarch and dispersed in kerosene, turned with the rotating cup when the suspending wire was touched by the bare hand, indicating response to the electrical charge maintained by the body. This gadget was an item taken to the U. S. Patent Office to show operability of the invention.

An electric relay was made consisting of a rotating brass cup containing a series of intermeshing discs like the plates of a variable condenser turning a shaft containing the contacts of a relay. When filled with an ER fluid, this relay closed contacts in response to the current through a vacuum type photocell connected directly to the relay. A sketch of this relay is featured in the original U.S. Patent #2,417,850: Method and Means for Translating Electrical Impulses into Mechanical Force.

A microscopic view of dispersed particles in an electrical field was made by pasting tinfoil electrodes on a glass plate and viewing a drop of ER fluid between the tinfoil electrodes. This experiment revealed a similarity between iron filings in a magnetic field and dielectric particles in an electric field.

A motor driven clutch generator demonstration set was built to see if the effect worked at higher rates of shear. These horizontal concentric cylinders with pulleys at both ends, plus motor and generator, were taken to the U.S. Patent Office then located in Richmond, VA, to demonstrate control of generator voltage through control of clutch coupling.

A clutch driven loudspeaker was made to gain some idea of speed response. A hand cranked inner cylinder was coupled to a concentric outer cylinder through a film of ER fluid. The rotary force of the driven outer cylinder pressed against the voice coil arrangement of a conventional loudspeaker. The sound of the speaker was surprisingly good when the output voltage of the radio receiver was biased with a DC voltage showing the fluid coupling was responsive to both positive and negative voltage. This modulation of a mechanical movement had promise where large volumes of sound waves are required.

An ER valve consisting of two concentric cylinders with a pipe fitting on each end, allowed ER fluid to be pumped and circulated between the cylinders. Pressures up to 60 p.s.i. were observed with applied AC voltage.

A drill press operated cone and cup clutch arrangement was made to give comparative tests of various ER mixtures of oil and solid particles. A drawing of this Electro-viscometer is shown in my article, "Induced Vibration of Suspensions" published in the Journal of Applied Physics, Vol. 20, No. 12, 1137 - 1140. December 1949. The arrangement worked so well that the entire setup of drill press, clutch, force measuring scale and variable voltage supply, was taken to Boulder, CO to demonstrate the Winslow Electro-viscous Clutch to the student branch of ACHS. Work with fluids soon revealed difficulty in duplicating fluids with similar characteristics

*Progress in Electrorheology*, Edited by K.O. Havelka
and F.E. Filisko, Plenum Press, New York, 1995

It was found that the use of metal organic additives such as manganese naphthenate, allowed higher concentration of solids giving stronger ER forces. This finding led to the use of surfactants.

An ER viscosity pump consisting of two concentric cylinders was produced by the inner rotating cylinder with square threads like grooves to drive the ER fluid in an axial direction between the stationary outer cylinder and the rotating inner cylinder. Pipe fittings at both ends of the outer cylinder allowed observation of fluid flow as well as pressure increased responding to voltage applied to both cylinders. Recirculating certain ER fluids through this pump often improved ER performance.

A shock absorber type of ER fluid test device consisting of a vertically mounted cylinder, into which was placed a movable cylindrical plunger, allowed measurement of rate of fall for various voltages applied to the cylinder and plunger.

While at the U.S. Patent Office to demonstrate operability of the invention, I visited with Jacob Rabinow of the Bureau of Standards who was interested in my invention, which had been offered to the Government for use in WWII. Later I designed and had built a drill press operated rotor and cup clutch that simultaneously utilized both electric and magnetic fields. In addition, a set of flat plates were made in which both electric and magnetic fields which were available. Tests revealed that most iron containing salts would decompose on heating to give an oxide responsive to both electric and magnetic fields.

Following issuance of the basic patent, WEFCO Winslow Electro Fluid Company was formed. Dr. Donald H. Menzel, a Harvard astro physicist, was instrumental in securing a contract with Warner Electric Brake and Clutch Company, in Beloit, Wisconsin. Warner, with the help of the Crystal Lake Labs of the Pure Oil Company, developed marketable Warner Electro Fluid. Of interest was the development of the Statigrip Chuck, consisting of a flat hard surface of high dielectric constant material on which high 3 phase voltage was applied to metal strips pasted on the underside. ER fluid was sprayed from a pressure can on the tope of the chuck and object to be chucked was placed on the fluid. During this contract WEFCO helped man a booth at the Design Engineering Show held in Chicago. A motor driven fluid clutch speaker attracted attention to manually operated sliding plate clutches. Postcards were given to prospective customers inviting their response.

Boeing's interest resulted in ER fluid controlled pistons operating shaker tables for spacecraft parts. The use of ER fluids allowed easy synchronization of multiple pistons.

An optional contract with Olin Industries called for the processing of more U.S. Patents. These patents featured the use of particles responsive to both electric and magnetic fields. Before issuance, the Patent Examiner set up an interference with GE which WEFCO lost for presumed lack of diligence.

An optional contract with IBM resulted in their referral to Arthur D. Little. The development of electric business machines over mechanical machines was given as the reason for rejection.

Early during my retirement from the Public Service Company of Colorado, my former college roommate also retired. We collaborated by mail on our mutual efforts toward fluid development, which resulted in the issuance of a U.S. Patent on starch ER fluids to Westhaver and the preparation of an unpublished paper dealing with the mathematical analysis of the force of attraction between particles in an electric field. During this work we first observed the permanent lock in between electrodes separated by a film of ER fluid when subjected to a prolonged electric field.

While consulting with Lubrizol, ER fluids were made from all liquid components by shearing a thin film of a mixture of polar and nonpolar liquids.

# OVERVIEW OF ER TECHNOLOGY

Frank E. Filisko

Materials Science and Engineering and
Macromolecular Science and Engineering
The University of Michigan
Ann Arbor, MI 48109-2136

## ABSTRACT

Electro-rheological materials (ERM) are materials whose rheological properties are strong functions of the electric field strength imposed upon them. ERM are typically fluids in the absence of an electric field but under constant shear stress at high enough fields, the materials can solidify into viscoelastic solids. In the liquid state during flow, these materials exhibit an apparent viscosity which can be increased by thousands of times by the application of an electric field. In their solid state, the materials are viscoelastic and characterized by complex modulii of which both the real and complex parts are strong functions of the electric field. Further, all field induced mechanical changes are virtually instantaneously reversible. The molecular mechanisms responsible for the phenomenon are however poorly understood as are the flow characteristics.

## HISTORY

The phenomenon which ultimately became known as electrorheology was first observed in the late 1800's by Duff [1] but it was not until the work of Winslow[2-5] that it's potential began to be recognized. Upon attempting to use these materials, it was soon realized that a serious obstacle existed which prevented their widespread use, i.e., the dispersed phase required significant amounts of adsorbed water [3,6-8]. Reasons for this will be discussed shortly. Attempts to replace the water with other substances such as glycerol [5,9] and silanol [10,11] resulted in ER activity which was substantially reduced. Recently reports have been made of particulate phases which produce ER active materials virtually free of or with substantially reduced quantities of water. These include the inorganic crystalline alumino-silicates [12], various polyelectrolytes[13], and certain polymeric semiconductors including phthalocyanine, polyaniline,and poly(acene-quinone radicals)[6]. In one case [14] an intrinsic mechanism is hypothesized due to free charge carriers intrinsic to the chemistry of the alumino-silicates whereas in the semiconducting polymers, the intrinsic mechanism would be electrons. Consistent with the elimination of water, most of the problems associated with use of ER materials have been rectified substantially and the practical use of such materials is now feasible.

## MOLECULAR MECHANISMS

One of the first models was proposed by Winslow[3] and follows from simple observations that particles in an ER fluid align between the electrodes under an electric field

*Progress in Electrorheology*, Edited by K.O. Havelka
and F.E. Filisko, Plenum Press, New York, 1995

in static conditions. He hypothesized that under shear, these chains would become distorted and break, but would reform again very rapidly. This could account for the increased stresses but does not address fundamental questions concerning the mechanism of interactions between particles, although a polarization mechanism is mentioned. Klass and Martinek [15] question this model since ER materials show activity at high frequency AC fields and these chains couldn't reform at such speeds, and Brooks et al.[16] reported a time scale for fibrillation of around 20 sec. which is much greater than the sub millisecond responses reported[15,17]. Other notable discrepancies with this model arise from the fact that chaining is a trivial consequence of polarization of the particles and therefore it would seem straightforward to increase yield strengths by using particles of higher polarizabilities. However as illustrated in figure 1, for an ERM containing $TiO_2$ particulates with a permittivity of 200, when dried, chaining still occurs but the fluid loses it's ER activity.

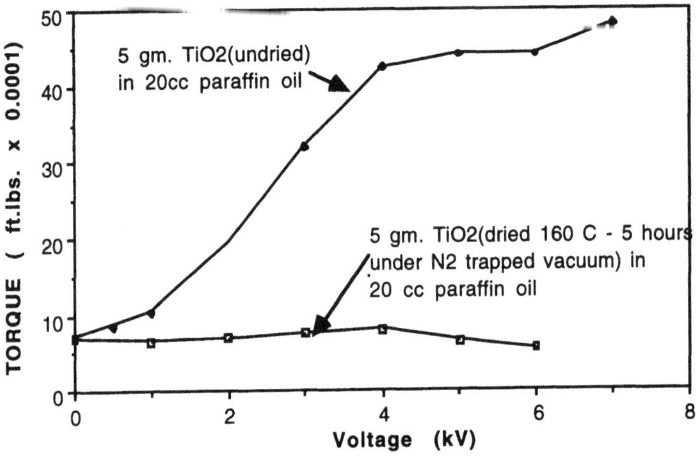


**Figure 1.** Rheological data for a titanium dioxide based ER fluid as -received andafter drying.

Another discrepancy arises from the fact that the strength of particle interactions due to polarization effects in chaining is related to the permittivity difference between the particles and liquid phase[18]. This would suggest that metal particles (if appropriately insulated) would result in the strongest interactions between particles. Although this probably occurs, it is not reflected in stronger ER activity. Although there are some reports of metal particle systems being ER active, the strengths are quite low. This as well as other information suggests mechanisms other then particle bulk polarizabilities must be involved in a major way in ER activity. An extension of this idea proposes that the particles, interacting Coulombically, flow as clusters ; but in static situations, will bridge the electrodes[19,20]. Neither of these addresses the basis for the particulate interactions on a molecular level.

The particles by themselves and/or in conjunction with the dispersing media must interact with the electric field in order for the particles to align, provide a yield stress, and hold the clusters together. The particles and liquids can interact independently with the field by virtue of their inherent electrical and dielectric properties; and/or the components can act cooperatively by virtue of the electrical double layer[21,22] which develops around colloidal particles in a dispersing liquid, and/or by virtue of interfacial polarization which develops due to mobile charges at the interface of the two materials[23,24]. It is the latter situations which are the most commonly considered as related to ER activity, but it is not clear to what extent these two are interrelated or in fact may be part of the same mechanism. Part of the confusion comes from the fact that although the basis for interfacial polarization is fairly well understood, theories related to electrical double layers [which are well developed for suspensions in electrolytic fluids[25], are poorly understood[26] in a non-conducting dispersing media of which ER fluids are an example.

Klass and Martinek[15,17] were the first to involve electrical double layers in their explanation of ER activity. They proposed that the diffuse portion of the double layers would become polarized under the influence of the electric field, and the electrostatic interactions of these distorted double layers require additional energy during flow, especially in concentrated suspensions where the layers overlap. This energy is required due to repulsion of the double layers so that the particles cannot simply move in a streamline but must have a transverse component which gives rise to the additional dissipation of energy, an idea similar to that suggested by Conway and Dobry-Duclaux[27] to explain the second electroviscous effect. They do not explicitly discuss the function of the adsorbed water even though without it, there would be no ER effect in these systems, yet double layers would still exist[26]. An interesting observation, based upon the relative permittivities of the systems of particles they used and the relative ER effectiveness, is that the bulk dielectric properties of the dispersed particles does not seem to play an important role. Interfacial and surface properties of the particles are much more important in ER activity. This finding is supported by others as well[7,15,17].

Schulman and Deinega[28] focus more on orientation of the particles and structures which may form in the electric field. They invoke electrical double layers and associate them with a surface conducting layer on the particles (i.e., water) in a non-conducting fluid where ion exchange with the fluid is negligible. In this case the mobile charges responsible for the Maxwell-Wagner-Sillars interfacial polarization also involve this water layer. The charge carriers can move along this conductive film under the influence of the electric field giving rise to an MWS polarization. The moisture here serves an essential function. Ion extension into the surrounding media, i.e., the dispersed double layer, may extend to various degrees depending among other factors on the degree of conductivity of this media. In reality it may be speculated that both mechanisms are probably involved in the ER phenomenon. What is certain; however, is that if either of the above are correct, then the surface charge conductivity introduced via the water certainly has a dramatic affect on the character of the double layer. This must actually be the case since it is a fact that the bulk conductivity of the systems increase many orders of magnitude for wet versus dry particles[14], thereby suggesting significant ion transfer to the media with water present versus without it.

Uejima[7] presented dielectric measurements which provided the most direct support for the above mechanisms. In these studies he followed loss factor and dielectric constant versus frequency for ER materials composed of cellulose particles with various amounts of adsorbed water. Specifically, he was observing the MWS interfacial dispersion which shifts to higher frequencies as the amount of water on the particles is increased. This is reasonable for this type of polarization[23,29] but has a number of other implications. The first is that the charge carriers involved in this dispersion are characterized by a relaxational spectrum of which the characteristic times, temperatures, and presumably distribution is strongly dependent upon the amount and type of water present[30]. Whether the MWS dispersion disappears as all the water is removed is an interesting mechanistic question since in these inherently heterogeneous systems, a MWS dispersion should still exist[23,29,30], but charge carriers may be of a different type.

Deneiga[25] also involving dielectric measurements and rheological measurements characterize this water layer further by suggesting that upon increasing temperature and field, there is a corresponding rise in ER activity as well as in the permittivity of the dispersions. However, these quantities peak at some point and beyond this peak the bulk electrical conductivity of the system begins to increase dramatically. They suggest a breakdown of the hydrate layer occurs with both temperature and field resulting in a lowering of the activation barrier for flow of carriers between particles. A very important point implied here is that the bulk conductivity may not be related to ER activity and the preferred situation is to contain charges on the particles by virtue of an infinite activation barrier between particles, if this is possible. This speculation is further supported by work of Deinega and Vinogradov[25], i.e., the relationship between ER activity, adsorbed water, and bulk conductivity.

Up to this point in time virtually all investigations continued to refine, with various modifications and extensions, the basic concepts of the electrical double layer extending into the liquid phase, a conductive surface layer of water (or other surfactant) on the particles giving rise to lateral mobility of ions, which are responsible for the classical Maxwell-

Wagner-Sillars interfacial polarization. All these imply the presence of a conductive layer on the particles, most commonly ions in the water, but none explained why the ER effect disappears when the water is removed even though the double layer still presumably remains as well as the MWS interfacial polarization.

A major advancement occurred with the recent reports of particulate systems which produced ER active materials without the need for adsorbed water or any water[14,31]. This is critical in resolving the model since either the same mechanism is operating both with and without water, or less likely that different mechanisms are operating. The implication here is very important since it suggests that the mechanism responsible for the ER activity can be an intrinsic characteristic of the chemistry and physics of materials and not due solely to extrinsic factors such as water.

The models proposed for this activity are similar to those previously discussed but are modified in that the electrical double layer is probably less dominant and the mobile charge carriers are not a consequence of an adsorbed electrolyte. In an article by Block and Kelly[32] more emphasis is put on the particle polarization which is really identical to the MWS interfacial polarization. Partially, support of this de-emphasis away from the double layer is a consequence of electrophoretic mobility (EPM) measurements on the actual fluids which indicate that the materials can be very active electro-rheologically yet show no significant EPM.

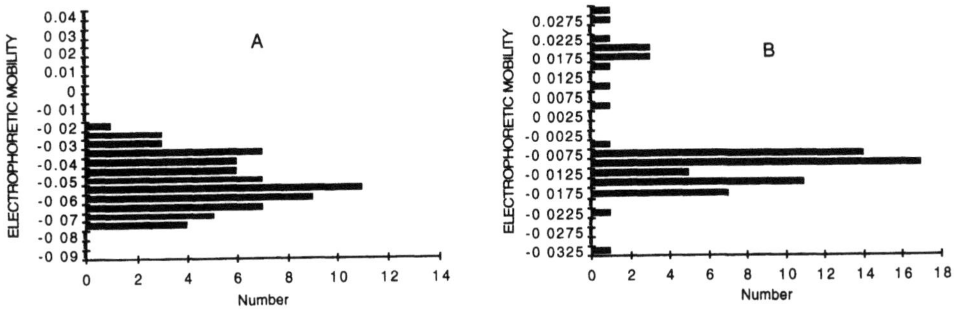


**Figure 2.** Electrophoretic mobilities for ER materials illustrating one with a net negative EPM(curve A) and one with a distribution around zero(curve B). (The lack of points at 0 is an artifact of the measurement technique since particles not moving cannot be detected). The best ER fluids have curves similar to B.

Further it appears that systems which show significant EPM are less active electro-rheologically. A suggestion here is that those mechanisms responsible for EPM, i.e., fixed surface charges and a diffuse ion layer, are different than those responsible for ER activity and in fact oppose each to some extent. Permanent fixed surface charges will cause attraction or drift of the particles toward one electrode and create a oil or particle free layer adjacent to the other electrode, thereby giving an apparent viscosity decrease. This also opposes formation of a particle mediated shear transfer between the electrodes which is necessary for ER activity. The mechanism responsible for ER activity, consistent with most others, are the presence of mobile charges (ions or electrons) on the surface of the particles which are able to move freely on the surface but cannot move off of it, i.e., a low activation barrier for surface migration but an infinite barrier for motion away from the surface.

The explanation for the activity of these dry systems is essentially based upon the presence of mobile charge carriers intrinsic to the molecular character or chemistry of the particles. This local mobility of the carriers on the particles is high, but mobility between particles should be very low. In the anhydrous materials of Block and Kelly[6,32], the carriers are presumably electrons as a consequence of the particle being semiconductors; while in the alumino-silicate systems the charge carriers are ions which are intrinsic to the chemistry of the particles and located on the surfaces[33,34]. Surface here includes the walls of the extensive interparticulate network of channels and cavities inherent to the morphology of the particles which can constitute more than 97% of the total surface area[35]. An important distinction between the two systems is that the bulk currents for the semiconductor systems are very high[6,32], presumably because electrons can more easily jump or tunnel between the particles and are all available to the outside surface of the particles by standard conduction mechanisms. However, in the Zeolite systems, ions on the outer surfaces have a much greater activation barrier to overcome in jumping particles, but more important, most are contained within the internal structure of the particles and cannot migrate to the outside surface, yet they are mobile within the internal labyrinth of channels and cavities afforded by the tremendous porosity inherent to the morphology of the materials. Apparent Maxwell Wagner Sillars interfacial dispersions are observed in all dried Zeolite systems as illustrated in figure 3. These all occur at very low temperatures and the corresponding Arrhenius plots (figure.4) produce activation energies between 12 and 16 Kcal/mol. These activation energies fall exactly in the range of values reported[36] for Zeolites but obtained by AC conductivity measurements. This suggests that the same carriers involved in their measurements (ions) on pure extensively dried Zeolites may be responsible for the dielectric dispersions on the Zeolite based ERM.

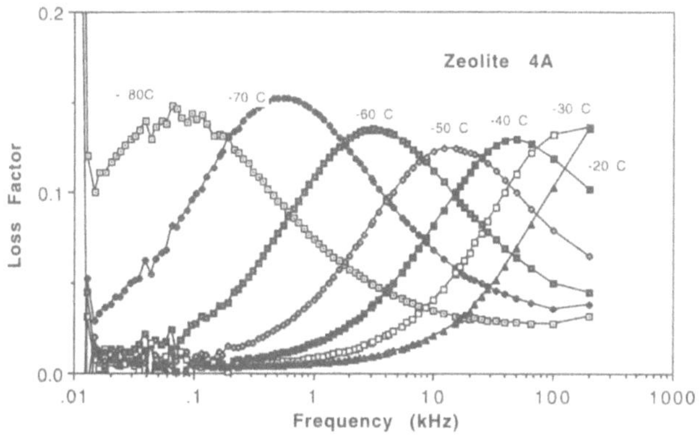


**Figure 3**. Dielectric dispersion due presumeably to MWS interfacial polarization for 4A Zeolite.

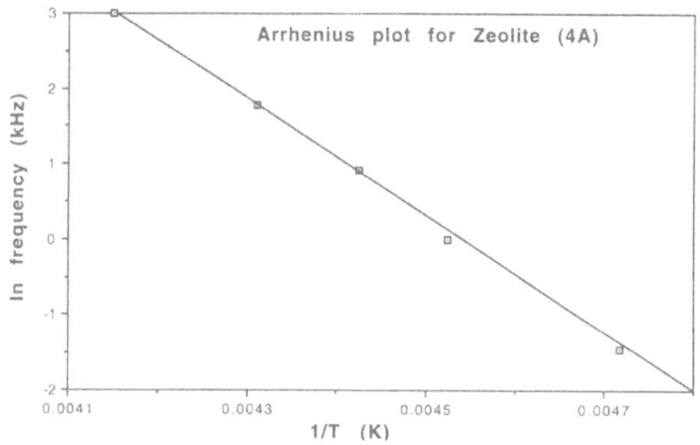


**Figure 4**. Typical Arrhenius plot from dielectric data as in figure 3.

Anhydrous polyelectrolyte systems are presumed to be ER active by virtues of locally mobile ions which can move within the environment of the chain coils or along a chain[37] but are not free to easily move between chains. It has been shown that ER activity is primarily a function of the pK of various polyelectrolytes[13].

## MATERIALS

Although the number of materials which can produce ER active suspensions is almost infinite, it is well understood that for most, the phenomena is a consequence of an extrinsic component, most commonly adsorbed water (or some other electrolyte) with various surfactants added and has nothing or little to do with the chemistry of the particles. There are of course properties of the materials which are beneficial to producing better ER materials such as particle porosity, high surface areas, and high affinity for water, but the ER mechanisms are not related to the particle chemistry. Such "wet" or extrinsic systems, most of which were summarized recently by Block and Kelly[6], were known for many years and it was as well realized that the water severely limited the potential application of ER technology. Some of the reasons are listed below.

1. Thermal Runaway- currents, although small, but at the high voltages required, cause $i^2R$ heating which drives off some water, which , in turn, increases the current which increases the $i^2R$ heating which drives off more water which increases the current, etc. until virtually all the water is off and the fluid no longer works.
2. Relatively high currents and therefore high powers needed.
3. Limited operating temperature range due to the freezing and boiling of the water.
4. Electrolysis
5. Corrosion of devices containing fluids
6. Instability of fluids with time and operation
7. Irreproduceability of different batches
8 Solid mat formation upon settling due to interparticulate hydrate bond formation.

Other important considerations certainly exist which will emerge as applications are developed such as cost, environmental acceptability, raw material availability, settling in low shear applications, plating of particles on one or both electrodes, sealing, effect on pumps, breakdown of particles upon shear for polymeric particles, adsorption of water, contamination, etc. The items above however are those associated directly with the presence of the adsorbed water.

With the recent discovery of water free or "intrinsic" systems and subsequent confirmation of these worldwide, a number of immediate improvements were realized, many having to do specifically with the water. Some of these are listed below:

1. Thermal runaway eliminated, since no water or other adsorbant required
2. Low currents . Currents $10^3$ to $10^6$ lower as a result of dryness, i.e. currents are in the range of micoamps/$cm^2$ or nanoamps/$cm^2$ instead of milliamps/$cm^2$ for alumino-silicate particulate systems. Drastic reductions in currents does not occur for the semiconductor systems.
3. Expanded temperature range . Zeolite base fluids operate from -60°C to 350°C when dispersed in silicone oil.
4. Electrolysis eliminated. Electrolysis does not occur being water is not present and needed.
5. Corrosion eliminated. Corrosion does not occur because water is not present.
6. Instability improved. A major cause of instability is loss of water due to operation or heating.
7. Irreproducibility substantially improved. Variability of water is a major source of difficulty in formulating water based fluids.
8. Solid mat formation and settling substantially reduced.

A very important additional consequence of these discoveries however implies that mechanisms responsible for ER activity can be associated with the basic chemistry and physics of the particles. Thus once these mechanisms are understood, materials can be synthesized specifically to optimize these mechanisms and improve ER properties in an intelligent manner.

It is realized that many continue to work on "wet" systems, mostly because it is relatively easy to improve properties to a limited extent. However we consider this a very limited and interim approach because prior attempts using this method to substantially improve properties in the 1950's and 1960's resulted in virtually complete failure. However it was also realized then that the water made the materials for the most part impractical.

An alternative approach with more potential is to try to understand the mechanisms responsible for the ER activity of the intrinsic systems and synthesize materials to enhance these. The approach is of course considerably more expensive and time consuming. For these reasons in the following we will concentrate only on 'dry" or intrinsic systems.

## Intrinsic ER Systems

Electrorheological fluids which operate without the need for adsorbed water on the particulate phase can be classified as follows:
1. Ionic conductors
2. Semiconductors
3. Polyelectrolytes
4. Solutions

## Ionic Conductors

The main particulate systems in this category are the alumino-silicates or Zeolites. The particles are highly porous, containing numerous cavities and interconnecting channels such that around 97% of the total surface area of the particles are contained within the particles, i.e. the walls of the cavities and channels. The dimensions of the cavities and channels can be varied by synthetic methods and by varying the aluminum/silicon ratio. The cationic charge carriers arise from the requirement for stoichiometry when some tetravalent Si atoms are replaced by trivalent Al without disrupting the crystal structure. Thus an Al at the center of a tetrahedron with oxygens at the vertices can bind to only 3 of these oxygens, leaving one unbonded with a net negative charge in the structure. This negative charge is balanced by the introduction of cations into the system. These cations however cannot fit into the close packed crystal structure and must therefore reside on the surfaces of the cavities and channels. Thus the cations are present as a consequence of the chemistry of the Zeolites ( not the presence of an electrolyte such as water) and their mobility is a result of their being on surfaces which are primarily internal and not confined within the crystal structure. Common uses of Zeolites are as molecular sieves due to the ability to synthetically control the channel dimensions, and as ion exchange materials due to the presence of unbonded cations which readily exchange with other cations in an aqueous suspension.

The intrinsic ER activity of these materials is associated with the presence of these cations which presumeably can move locally under the influence of an electric field. Such materials are susceptible to modification partly by varying the Si/Al ratio, by incorporating atoms other than aluminum, and by varying the types of cations. These materials have been available commercially for years as molecular sieves.

## Semiconductors

Most of the work in this area has been performed by H. Block and associates using various polyacene quinone radicals (PAQR) and recently polyaniline. The PAQR are not available commercially but can be prepared by the method described by Pohl[19]. The mechanism of activity for these materials is presumeably associated with the electronic charge carriers which can move locally under the influence of an electric field. A characteristic of these materials, presumeably associated with the electron mediated ER activity, are relatively high bulk currents due to the relative ease with which electrons may jump or tunnel between particles.

Although there are many types of semiconductors which can be used to make ER materials, many are ineffective or only slightly effective when dried. The reasons for this may be related to the size of the energy band gaps, the charge mobility, and/or charge concentration, although I know of no studies reported in this regard.

Photoconductors represent an interesting group of semiconductors which can be used to make ER materials. In this instance, many fluids which are inactive or weakly active can show a much enhanced ER activity when exposed to the correct frequency light. Phenothiazine demonstrates this rather dramatically even when dried. Photo-electrorheological materials (PHERM) represent the clearest proof of the relationship between charge mobility and ER activity, since exposing such a materials to light produces a tremendous increase in the number of free electronic charge carriers.

### Polyelectrolytes

Although many types of polyelectrolytes have been used in ER materials, poly lithium methacrylate being the most well documented, most require adsorbed water to function, presumeably to dissociate the cations from the macroions. Treasurer[13] evaluated various polyelectrolytes commercially available as ion exchange resins and reported that many function with greatly reduced amounts of water. These were all dried at 120 °C under vacuum for 4 days but retain between 0.1% and 2 % water. No correlation exists between the ER activity and residual water, however a strong correlation exists between the dissociation constant, pK, and ER activity. Another interesting observation was that some systems were ER active at 23 °C and 100°C, some were inactive at both temperatures, and some were inactive or weak at 23 °C but showed a much enhanced activity at 100 °C. Materials with high and low pk's demonstrated activity at both temperatures. Materials with intermediate pK's showed partial or no activity, while materials which were acidified, i.e.contained no cations, showed activity at 100 °C but not at 23 °C. A common feature of these latter materials was that they all contained quartenary ammoniums, but no reason is given as to how this is related to ER activity.

The mechanism for ER activity in these dried materials is presumeably due to cations which, even in the absence of an electrolyte, are able to move locally within the confines of a chain coil in the presence of an electric field but cannot move outside the coil into the surrounding liquid[37]. Direct evidence for this has not been obtained, but dielectric dispersions associated with interfacial polarization have been detected and are presumeably due to these locally mobile ions.

Systems containing polymers as the dispersed phase are very popular because of the softness of the particles and therefore reduced abrasion, because of the relatively low density which will aid the settling problem, and because of the vast body of knowledge on latex suspensions. Additionally, however, polymers represent an almost infinite range of systems which can be chemically customized for ER materials, once the basic chemical mechanisms associated with ER activity in "intrinsic "systems are better understood.

### Solutions

Recently reports have been made of true homogenous solutions which are ER active and presumeably operate free of water. One is that of solutions of poly -γ-benzyl L glutamate (PBLG) in various solvents[38], and the other is that of poly( hexyl isocyanate)(PHI) in various solvents[39]. Difficulties encountered with the PBLG systems include getting to high concentrations before gelling occurred, and the better solvents were polar thus resulting in high currents. Nonetheless, these solutions showed very significant increases in viscosity upon application of a field. Further, the effectiveness increased significantly with temperature, the limit being the boiling point of the solvent used.

The PHI systems on the other hand are soluble to much greater concentrations and in nonpolar solvents. The ER activities are also significantly higher.

The discovery of ER active solutions represents another very significant advancement in the field of ER. One reason being that it would resolve the problem of settling which has remained a major concern in some device design. These however will have unique disadvantages such as the greater toxicity and aggressiveness of the solvents, limited upper operating temperatures due to solvents and thermal degradation of the polymers, and costs will generally be higher. What is most important however is that this discovery emphasizes the enormous versatility in the compositions of ER active materials as well as the complexity in attempting to describe the behavior to a single mechanism.

## MECHANICAL PROPERTIES OF ER MATERIALS

The rheological behavior of ER materials under the influence of an electric field is commonly characterized by observing their properties during steady state flow[15,17,28,40].

Under these conditions, the flow properties of ERM can be adequately described as Bingham bodies and their flow or shear stress - shear rate behavior is qualitatively illustrated as show in figure 5.

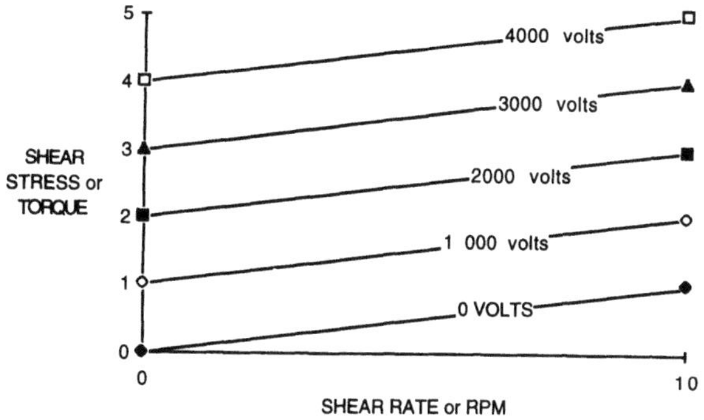


**Figure 5.** Bingham Body illustration of rheological behavior of an ER material.

ER materials are usually fluids when subjected to unidirectional shearing and under zero field conditions. However, when shearing conditions are maintained constant, the shear stress will increase with increasing applied electric field strength. The shear stress dependence is commonly reported as being proportional to the field squared[3,15,17,28,48], but many other types of behavior are observed[8,32]. According to ideal Bingham behavior, the ER materials are fluids under zero field, but under a non-zero field are solids up to a certain critical shear stress ($S_c$) and liquids at shear stresses above $S_c$. This model, although adequate in steady flow situations where transient or "start up" effects are neglected or unimportant, is not applicable in situations where the transient behavior is important or under dynamic loading (i.e., rapid or impact stresses or in damping applications). In these situations, the Bingham model completely overlooks the properties of the materials at stresses less than $S_c$[41]. A more complete description of the behavior of ER materials may be illustrated by a plot of stress versus strain as illustrated in figure 6.

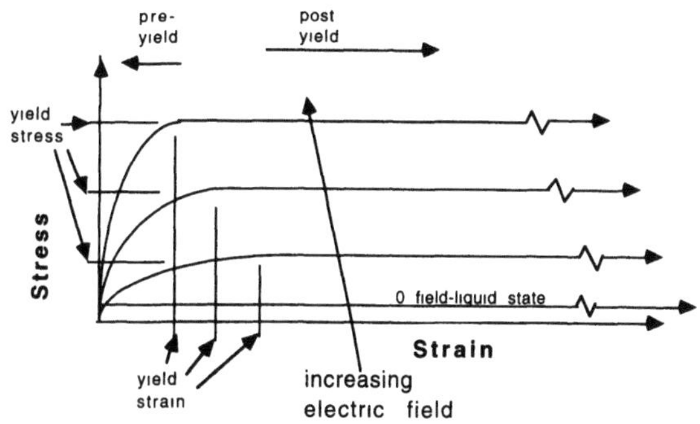


**Figure 6.** Illustration of more correct rheological behavior of ER fluid.

Under these deformation conditions, ER materials can be described as viscoelastic solids below a certain critical yield stress or yield strain and as viscous liquids for stresses at or above $S_c$ and strains greater than the yield strain. If characterized in this manner ER materials can be described in terms of their overall rheology as viscoelastic-perfectly plastic materials in which $S_c$ and the yield strain are strong functions of the electric field strength. The yield stress $S_c$ is highly variable and is strongly dependent upon numerous factors including the ER material composition. Furthermore, the energy dissipation mechanisms in the pre-yield region are different than those mechanisms present in steady flow.

In the simplest case, the rheological behavior of ER materials in the pre-yield region can be characterized by a modulus and a yield stress ($S_c$), in contrast to the post-yield region (i.e., liquid state) in which the material is characterized by an apparent viscosity ($\eta_a$). Furthermore, the behavior of the ER material is linear viscoelastic when deformation is restricted to the pre-yield state and the ER material is characterized by time constants and damping factors which are complex functions of the field strength.

## ERM in Steady State Flow (Post-Yield Behavior)

In steady state flow at a shear rate of " $\dot{\gamma}$ ", ER materials are characterized by an apparent viscosity "$\eta_a$" which is defined as:

$$\eta_a = [S_c(E) + S_o(\dot{\gamma})]/\dot{\gamma}_a$$

where $S_c$ is a function of the electric field strength (E) while $S_o$ is a function only of the shear rate and temperature, and is material specific.

Two common criteria for evaluating ER materials in flow include: 1) the magnitude the viscosity can be increased to and 2) what factor it can be increased by. The second criterion is more important since it relates how effective an electric field strength is on the rheology of the material. Regarding the latter point, we can define a fluid effectiveness factor, K, as:

$$K = \eta_a(E)/\eta_o = [(S_o+S_c)/\dot{\gamma}]/(S_o/\dot{\gamma}) = 1 + S_c/S_o$$

where $S_c$ is essentially a constant at a certain field strength but $S_o$ increases continuously with increasing shear rate. Therefore, this suggests that K decreases toward "1" as the shear rate increases. This is an important first order relationship for understanding the characteristics of ER materials under flow since, it implies that to make a more effective fluid requires making $S_c$ as large as possible while keeping $S_o$ as small as possible. The second parameter, $S_o$, is a function of the ER material composition and flow conditions. Thus, ER materials of low or high viscosity can be made by varying the solid concentration or the viscosity of the dispersing liquid. However, though the maximum shear stresses can be increased by making the zero field materials thicker, the K factor can become extremely small such that the field induced change in $S_c$ will become insignificant. Furthermore, it has been reported that $S_o$ is a much stronger function of concentration than $S_c$[42], and it appears that $S_c$ is a linear function of concentration[25]. Therefore, understanding the basic parameters (chemical, physical, and rheological) and controlling the magnitude of $S_c$ is essential to understanding the electrically induced properties of these materials.

## ERM in Oscillatory Shearing (3 Rheological Regions)

The response of ER materials to dynamic loadings can be discussed in terms of three distinct rheological regions: a pre-yield, yielding, and post-yield region. In the previous section, discussion was limited to conditions of steady state flow in which the transient effects of the pre-yield region were not considered. The pre-yield region can be effectively studied when oscillatory stresses are applied to the ER material such as may occur in vibration damping[43-45]. Under these straining conditions, the amplitude of the shear stress response is a strong function of the applied field strength. However, there exists a limiting shear stress value beyond which the shear stress response no longer follows the shape of the shear strain function, but becomes "cut off" or truncated[43,45]. The value of the shear stress at the onset of truncation is a function of the field strength and is also related to $S_c$. Rheologically, the appearance of the cut off is an indication that the material is beginning to flow. During an oscillatory shear strain, the ER material will deform as a linear viscoelastic solid over part of the deformation cycle and as a liquid over the other part.

A representative series of stress responses for the ER material when subjected to a sinusoidal shear strain is presented in Figure 7. Curve A is the applied sinusoidal shear strain of frequency 15 Hz and amplitude of 0.25. Curve B is the shear stress response when the material is subjected to a zero strength electric field. The shear stress response appears sinusoidal and the phase angle between the applied shear strain and shear stress response is 90°; suggesting that the material is deforming as a viscous body. Subjecting the ER material to an electric field of strength 1.0 kV/mm yields Curve C; the shear stress response amplitude increases and the phase angle decreases. Thus, when the ER material is subjected to a nonzero electric field, the material behaves as a viscoelastic material. If the strength of the

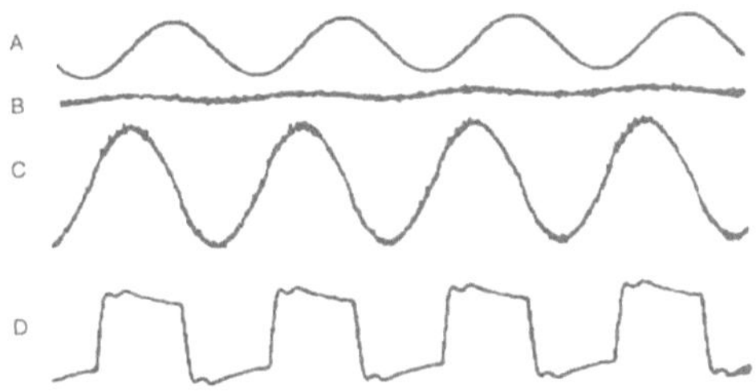


**Figure 7.** Shear stress response to a constant strain amplitude at various electric fields. Curve A is the strain, B the shear stress with E=0, C the shear stress with E=1 kV/mm, D the shear stress with E=2.5 kV/mm.(D is 1/4 the scale of the other curves)

electric field is increased to 2.5 kV/mm, the shear stress response deviates from sinusoidal behavior (Curve D). A nonsinusoidal response suggests that the material is behaving as a nonlinear viscoelastic material. In addition, as the material is subjected to a 2.5 kV/mm field, the fundamental harmonic of the shear stress response increases and the phase angle between the fundamental harmonic of the shear stress response and the applied shear strain decreases. Thus, as the strength of the applied electric field is increased from 0.0 kV/mm to 2.5 kV/mm the ER material transforms from a viscous to a linear viscoelastic to a nonlinear viscoelastic body. Moreover, the energy storing and energy dissipating properties of the ER material are strong functions of the applied electric field strength.

The linear viscoelastic parameters, shear storage modulus (G') and shear loss modulus (G") are strong functions of the applied electric field, strain amplitude, strain frequency, and material composition[16,25,44]. In addition, it was shown[25,44] that the shear storage modulus is a stronger increasing function with increasing electric field strength as compared to the shear loss modulus. Furthermore, it is of particular interest to note that ER materials become greater energy storing bodies as they simultaneously become greater energy dissipating bodies.

A second technique to observe the effect of the electric field under cyclic loadings is to observe shear stress - shear strain loops (hysteresis loops) for these materials at a constant strain frequency and amplitude, while varying the strength of the electric field. A sequence of hysteresis loops generated when the ER material was deforming as a linear viscoelastic body is shown in Figure 8.

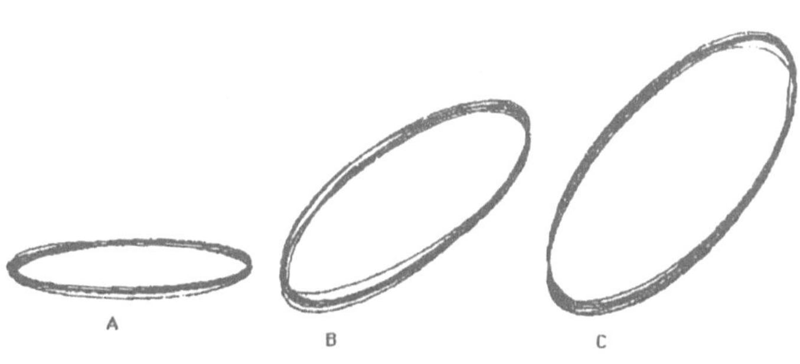


**Figure 8.** Hysteresis loops for an ER material subjected to a strain of 0.001 amplitude at 300 Hz. A is at E=0, B at E=1 kV/mm, C at 2 kV/mm.

As the strength of the electric field increases, both the area within the hysteresis loops and the angle the major axis of the hysteresis loop makes with the abscissa increase. The hysteresis loops are all elliptical in shape which is indicative of linear viscoelastic material response. The viscous component (energy dissipated) is determined by the area within the loop and the elastic component (stored energy) is determined by the major axis inclination.

The existence of a deformation transition limits the applicability of linear viscoelastic mathematics to quantify the energy storing and energy dissipating properties of an ER material. The amount of energy dissipated by the ER material during one deformation cycle, irrespective of linear or nonlinear viscoelastic material response, can be obtained by generating a hysteresis loop. The energy dissipated by an ER material is found by calculating the area within the loop. The recorded hysteresis loop for an ER material subjected to a strain of moderate frequency and moderate amplitude, and zero strength electric field is elliptical and the major axis of the hysteresis loop is parallel to the abscissa; this response is indicative of a viscous material (figure 9A). When the ER material is subjected to a field strength of 1.0 kV/mm the area within the hysteresis loop increases and the angle between the major axis

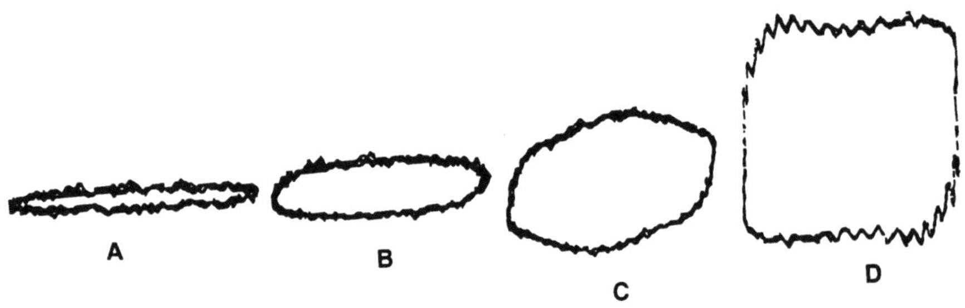


**Figure 9.** Actual hysteresis loops recorded for an ER material under various electric fields. A: E=0, B: E=1 kV/mm, C: E=2kV/mm, D: E=3kV/mm.

of the hysteresis loop and the abscissa increases (figure 9B). The increased area within the hysteresis loop suggests that the ER material dissipates more energy while the increased angle is related to the energy storing properties of the ER material. Continuing to increase the strength of the electric field yields hysteresis loops which encompass greater areas, suggesting the material dissipates more energy (figures 9C and 9D). However, the loops are no longer elliptical and thus the ER material is behaving as a nonlinear viscoelastic material. Moreover, the energy storing properties of the ER material cannot be calculated when nonelliptical hysteresis loops are recorded. Thus, if the amplitude of the imposed impulse is large enough, nonelliptical hysteresis loops are recorded which indicates nonlinear viscoelastic response. This occurrence suggests that the liquid or flow regime has been encountered.

## MECHANICAL MODELS

Presently the two most serious impediments to the intelligent and inovative design of various types of devices employing ER materials is the lack of a good mechanical rheological model and corresponding mathematical equations which adequately simulate, at least qualitatively, the behavior of ER materials, and the lack of quality engineering design data. The latter point is relative since, as mentioned previously, the composition of the materials is virtually infinite and they will ultimately be customized to optimize desireable properties and minimize undesireable properties for specific purposes. However there is sufficient data in the literature to allow reasonable levels of properties to be estimated and inserted into appropriate equations. However, an adequate rheological model is presently nonexistent.

In one case a model by Bullough and Foxon[46] is proposed to simulate only damping and is then only linear viscoelastic so it can be treated mathematically. It is unrealistic in regard to three obvious deletions: First, it incorporates no flow element; second, it incorporates no Coulomb damping term; and third, it is completely recoverable. It was not however intended to simulate the behavior of ER materials. The only other model by Shul'man[47] suffers from the same problems and that of Bullough and Foxon, but again it was intended only to simulate electric field control of damping.

A current model proposed by Gamota and Filisko[43] shown in Figure 11, correlates qualitatively with the overall observed behavior of ER materials (i.e., pre-yield, yielding, post-yield regions) and has been tested quantitatively for experimentally observed pre-yield behavior[44].

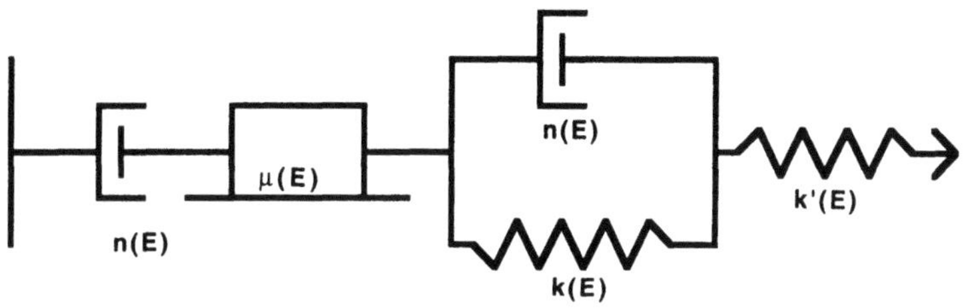


**Figure 10** . Mechanical model simulating behavior of an ER material.

Element "1" is a dashpot which is essentially field independent but governs the slope of the shear stress versus shear rate data, i.e., the slope of the lines in figure 5. Element "2" is a Coulomb friction element which is strongly field dependent and controls the magnitude of the ER response under steady shear, i.e., the yield stresses or $S_c$. Elements "3 and 4" combined in parallel form a Voigt element. Both elements are strongly field dependent and are responsible for the transient response and damping behavior under impact or vibration, i.e., both the angle of inclination and areas within the ellipses in figure 8. Under low amplitude it may be only this portion of the model which is deforming. Element "5" is a spring which is strongly field dependent and in conjunction with the Voigt element governs the dynamic response.

Notice that under constant strain rate, elements "3,4, and 5" rapidly reach an equilibrium extension and contribute nothing in steady state flow. Element "1" simply determines the slope of the stress - strain rate data which is essentially field independent, and element "2" which is field dependent determines the yield stress. However under dynamic loading, at low amplitudes, the block (friction element) may not move and then only elements "3,4, and 5" are determining the material characteristics. In more complex loading situations, all elements may be contributing to various degrees to the rheological behavior of the electrorheological materials.

## ACKNOWLEDGMENT


Much of our work on ER was supported by the University of Michigan Industrial Affiliates Program in Electrorheology, which includes the following sponsors: Aeroquip Corporation, Ford Motor Company, Monroe Automotive Equipment Company, Naval Underwater Systems Center, Nissan Motors, BASF Corporation, Chrysler Corporation, Tremec Trading Company, and Ethyl Corporation.

Also, special appreciation is extended to the American Society of Mechanical Engineers for permission to reproduce for this article portions abstracted from "Recent Advances in Non-Newtonian flows" ASME-AMD(153), ASME-FED(141),75-87(1993).

## REFERENCES


1. Duff, A.W.(1896), Phys. Rev.,4,23.

2. Winslow, W.M.(1947),"Methods and means for transmitting electrical impulses into mechanical force", (U.S.Pat.), 25 Mar.1947, No.2417850.

3. Winslow, W.M. (1949), "Induced fibration of suspensions", J. Appl. Phys. 20:1137-1140.

4. Winslow, W.M. (1953)," Field controlled hydraulic device ", (U.S.Pat.), 8 Dec. 1953, No.2661596.

5. Winslow, W.M.(1962)," Field response force transmitting compositions", (U.S.Pat.), 1962, No. 3047507.

6. Block, H. and J.P. Kelly(1988). "Electrorheology", J. Phys. D: Appl. Phys. 21:1661 1677.

7. Uejima, H. (1972). "Dielectric mechanism and rheological properties of electro-fluids," Jap. J. Appl. Phys. 11(3): 319-326.

8. Sugimoto, N.(1977), "Winslow effect in ion exchange-resin dispersions", Bull. Jap. Soc. Mech. Eng. 20(149),1476.

9.Trapeznikov, A.A., Petrzhik, G.G. and O.A. Cherlkova (1981), "Electrorheological properties of nonaqueous dispersions of titanium dioxide and silicon dioxide in relation to concentration and moisture content of filler", Kolloid Z. 43(6), 1134.

10. Goosens,J.(1987), "Electroviscous fluids", (U.S.Pat.), 27 Oct. 1987, No. 4702855.

11. Klass, D.L. and T.W. Martinek (1966), "Preparation of silica for use in fluid responsive compositions," (U.S.Pat.), 10 Oct., No. 3250726.

12. Filisko, F.E. and W.E. Armstrong (1988). "Electric field dependent fluids", (U.S.Pat.), 17 May 1988, No. 4744914.

13.Treasurer, U., L. H. Radzilowski and F.E. Filisko (1991). "Polyelectrolytes as inclusions in water free electrorheological materials: chemical characteristics," J. Rheol. 35(4).

14. Filisko, F.E. and L.H. Radzilowski (1990). " Intrinsic mechanism for activity of alumino-silicate based electrorheological fluids", J. Rheol. 34(4): 539-552.

15. Klass, D.L. and T.W. Martinek (1967a). "Electroviscous fluids I. Rheological properties," J. Appl. Phys. 38(1): 67-74 .

16. Brooks, D., J. Goodwin, C. Hjelm, L. Marshall, C. Zukowski (1986). "Viscoelastic studies on an electrorheological fluid." Colloids Surf. 18:293.

17. Klass, D.L. and T.W. Martinek (1967b). "Electroviscous fluids II: Electrical properties," J. Appl. Phys. 38(1): 75-80 .

18. Gast, A.P. and C.F. Zukoski (1989),"Electrorheological fluids as colloidal suspensions", Adv. Colloid Int. Sci.,30, 153.

19. Pohl, H.A. (1951). "The motion and precipitation of suspensions in divergent electric fields", J. Appl, Phys. 22(7): 869-871.

20. Voet, A. (1947) "Dielectrics and rheology of non-aqueous dispersions," J. Phys. Colloid Chem., 51:1037-1063.

21. Davies, J.T. and E.K. Rideal (1961).Interfacial Phenomena, Ch.2, Academic Press,NY.

22. Lyklema, J.(1985)," Interfacial Chemistry of Disperse Systems", J. Matls.Ed. 7(2),211.

23. VonHippel, A.R. (1954). Dielectric and Waves, pp. 228-234, John Wiley & Sons, Inc., New York.

24. Dukhin, S.S. and V.N. Shilov (1974). Dielectric Phenomena and the Double Layer in Disperse Systems and Polyelectrolyte, translated from Russian by D. Lederman, Keter Publishing House Jerusalem Ltd.

25. Deinega, Y.F. and G.V. Vinogradov (1984). "Electric fields in the rheology of disperse systems," Rheol. Acta 23:636-651.

26. Kitahara, A. (1984). "Nonaqueous systems", in Electrical Phenomena at Interfaces: Fundamentals, Measurements and Applications, eds. A. Kitahara and A Watanabe, pp. 119-143, Marcel Dekker, Inc., New York .

27. Conway, B.E. and A. Dobry-Duclaux ( 1960 ),"Viscosity of suspensions of electrically charged particles and solutions of polymeric electrolytes", in Rheology:Theory and Applications, Vol.3, ed. F.R.Eirich ,Academic Press, NY.

28. Schul'man, Z.P., Y.F. Deinega, R.G. Gorodkin, and A.D. Matsepuro (1971). "Some aspects of electrorheology," Prog. in Heat and Mass Transfer 4:109-125 .

29. Maxwell, J.C. (1892). Electricity and Magnetism, Vol.1. 452, Claredon Press, Oxford .

30. Hedvig,P. (1977) , Dielectric spectroscopy of polymers, John Wiley & Sons, Inc., New York, pp.282-296.

31. Block, H. and J.P. Kelly (1985). Proc. IEE Colloq. 14:1.

32. Block, H., J.P. Kelly, A. Qin, and T. Watson(1990). "Materials and mechanisms in electrorheology", Langmuir 6(1):6-14.

33. Breck, S.W. (1974). Zeolite Molecular Sieves, John Wiley & Sons, Inc., New York.

34. Denkewicz, R.P.(1987). " Zeolite science: an overview", J. Matls. Ed. 9(15): 519-585 .

35. Barrier, R.M. (1959), Brit. Chem. Eng. 5:1.

36. Freeman,D.C. and D.N.Stamires (1961), J. Chem. Phys.35,799.

37. Oosawa, F. (1971). Polyelectolytes, Marcel Dekker, Inc., New York.

38. F.E.Filisko(1993),"Materials aspects of ER fluids', in Electrorehological Fluids: A Research Needs Assessment, Department of Energy Report DOE/ER/30172, May.

39. Yang,I-K and A.D.Shine(1991), "Electrorheology of Poly(n-hexyl isocyanate) solutions", presented at Soc. of Rheology meeting, Oct.20-24,1991(Rochester NY.

40. Conrad, H., M. Fisher, A.F. Sprecher (1989). "Characterization of the structure of a model electrorheological fluid employing stereology." Electrorheological Fluids- Proceedings of the Second International Conference on ER Fluids (pub. in 1990), Raleigh, North Carolina, USA, August 7-9, Technomic Pub. Co., Inc.

41. Sprecher,A.F., Carlson,J.D., and H. Conrad(1987),"Electrorheology at small strains and strain rates of suspensions of silica particles in silicone oil", Matl. Sci. and Eng. 95, 187-197.

42. Frisch, H.L. and R. Simha (1956). "The viscosity of colloidal suspensions and macromolecular solutions" in Rheology: Theory and Applications, Vol.1, ed. F.R.Eirich, chap.14, Academic press, NY.

43. Gamota, D.R. and F.E. Filisko(1991a), "Dynamic mechanical studies of electrorheological materials: Moderate frequencies", J. Rheol. 35(3), 399-426.

44. Gamota, D.R. and F.E. Filisko(1991b), "High frequency dynamic mechanical study of an aluminosilicate electrorheological material", J. Rheol. 35(7), 1411-1426.

45. Stanway, R., Sproston, J. and R. Firoozian(1989)," Identification of damping law on an electrorheological fluid: a sequential fitting approach", J. Dynamic Systems: Measurement & Control 111, 91-95.

46. Bullough, W. A. and M. B. Foxon (1978). "A proportionate Coulomb and viscously damped isolation system." J. Sound & Vib. 56(1): 35-44.

47. Shul'man, Z. P., B. M. Khusid, E.V. Korobkov and E.P. Khizhinsky (1987). "Damping of mechanical-system oscillations by a non-Newtonian fluid with electric-field dependent parameters." J. Non-Newtonian Fluid Mechanics 25: 329-346.8

48. Saito,N. and T.Kato(1957),"Viscoelasticity and complex dielectric constant in the presence of an electric field and shear laminar flow in macromolecular solutions", J. Phys. Soc. Japan 12(12),1393-1397.

# RECENT DEVELOPMENTS IN ER FLUIDS

Harry Block and Paul Rattray

School of Industrial and Manufacturing Science
Cranfield University
Cranfield
Bedford, MK 43 0AL
UK

## INTRODUCTION

Electrorheological (ER) or smart fluids are materials which have the ability to reduce their flow when strong electric fields are present, in the manner illustrated in figure 1. There is an extra shear stress ($\Delta\sigma$) needed to maintain the flow of such a liquid against the action of the field or the flow is reduced and in some circumstances, may cease altogether. In the latter event the material has converted from liquid to solid. Such dual behaviour is reminiscent of a Bingham plastic and the ER phenomenon can be considered as invoking electrically induced Bingham body behaviour. Since removal of the field causes the viscoelastic moduli of such a fluid to revert to their original state an ER 'fluid' still under stress will recommence flowing when that field is removed. Thus, ER fluids are amenable to switching by electric field and this, together with the rapidity ($\approx$ ms) of at least parts of these interconversions for many ER fluids, renders them of practical importance. Electrically enhanced stresses in the kPa range for electric fields of $\approx$ kVmm$^{-1}$ can be achieved with a wide variety of fluids. The phenomenon occurs with ac as well as dc fields provided that the field frequency is not too high.

Most ER fluids comprise dispersions of powders, the ER active ingredient, in insulating oils. Another group of materials showing ER effects are liquid crystal systems which are discussed more fully below. In the early days of ER technology, the dispersions were found only to be effective if a third component, a promoter was present. These additives, the most common of which were water or water based surfactants, ended up largely in the powders rather than the dispersing oil and had their activity there. The view gained ground that these promoters were essential for the effect. However, their presence often led to practical problems because of the high currents drawn by such damp fluids under the large operating fields, and even more so, because the temperature coefficient of this conductance was large and positive, necessitating costly external cooling to avoid thermal runaway. Drying out of the fluid or ice formation were other sources of difficulty for water promoted fluids. However modern fluids are not limited to ones containing water since the discovery of ER active substrates which did not require promotion[1-3]. This has led since the mid 1980's to increased interest in the ER phenomenon and to a diversification of ER fluids.

*Progress in Electrorheology*, Edited by K.O. Havelka
and F.E. Filisko, Plenum Press, New York, 1995

It is known that as part of their ER action, the particles experience field induced forces and unless flow prevents it, these cause chaining of the particles in the direction of the field to form fibrils and then columns, which finally span the gap between the electrodes. These same forces are also believed to be responsible for the field induced enhancements in viscoelastic moduli, whether in the solid form or in flow. As discussed below, the way these ER structures grow and decay, their extent in flow, and the magnitudes of the viscoelastic moduli which result are areas of considerable current interest and study.

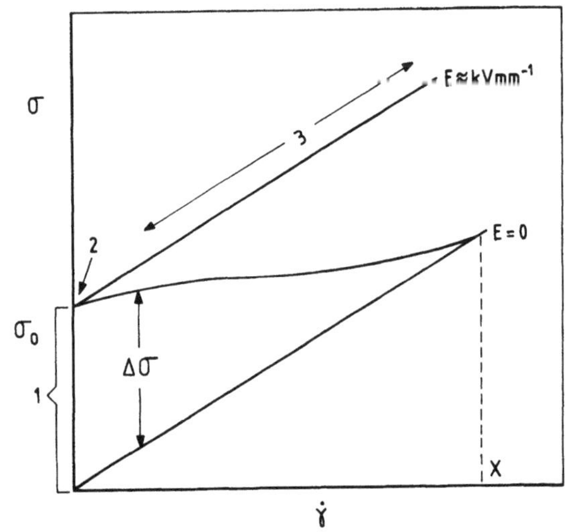


**Figure 1.** An illustration of the shear rate ($\dot{\gamma}$) - shear stress ($\sigma$) behaviour of ER fluids under a field E. For simplicity the flow is assumed Newtonian when E = 0 and shows induced Bingham behaviour under field with a solid state region 1, the onset of flow at the static yield point 2 (static yield stress $\sigma_o$) and in an ideal case, the retention of an enhanced field induced stress in all regions of flow 3. A more common situation discussed in the text is some reduction of the enhanced stress ($\Delta\sigma$) which can be so extreme as to lead to a collapse of the ER effect at a higher shear rate marked X.

In discussing the ER phenomenon, it is important to appreciate that the rheology of ER fluids is very complex as figure 1 illustrates. Under certain levels of shear stress ($\sigma < \sigma_o$, the static yield stress) the material shows at least some aspects of solid behaviour with a structured morphology of particles. At sufficiently high shear rates ($\dot{\gamma}$) such structures must degrade, either to partial structures such as broken fibrils, or more fully. However, in order to maintain an enhanced shear stress ($\Delta\sigma$), some level of particle-particle interactions must persist in the flow. Failure fully so to do will cause a drop in $\Delta\sigma$ as illustrated in figure 1, whilst a total loss of force field in shear flow would degrade $\Delta\sigma$ to zero, destroying the ER effect. Decreases in $\Delta\sigma$ are common and total loss of $\Delta\sigma$ is not unknown[4-6]. As Gamota and Filisko[7] have pointed out, the full rheological picture can be discussed in terms of three regions: the solid state, yielding of that state, and the liquid state in flow. This is a useful conceptual division, provided that it is appreciated that the transitions between them are not sharp.

This in-brief outline, sets out the more salient features of the ER phenomenon. For more detail the reader is referred to a number of reviews[4-6,8-17] and conference reports[18-20]. Here we will limit our discussions to more recent advances, where in the authors opinion there are particularly novel aspects of the phenomenon or where questions remain in relation to our understanding and use of ER. This will be attempted within sections devoted to theories for ER, techniques for probing the ER mechanism, ER materials, and applications.

## THEORIES FOR THE ER EFFECT

It appears to be the opinion of most researchers that the ER effect originates when the electric field polarizes the particles to induce dipoles, and it is their interaction which provides the forces leading to both changes in rheology, and finally structuring if flow permits. This scenario is attractive, not least because it has a magnetic counterpart in the behaviour of magnetorheological (MR) fluids[21-23] which show an analagous property following magnetization of magnetizable particles in dispersion. What is less clear is what further, if any, specific aspects of polarization are necessary for the ER effect to be operative and how precisely such polarization explains all facets of the phenomenon. Particularly lacking is a quantitative theoretical framework which covers all the three regions defined in the introduction. There is not even universal agreement as to appropriate dielectric parameters for modelling ER.

The approach which until recently was mainly followed was to model and correlate ER effects in terms of the Mason number ($M_n$), following the work of Mason *et al*[24-27]. These workers analysed the behaviour of a pair of interacting dielectric spheres of relative permittivity $\varepsilon_2'$ dispersed in a liquid of relative permittivity $\varepsilon_1'$ and viscosity $\eta_l$ and did so in terms of the dimensionless variable $M_n = \eta\dot{\gamma}/2\varepsilon_0\varepsilon_1'(\beta E)^2$ obtained* by ratioing the viscous to electrostatic forces set up under a shear rate $\dot{\gamma}$ and dc electric field E. In this expression $\varepsilon_0$ is the permittivity of free space and $\beta = [(\varepsilon_2' - \varepsilon_1')/(\varepsilon_2' + 2\varepsilon_1')]$ is the internal field correction term for spheres. However, the strict reliance on the above form of the Mason number neglects a number aspects of polarization of particles in flow, some of which can be particularly relevant to ER. They are:

1. An assumption that the polarization is entirely due to a mismatch in the permittivity between particle and dispersing fluid, with no account being taken of any differences in conductivity between the two phases.
2. The assumption that any time taken for the polarization to be established has no relevance because the field is dc.
3. That the form of ß is correct in that spheres are involved whose cavity field is well represented by the Clausius Mosotti relation.
4. That the polarization is proportional to field strength.

It is the first two of these assumptions that are primarily suspect in the use of the simple Mason number in analysing ER. The reasons have been stated in the past[4-6,29] with more recent quantitative analyses of the implications of a finite conductivity given by Anderson[30]. To appreciate the thrust of the arguments relating particularly to assumptions (1) and (2) it is informative to pose the question as to what type of fluid components might lead to the situation envisaged where the Mason number would be appropriate? They would involve non-conducting spherical particles whose polarization would be classical in the sense that the orientation of molecular dipoles would be involved, and these would be dispersed in a non conducting dielectric liquid with a lower relative permittivity. With the exception of

* There are alternative scalings for $M_n$ depending on the units used[28].

electrets or the particle-free liquid crystal systems, such orientation polarization has a low strength (low relative permittivity) because of thermal randomization of the molecular dipoles. Electronic plus atomic polarizations again only make small contributions. In consequence ER effects do not occur or are very small. One way of inducing large polarizations into particles and thereby increasing the interactive forces is to introduce charge carriers which are mobile within or around the particles. If their transport out of or between particles are prevented or attenuated by the particle - oil interface then a polarization will build which can be very much larger than that achievable by a simple dipole orientational polarization. The process being described is interfacial polarization in its most general form, and because of its large magnitude, is the one most commonly involved in ER. As discussed by Davis[31-33] there exist levels of conductivity *vis-a-vis* permittivity mismatch where the Mason number approach fails and where developments based on interfacial polarization theories such as those of Maxwell - Wagner[34,35] or for reasons elaborated below, even Sillars[36], are more appropriate. Recently Foulc, Felici and Atten[37,38] have proposed that the disparity in conductance takes a very dominant role in the ER phenomenon, but how their theory differs in detail from those of Hill and diverge from any of the models based on interfacial polarization are not discussed.

The second problem with using the Mason number in some flow situations is that it assumes that the induced particle polarization is independent of $\dot{\gamma}$, a statement tantamount to assumption (2). The cause stems from the rotational nature of most regions of shear flow, where the effect of shear is to rotate either the individual particles, or any structured or interacting groups of particles, unless prevented by other counteracting forces[39]. However, the global direction of the electric field is not changed by the flow so that within the framework of rotating particles or clusters, the direction of the field does change periodically with time. The effect on the particle polarization is that the nominally dc field acts at the particle as if it were a time varying field, causing the polarization to become shear rate dependent[40-42]. At too high a shear rate the polarization cannot be established because the charge redistribution mechanism is too sluggish, whilst at much lower shear rates the polarization is always established and co-linear with the field. The parameter delineating this rate condition is the critical frequency for the polarization which, if it involves interfacial polarization of the Maxwell-Wagner-Sillars type, is governed by mismatches in the conductance of the phases as well as differences in permittivity, with the former often taking the dominant role. If the shear rate at which an ER fluid is required to operate is such that the polarization mechanism for inducing dipoles is starting to relax out, then the ER effect will start to be attenuated and at much higher flow rates the ER effect will not operate (illustrated by curve 3 in figure 1). On this basis we predicted[4-6,9,29] that a lower limit should exist for the relaxation frequency for effective ER fluids giving a lower limit to the conductance of the material being dispersed as particles if interfacial polarization is involved. Empirically we found these to be $\approx 10^2$Hz and $\approx 10^{-7}$Sm$^{-1}$ respectively.

Materials which polarize rapidly in relation to the rotation rate can of course fully maintain the polarization, which then is always co-linear with the field. This provides forces attempting to chain the particles into the direction of the field but the shear gradient introduces motion tending to counteract this by rotating incipient chains into the flow lines. In most ER situations this means turning these across the field direction and as that proceeds, the induced dipoles would increasingly realign towards a side-by-side orientation which rather than being attractive, is repulsive. The structuring weakens and rearranges resulting in a complex orbital motion overall. We hypothesized[4-6,29] that there would be a most advantageous ER situation where both the polarization and field vectors are not colinear (a situation possible in particles polarizing in flow[40-42]) and where there could be some permanence of the structuring. On that basis there should be an optimum range in the dielectric relaxation frequency for ER fluids, above which particle association due to electrostatic forces is transient and below which electrostatic forces are too weak due to poor polarization. For interfacial systems there would be an equivalent upper level for the conductance. Our empirical estimates for this is $\approx 10^{-2}Sm^{-1}$ or in relaxation frequency terms $\approx 10^6$Hz.

To the authors' knowledge, these predictions have been upheld and even appear to apply to situations of no flow in that $\sigma_0$ levels correlate with the conductance of some semi-conducting polymers[4-6,43]. However, it should be pointed out that the experience that dispersions of good conductors tend to show disappointing ER effects, particularly as regards the level of $\sigma_0$, may be due to another cause. In such cases what is normally taken as the field strength (voltage/electrode separation) may not represent the polarizing field experienced by all the particles. There will be screening of parts of the fluid by a conducting network reducing the local field and the current drawn by the columns will also degrade the field. Some of these effects will persist in flow in so far as any clustering of good conductors will effect the field strength local to them (increased at high curvatures within the field lines and screened in regions adjacent to the cluster). As far as the authors are aware, these complexities have not been quantitatively assessed, but their possibility does raise some doubt on the validity of field level estimates based on the ratio of applied voltage to electrode gap, particularly when the dispersed particles have a significant conductivity.

Work by Hemp[39] adds a further dimension to the complexity of the polarization in flow when high fields are present. His theoretical study of the polarization and rotational motion of an isolated spherical semi-conductor has shown that if more electrical energy is dissipated to maintain the circulating current within the particle as it rotates (essentially a loss current) than would be gained in an adjustment of the viscous flow so that the particle stops rotating, then the particle locks. The permittivity and conductance of the particle and the viscosity of the base oil are the parameters which determine the levels of E and $\dot{\gamma}$ when this occurs. Admittedly, the behaviour of an isolated particle is not the full ER effect, involved as that is with many bodied interactions, but the same locking of particles may still apply. It is interesting to speculate that if such locking of the spin of isolated particles or their clusters occurs in concentrated dispersions, then that might aid fibrillation (*qv* the section on dielectric measurements).

For some of the materials dispersed in ER formulations there exist possible alternative mechanisms to interfacial polarization as the basis for the phenomenon. For example, in ER active fluids based on ferroelectrics the implication is often drawn that the activity is due to mismatch in permittivities that is large because of the large permittivity of such ferroelectrics as barium titanate (see however discussion below). Even in such situations, the use of Mason number may be inappropriate because of the caveats listed under (3) and (4) above. With large mismatches in permittivity the simple internal field approximation involved in ß is likely to be wrong and there is also the possibility that with electrets $\varepsilon_2'$ will be field strength dependent.

Although polarization is considered as the first stage in the ER effect, there has been a recent suggestion[44-46] that an older theory involving water bridging[47] might operate in a modified way, once structuring has occurred. As distinct from the model of Stangroom[47], the driving force for water redistribution is taken to be electrostatic in origin, with the 'condenser' regions between particles when filled with water. This model is put forward to explain the maximum in the water content for optimizing some ER fluids[48] but as explained above, that could also be attributable to other causes relating to conductance and polarization rates. However, whether water bridging does cause enhanced ER effects in promoted fluids is a subject worthy of further investigation.

The polarization models for ER fluid behaviour outlined have not had all their ramifications fully implemented into any global theory for the ER effect. However, in recent years in particular there has been considerable effort put into more quantitative modelling. This is having its greatest success in predicting the structure, and to some extent the strength, of the solid Bingham state. Minimum energy computations[32,49-56] have explained many of the features of the observed structuring. Single fibrils have been shown to be energetically unstable with calculations showing that the lowest energy packing of these into columns involves body-centred-tetragonal lattices. Structural variation, including the observed branching, is also explicable in terms of these computations. There have also been simulations of the dynamics of structure formation when flow is absent[57,58], including the development of power laws relating a characteristic time scale to the field strength, particulate concentration and base oil viscosity[58].

Various dynamic simulation of ER fluids in flow have been undertaken[28,59-68]. These use computer simulation on a position lattice occupied by a limited set of particles and develop the structures in real time, using a combination of electrostatic and viscous energy terms. Although these show the formation of fibrils, they appear to fail in closely predicting dynamic moduli. Generally, Mason numbers are used for scaling but without any cognizance being taken of the shear dependence of the extent and direction of polarization in relation to the rotatory motion of particles or clusters. Permittivity rather than conductivity mismatch as the cause for the dipolar forces are generally assumed in these computations.

## TECHNIQUES FOR PROBING ER

### Rheology

Although studying the rheology of ER fluids under field is the most direct method of obtaining data on the ER effect, correlating this to the physico-chemical attributes of the fluids is much more difficult because of the paucity of quantitative theory relating cause to effect. Our ability to fully understand and correlate parameters such as base oil viscosity, particulate concentration, shape, size distribution and particulate nature to the rheology of concentrated dispersions in the absence of field is difficult enough. The influences of such variables on ER when fields are also acting is even more so. The most impressive progress seems to have been made in studying the rheology of ER fluids under oscillatory shear where evidence of both elastic and viscous terms appear in the complex modulus. The strain imposed is small so that major flow disruption is avoided and the influence of strain rate, field strength, field frequency and particulate loading has been studied for a number of fluids[7,69,70].

As a method of physical investigation rather than as an essential component of fluid use, simple rheometry on ER fluids has, in recent years, not been as actively pursued as in the past. There has been some work using pulsed electric fields, when the lag or otherwise in rheological response can become evident[43,71]. An example of the type of information that can be obtained is shown in figure 2 where the response of a controlled stress Couette rheometer to unidirectional rectangular pulsed fields of various periods $\tau$ are shown. At low frequencies (long period) the system is switching between a free and field restrained rotation rate under preset levels of applied driving torque (two different torques are shown). As the frequency of switching increases, and the consequent period of field application decreases, the inertia of the rheometer and spinning fluid come into play, so that the response of the bob progressively fails to discriminate between the field 'on' and 'off' state. Finally, only an average response is achieved, which may (case I) or may not (case II) differ from the angular velocity for the zero field state. These differences in mean flow level show that for the fluid under test, a $\dot{\gamma}$ of 1350s$^{-1}$ is too high for the ER effect to be maintained with trains of pulses of period $\leq 5\times10^{-3}$s (case II), whilst at the lower $\dot{\gamma}$ of 730s$^{-1}$ the fluid was still able to respond to the pulsating field, even when the pulses are as short as $10^{4}$s (case I). The

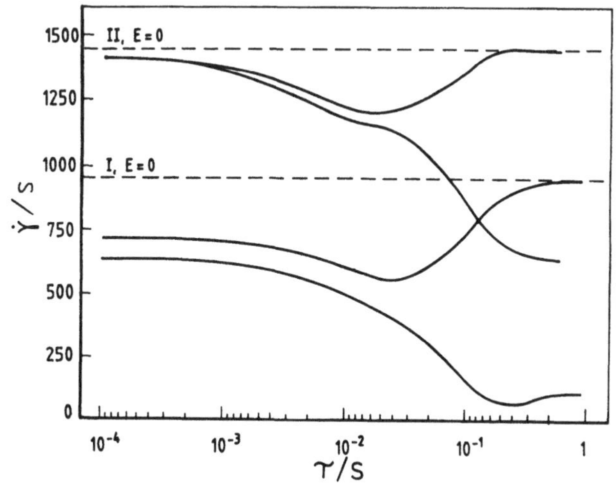


**Figure 2.** The effect of trains of 2 kVmm$^{-1}$ square unidirectional pulsed fields of period $\tau$ on the $\dot{\gamma}$ developed by a controlled stress rheometer containing an ER fluid (35% v/v PAnQR in 100 cS silicone oil). The response under two preset torques of 25 mNm (I) and 50 mNm (II) are illustrated.

cause is the inability of the particles to develop the polarization forces in flow at the higher shear rate. Dielectric measurements confirm this general view in that under the applied field of 2kVmm$^{-1}$ the experimentally observed dielectric relaxation frequency is ≈700Hz, so that it is not surprising that pulsing at $\tau < 1.4 \times 10^{-3}$s progressively attenuates the ER response, even though the pulses are unidirectional. There is also shear induced rotation which in its effect rotates the direction of the field. Experiments such as this can lead to estimates of the critical time scale for structuring under flow even when the inertia of the rheometer prevents full on-off switching. That the inertia can override the ability of a fluid to switch even when its response is fast should also be borne in mind when applying ER.

## Optical Techniques

The structuring of quiescent ER fluids under fields was first discovered[72] by direct microscopic observation and this type of visualization has been repeated with a variety of fluids. On occasion the structural details of the dimensions of the columns and their spacing have been analysed by direct visual observation as has the fracturing of the solid state using video recording[73]. The difficulties with all such direct observations, and with the alternative structural probe of light scattering, are that the majority of ER fluids are very opaque. To view the effects of applying fields, experiments with such opaque fluids have to be done using thin sections or employ fluids containing low volume fractions of particles. These conditions are atypical for the ER effect as normally employed. In particular, using thin sections may cause artefacts because of the greater dominance of non-uniform field gradients at electrode edges which can result in the dielectrophoresis of particles and in ill defined non-linear fields.

Until recently attempts to study the structuring of ER fluids by light scattering always failed because the fluids were too turbid. Recently however this problem has been by overcome by using silica or glass micro-spheres and adjusting the base fluid's refractive index so as to achieve only a small refractive index mismatch[52,74,75]. ER activity was induced by using damp particles, but the level of the ER moduli are presumably rather low. Nevertheless, there are no grounds for believing that the nature of the structuring would be significantly different then that expected for other stronger and more turbid fluids. Indeed the investigations generally confirm the theory in that the columns seem to be based on body-centred-tetragonal packing of the spheres but with considerable disorder to the lattice. The scattering theory to be applied should take into account that the scatter from diffracting spheres in these structures involved multiple lens effects and are not adequately described by point scattering theory[74,75]. For more opaque systems, diffuse light scattering has provided information on the time scales for structuring[76].

Light scattering from index matched ER fluids in flow could prove even more informative, particularly on the extent and development of structures in flow and their motion.

**Dielectric Measurements**

Since polarization is involved in the ER phenomenon, the study of the dielectric properties of ER fluids is bound to be of relevance. A knowledge of the impedance of ER fluids is of importance to the user who needs to supply the fields and switch these with the required rapidity. However, this is not the only reason why the dielectric spectroscopy of ER fluids is of interest. For the reasons discussed above, the polarization magnitudes and rates are fundamentally involved in the ER phenomenon and any method of studying them under the conditions when ER effects are present must be fruitful. Polarization rates and structural changes of ER fluids can be probed by means of dielectric measurements and as we have shown these can be done not only on quiescent fluids but also on fluids under shear, dc field or these in combination[4,5,43,71,77].

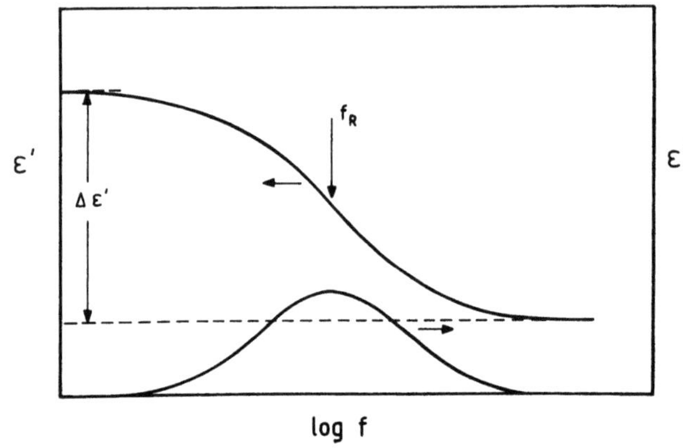


**Figure 3.** The frequency (f) dependence of relative permittivity (ε') and dielectric loss (ε'') of a material, such as an ER fluid, showing a dielectric relaxation at the frequency $f_R$. The intensity of the polarization relaxing is proportional to $\Delta\varepsilon'$, also shown.

Dielectric spectroscopy involves the frequency dependence of the complex permittivity $\varepsilon = \varepsilon_0(\varepsilon' - i\varepsilon'')$ in which $\varepsilon_0$ is the permittivity of free space, ε' the relative permittivity and ε'' the dielectric loss. ε' and ε'' are directly measurable in terms of the capacities and conductances of materials (in this case the ER fluid) and their frequency variation in the radio frequency region (which is of interest here) can be readily measured, for example by using analogue bridges operating at low voltages. It is then quite common, and in ER fluids we believe inevitable, to observe dielectric relaxation processes as illustrated in figure 3 with a characteristic relaxation frequency $f_R$. At frequencies $f \ll f_R$ the polarization due to the process occurring at a mean relaxation time of $\approx(2\pi f_R)^{-1}$ is fully established, whilst it is progressively attenuated as f traverses through the relaxation region until at $f \gg f_R$ the polarization for that specific process ceases to contribute.

Such simple dielectric measurement are of interest as an adjunct to ER[78] but as already indicated much more than this is possible: dc fields can be imposed and measurements still made with conventional equipment by using blocking capacitors between the dielectric cell and a low field impedance measuring instrument. Dielectric spectroscopy can also be done on flowing liquids, where the technique has been termed flow modified permittivity (FMP)[42]. FMP cells involve rotating couette capacitors with the advantage that geometry ensures uniformity of both electric and shear fields.

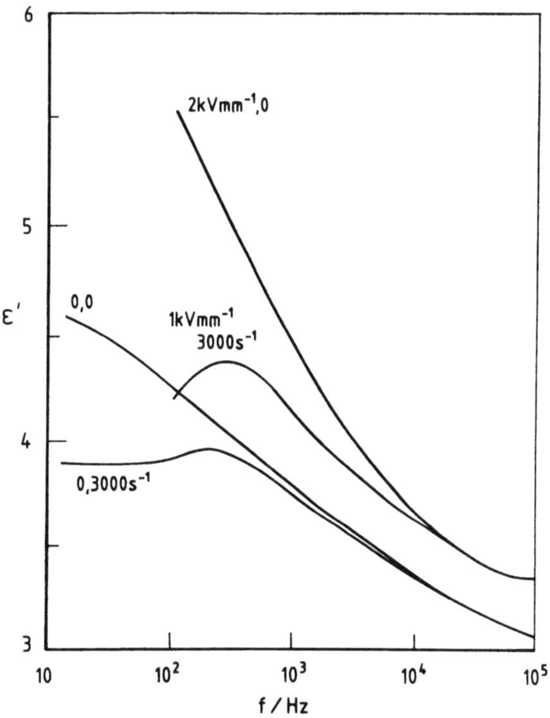


**Figure 4.** The frequency (f) dependence of relative permittivity (ε') of 20 vol% PTQR in 100 cS silicone oil under various fields and shear rates as marked. The FMP resonances under flow are at the maxima of the curves

Recently we have carried out experiments with combined shear and strong electric fields on ER fluids with these fields also acting separately[43,71]. With such studies, a number of features reflecting aspects of the ER phenomenon come to light. Some are illustrated in figure 4 which shows the dielectric behaviour for a poly(p-terphenylanthraquinone radical) (PTQR) based ER fluid. In the absence of field and flow the ER fluid shows a typical

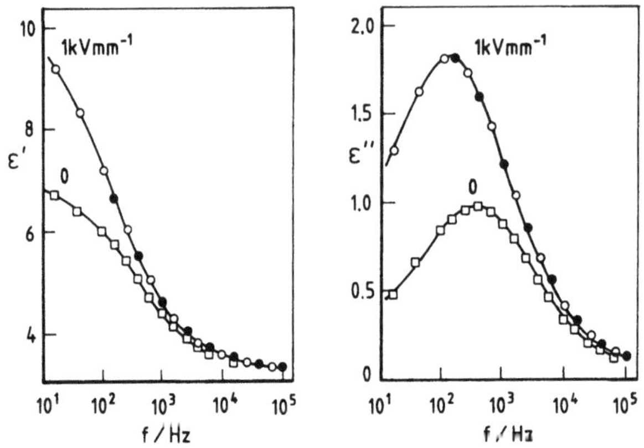


**Figure 5.** The persistence of structures once they have been formed as evidenced by dielectric spectroscopy. Closed points refer to data obtained in the presence of a 1kVmm$^{-1}$ field, and open points to the measurements after removal of the field but without mechanically disturbing the fluid. Permittivity data for the 20 vol% PAnQR/silicone fluid before the field was applied is also given to show the enhancing effect of field on the permittivity.

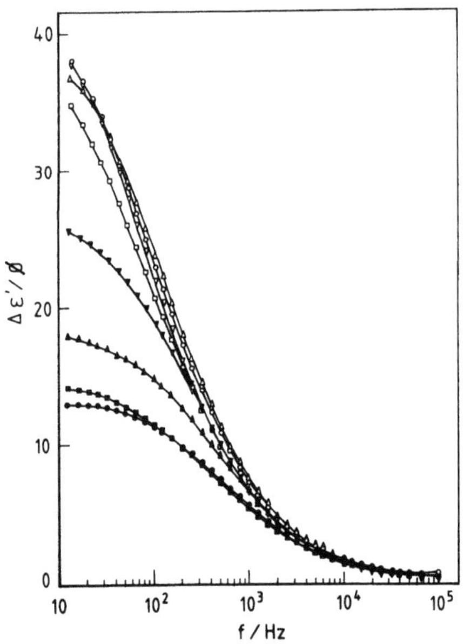


**Figure 6.** The effect of volume fraction (ϕ) of PAnQR on the permittivity increment (Δε'/ϕ) for 100 cS silicone based fluids before and after being structured by a 1 kVmm$^{-1}$ field. Solid points show data prior to field application and open points to data thereafter; ○ = 5, □ = 10, Δ = 20 and ∇ = 35 vol%.

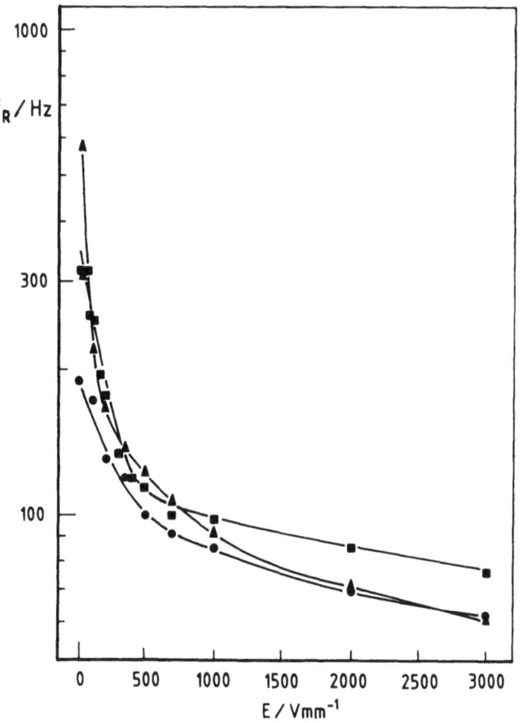


**Figure 7.** The changes in relaxation frequency ($f_R$) as a function of field strength (E) for 5, (▲); 20 (■) and 35 (●) vol% PAnQR in 100 cS silicone oil.

dielectric relaxation (shown in terms of ε') which is due to interfacial polarization of the dispersed semi-conducting polymeric particles. Under a 2kVmm$^{-1}$ field but without shear there is a dramatic increase in ε' and therefore in the polarization of the solid state. On the other hand, shearing this particular fluid at 3000s$^{-1}$ in the absence of field reduces ε' at low frequency because PTQR is a sluggish polarizer (conductance ≈10$^{-8}$Sm$^{-1}$). The curve also shows the forced FMP resonance at 240Hz, a feature arising because of a match between the particles spin and the field frequency[40-42]. If the ER fluid is subjected to both field and flow then there is an increase in ε' which, however, does not match the increase shown in the absence of flow. This behaviour, given here in outline, can be further investigated by more dielectric experimentation. For example, as figure 5 shows for a fluid based on dispersed poly(anthracenequinone radical) (PAnQR), the dielectric spectrum of fluids remain unchanged after applying and then removing the field, provided the fluid is left undisturbed[77]. As with PTQR, ε' is increased and $f_R$ decreased and since these changes remain after the field is removed, they reflect the structuring rather than field effects on the polarization. Such effects on ε' and $f_R$ are predictable for interfacially polarizing columns aligned in the field direction[36]. The effect of particle concentration shown in figures 6 and 7 is particularly informative. For a fluid with randomized particles (E = 0) the relative permittivity increment for the interfacial polarization (Δε'/volume fraction) and $f_R$ are still concentration dependent, showing coupling of the polarization between neighbouring particles, but after such particles are collected into columns, neighbour interactions now involving columns is much reduced because the separation between the columns is much larger than between randomized particles.

Since there is persistence of structure to an undisturbed fluid after a field has been applied, application of a rectangular pulse can be used to look at the kinetics of the structuring. By analysing each individual dielectric spectrum obtained after applying a single pulse of time $\tau$ to an unstructured PAnQR based fluid, Block and Rattray[71,77] followed the polarization and hence fibrillation in real time. Figures 8, 9 and 10 show the effect of field strength, particle concentration and base oil viscosity on that polarization. The

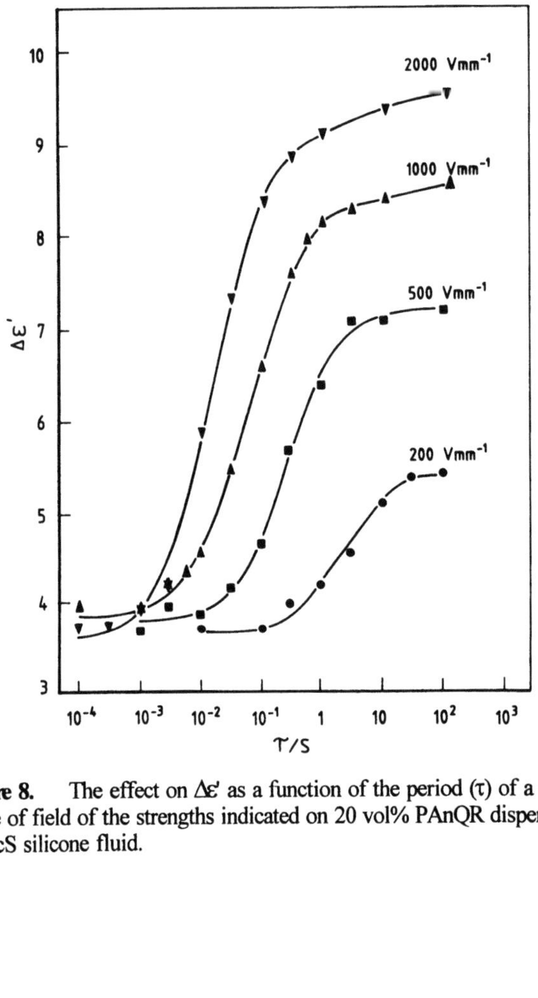


**Figure 8.** The effect on $\Delta\varepsilon'$ as a function of the period ($\tau$) of a single pulse of field of the strengths indicated on 20 vol% PAnQR dispersed in 100 cS silicone fluid.

combined results fit a very simple master curve (figure 11). This is based upon the development of columns being reflected by the fractional extent of the polarization enhancement, and the rate at which this occurs being determined by the Stokesian limited coalescing velocity under the influence of the induced dipole - dipole forces over the period $\tau$.

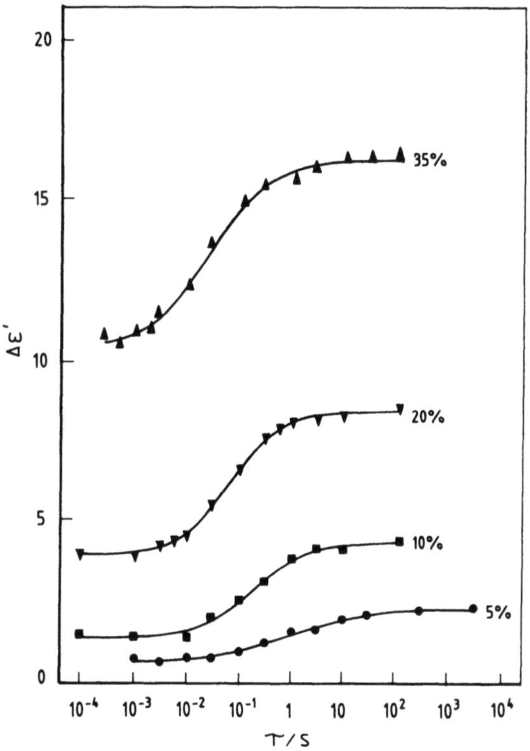


**Figure 9.** The effect on $\Delta\varepsilon'$ as a function of the period ($\tau$) of a single 1 kVmm$^{-1}$ pulse on PAnQR at volume fractions indicated dispersed in 100cS silicone fluid.

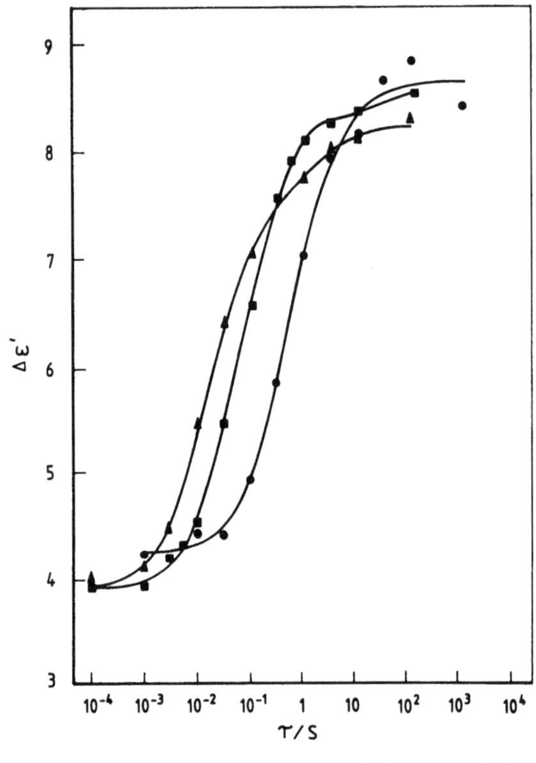


**Figure 10.** The effect on $\Delta\varepsilon'$ as a function of the period ($\tau$) for a single 1 $kVmm^{-1}$ pulse on 20 vol% PAnQR dispersed in silicone fluids of various viscosities: ▲ = 20 cS; ■ = 100 cS and ● = 500 cS.

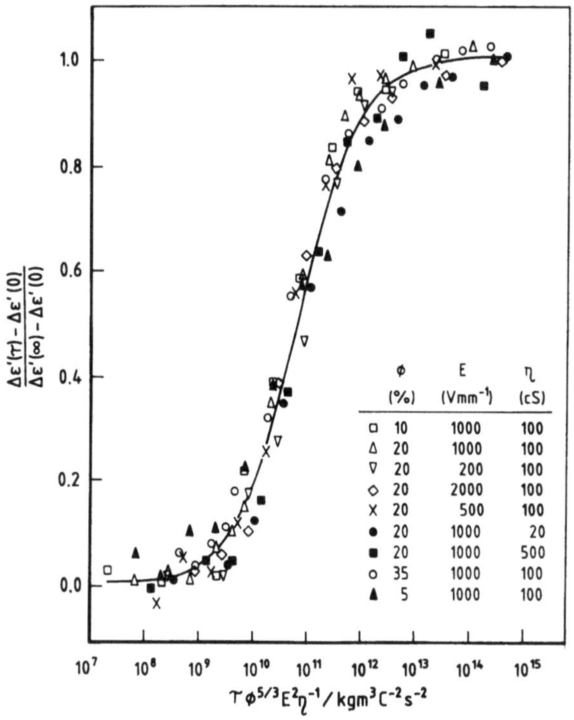


**Figure 11.** The master curve based on the data of figures 8 - 10.

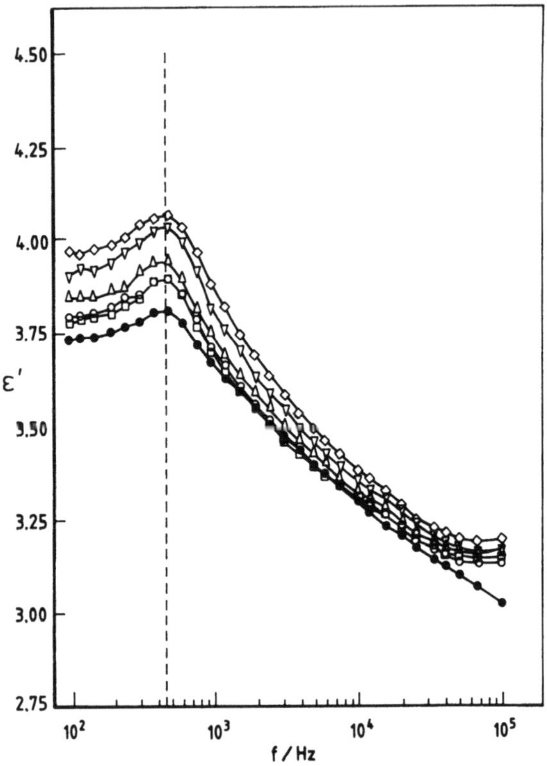


**Figure 12.** The FMP resonances observed in the flow of 10 vol% PTQR in 100 cS silicone fluid sheared at 6000 $s^{-1}$ under different fields: (●) = 0; (○) = 0.2; (□) = 0.4; (Δ) = 0.6; (∇) = 0.8 and (◇) = 1.0 $kVmm^{-1}$. The theoretical position[40-42] of the FMP resonance for unhindered rotating spheres at 477 $s^{-1}$ is shown by the broken line.

Measurements with a combined flow and electric field are continuing but one interesting facet has come to light in the study of a PTQR based fluid which shows an FMP resonance. Figure 12 shows that the resonance position under $\dot{\gamma} = 6000$ $s^{-1}$ is not significantly shifted from its theoretical no field level of 477 $s^{-1}$ and similar absences of any shifts are shown at lower shear rates[77]. This is an unexpected result since the field might have been expected to shift the resonance position down in frequency, if the spin rate of the particles or clusters were reduced. One explanation is that the forces of interaction are insufficient to cause shifts in this rather weakly interacting ER system. Another possibility is that the interactions do occur but rather than a gradual shift in the resonance peak position, the change is discontinuous between some of the particles becoming locked (hence no resonance) whilst others rotate with an unchanged velocity. As discussed above, the analysis by Hemp[39] provides some support for this possibility but whether such effects do occur and to what extent requires further experimentation and also analysis of the motion of interacting polarizable particles under the more concentrated conditions applicable to ER.

## Other Techniques

As far as the authors are aware, very few other techniques have been used to probe ER behaviour. There have been a few ultrasonic[79-82] studies of the solidified ER state.

Propagation of MHz compression waves across the structures attenuates the intensity but leaves the velocity unchanged, whilst there was no attenuation propagating along the columns, although a small increase in velocity was observed[79-81]. Bellin *et al*[82] have used shear wave reflectometry at 3MHz and measured the sound velocity and complex shear moduli of three ER fluids at 0 and 1.5 $kVmm^{-1}$. Such studies should be extended to include variation of field strength, changes in particle concentration and, if at all possible, should also be done under conditions where the fluid is subjected to shear flow.

Probing structures by magnetic resonance techniques is another possible method of investigating ER behaviour and could include the effect of flow. There has been one esr study[83] using esr probes, but this was narrow line and targeted to investigating the role of water as an ER promoter. The use of broad line or imaging techniques, particularly nmr but also esr if a free electron containing semi-conducting ER substrates is present, should prove a useful tool in studying structure formation. As far as the authors are aware such studies have not been done.

## ER MATERIALS

ER fluid formulations are continuously being diversified and added to, but in this article we shall restrict the discussion to ER fluids which show attempts to produce particular attributes or where the materials might produce the phenomenon by polarization mechanisms other than interfacial.

If an ER fluid is to be considered for the front running commercial applications in dampers, engine mounts or to a lesser extent clutches and latches, then consideration of a number of properties are important. These include:

i. A high induced yield together with a low no-field viscosity.
ii. Persistence of the field induced stress enhancement at high shear rates.
iii. An ability for the electrically induced stress to be rapidly turned on and off by the field.
iv. Low currents drawn by the fluid under field.
v. A wide temperature range of operation including retention of the properties listed under i to iv.
vi. The fluid to have long term stability against sedimentation and in some applications, centrifugation.
vii. More generally, the fluid should not degrade in storage or use.
viii. The particles within the fluid should not be subject to electrophoresis and not plate out under dc fields.
ix. The fluid must not cause mechanical abrasion in flow.
x. The fluid to be non-toxic, of low flammability and easily disposed of: it must be environmentally friendly.
xi. The cost must not be prohibitive.

The precise requirements and levels which are necessary under these headings depend upon the individual application. Even then there is some diversity of opinion as to the acceptable conditions, partly because some of the difficulties can be overcome by engineering design of the devices, but not to the extent that they become so complex and costly as to make the application commercially non-viable. The reader is referred to an number of articles discussing the viability of ER technology and the fluid requirements[6,8,10,13,14,16,84-92], but it is useful to bear in mind ball-park figures when considering particular fluids. Electrically enhanced stresses of several kPa are really necessary and, if possible, need to be achieved with a field strength $\leq 3kVmm^{-1}$. The enhanced stresses have to persist in flow up to shear rates in excess of $10^3s^{-1}$ and the response time to switch this level should not exceed a few ms. It is of course possible to enhance the stress level by loading the fluid with more active particles but doing so can raise the no field viscosity very dramatically; the discrimination between on and off is then lost. Thus, in terms of the shear stress - shear rate behaviour illustrated in figure 1, an ideal ER fluid would have a very small slope of the E = 0 line with a large difference between it and the fielded response line, which should not diminish with shear.

There might well be considerable viscous heating during the shearing of an ER fluid which carries a high loading of particles. In consequence the bulk conductance of the fluid could also be adversely effected. Amongst many problems, such loading effects could produce not only high power demand, but also difficulties in controlling the temperature if the positive thermal coefficient of the bulk conductance of the fluid is large. Indeed the current drawn and its variation with temperature are important fluid properties. The current density should be small ($<$ $100\mu Acm^{-2}$) and remain so over the temperature range of operation and not result in such electrical heating as to cause a thermal run-away. The temperature range over which dampers, engine mounts and clutches are expected to operate in an automotive applications are often quoted to be as wide as -40 to 120°C.

As the above list illustrates, these parameters are not the only criteria in judging the fluids. Particularly important is the question of sedimentation of particles. Obviously large density mismatches between the material of the particles and the base oil are undesirable, but attempts to match densities at all temperatures would prove very difficult if not impossible. Small particles, high loadings and high viscosities of base oil would all help to reduce the rate of sedimentation but as already stressed, there are strong counter indications in the need to have as low as possible a no-field viscosity. Any tendency for the particles in an ER fluid to flocculate would increase sedimentation rates and also might make redispersal by agitation difficult. This is an important consideration because in applications such as the damper or engine mount, an initial agitation on start up is often expected to induce ready redispersion of any sediment formed due to standing.

The no-field viscosity of dispersions is strongly influenced by the same interactive forces responsible for flocculation and thereby the rate of sedimentation and redispersal. This together with the need to reduce the electrical power demand has led to considerable interest in the coating of ER particles. Using an insulating coat can reduce power demand and if that layer also acts as a steric stabilizer, then its presence might reduce flocculation and the rate of sedimentation and act in a flow enhancing role. Poly(aniline)[5,16,43,93,94] and poly(o-toluidine)[95], which if doped at the correct level can act as very effective ER substrates, have the interesting property that their hydrochlorides initiate vinyl polymerization, thereby making it possible to surface graft insulating layers to them. That has been shown to lead to a large reduction in the current density of such fluids[96]. Choice of monomer in relation to the base oil for dispersion should also make it possible to attach steric stabilizer coatings to such particles.

## Commercial fluids.

Commercial fluids have appeared and it has been reported[92] that the cost for one of them is currently running at $700/dm³ with the potential to drop to $50/dm³. It is water promoted and based upon poly(lithium methacrylate) as the active component. Bayer have reported[97] the use of a cross linked polyethylene oxide doped with inorganic salts but without water or promoter. Presumably the cations bind to the ethylene oxide chains leaving the anions more mobile, a mechanism believed operative in the battery application of salt loaded polyethylene oxide systems. Semiconductors are the basis of the Bridgestone[98] and Fujikawa[99,100] fluids. In the former, carbon with conductance in the range $10^4$ - $10^{-7}Sm^{-1}$, derived from pitch or an acrylate resin by controlled pyrolysis, is use as the ER active substrate. The particles are also treated to produce an insulting coat so as to reduce bulk conductance of the fluid. The fluid described by Fujikawa has a very low quoted bulk conductance. It is based upon polystyrene or polyacrylate microbeads on whose surface are located sub-micron particles of ER active inorganic semi-conductors. Presumably the semi-conductors are insulated by the polymer or separated on its surface, at least in some places, so as to break any conducting chains when fibrillation sets in. Another claimed advantage of the Fujikawa fluid is that the buoyancy of the polymeric microbeads reduces the effective density of the compound particles.

As well as these commercial or commercially targeted systems, recent research into a number of other ER fluid systems deserve comment.

**Barium titanate based fluids**

The use of barium titanate or other ferroelectric materials as ER active substrates has been known for some time. Recently interest has again turned to whether these materials act as ER substrates because of their ferroelectric character. Thus Ginder and Davis[101] have looked at barium titanate based ER fluids and discuss the ER effects in terms of permittivity mismatches, which, because of the high permittivity of the ferroelectric state, are presumed to be large. However, some aspects of the ER behaviour of such ferroelectric dispersions are puzzling. It has been reported[102] that barium titanate as an ER substrate is very strongly water promoted and Ginder and Davis also find it a rather weak ER material when dry, especially under dc fields, with stronger effects seen under ac. It is also known that barium titanate based ER fluids can be quite conductive and it has been suggested that surface conduction may occur in powdered barium titanate[103]. Furthermore, if the ER effects are electret in origin then we suggest that such a system should show unusual field dependencies. There should also be interesting temperature variations for ER fluids based on ferroelectricity, particularly near any accessible Curie temperature.

Because a ferroelectric based ER effect would be novel, such fluids merit further studies and these should include dielectric measurements so as to establish the permittivities of the particle phase. To what extent domain structures and high permittivities are present in the barium titanate powders used for making ER dispersions is also questionable.

**MR/ER fluids**

There have been a number of attempts to produce fluids which combine ER and magnetorheological (MR) effects. Two approaches are possible: dispersing separate ER and MR particles or the production of some form of compound particle capable of acting under either magnetic or electric fields. Fujita *et al*[104] have studied an example of the former involving submicron dispersions of magnetite and barium titanate in kerosene. They subjected such a fluid to magnetic and electric fields (50 Hz ac) acting in parallel and showed that the enhanced stress levels caused by the fields were additive. By comparing the effects individually, using pure magnetite or barium titanate based fluid of similar concentrations, they conclude that the electric field was more effective than the magnetic field for these system. They make the comment that the magnetic field helps to stabilize the compound fluid against sedimentation because of the levitation induced on the magnetite particles by the magnetic field.

Reports on the preparation of compound particles have appeared but some difficulties are being encountered in producing such a fluid with a sufficiently high electrical resistance[105]. As far as the authors are aware, no rheological data under combined electric and magnetic fields involving compound particles have been published. It is our opinion that it would stimulate such work if the scientific interest or practical advantages of such combined ER/MR fluids of either type were outlined by those advocating their preparation.

## Lyotropic liquid crystals as ER fluids

All of the ER fluids discussed above are based upon dispersions and that can lead to problems, particularly sedimentation. The use of a single phase liquid would remove these. It has been known for some time that thermotropic liquid crystals show changes in their rheology under electric fields with the Leslie coefficients becoming enhanced due to the tendency to align domains under field. In comparison to the ER effects in powerful ER dispersions, the changes that similar fields make to the flow of thermotropic liquid crystals are small[106-108] and there is no induced Bingham body behaviour. Cost, sensitivity to impurities and the limited temperature range for liquid crystal phases would also limit their industrial use in an ER role. However, recent reports showing substantial ER effects with certain lyotropic liquid crystal systems are very interesting and might have technological applications[109]. The ER fluids here are solutions of poly(n-hexylisocyanate) at a sufficient concentration to form homogeneous liquid crystal phases. These macromolecule have chain conformations which tend to give them an extended helical architecture, which persists even in solution, and this is the cause for their liquid crystal forming tendency. The same configurational attributes impart an along chain permanent dipole contribution per residue. In high molecular mass polymers this leads to a large total dipole per macromolecule[110,111], large enough, under fields ≈kVmm$^{-1}$, to be able to reach alignment energies ≈kT. Using a 26 wt% poly(hexylisocyanate) solution in xylene, Yang and Shine[109] have shown a thirty-five fold increase in viscosity under field of 1.5 kVmm$^{-1}$ when $\dot{\gamma} = 0.4s^{-1}$. Whether there is any Bingham state is not recorded but even on scientific grounds alone, such an intense phenomenon in a liquid rather than a dispersion is of interest. The potential for application may also be there, although at 26 wt% the viscosity without field of 63Pas is undesirably high. Other factors such as cost and possible limits in operating temperature (to avoid possible transition to the isotropic state) may also place limits on application. It is also of potential interest to study the MR effects of such liquid crystal systems since magnetic ordering of lyotropic liquid crystals can occur. If effective these would be alternative classes of combined MR/ER fluids.

## NOVEL APPLICATIONS

The traditional applications of ER are concerned with applying the effect for switching flows or interconversion between solids and liquids. Potential uses in this mode are varied and range in their mechanical demand from the control of large to quite small forces. The implication of that and the types of devices which might employ the ER effect have been reviewed[84-92,112], but in essence, the current perception is that the costs of starting the commercial production of an ER fluid require a large scale market for the fluid. This, together with practical considerations relating to the viability and reliability of devices, places vibration damping in engine mounts and suspensions at the forefront of applied research. Applications in clutches, locks and valves for robotics currently have less dominance. There are of course a large variety of other rheologically based uses for such fluids and many of these could become viable as standard fluids become commercially available. As already indicated above, this situation seems to be at hand.

The phenomenon involved in ER is also of potential applicability outside the switching of fluid flow. However, examples are much less numerous or well known. One of the earliest described by Hollmann[113] makes use of the chain forming property of dispersed graphite in silicone to produce a field variable conductor, which he showed could also be made to act as a rectifier. These effects do not seem to have been developed into an application, but the concept of applying an electric field to chain particles in order to modify the properties of some composite system, be it a dispersion or even a solid, has caused some recent interest. It is not even necessary to limit the changes to mechanical ones. For example, Randall *et al*[114] extended this concept to the alignment of ferroelectrics, semiconducting, metallic or other particles dispersed in various resins. Electric fields were applied during cure and when the electrodes were laid down in a pattern on the surface, then that pattern was reproduced by the columns of particles concentrated underneath. Cure then locked these structures into position. They aligned aluminium particles in a silicone elastomer under imaging electrodes to produce transparent patterns in an otherwise opaque sheet. The technique was also applied to chaining and encapsulating silver coated acrylic fibres or powdered barium titanate into silicone elastomer and the suggestion was made that these and analagous structured composites could be used for novel conductive connectors, non-linear I-V varistors, piezoelectric transducers, positive temperature coefficient resistors and even superconductive composites. The possibilities are obviously legion, particularly when it is remembered that in such applications the fibril strength or mechanical interactions may not be very important, provided structuring occurs. There are many classes of materials which fibrillate under field but are not effective ER substrates.

However using substrates which would be ER active in a fluid dispersion and compounding these into an elastomer does provide ER properties to the rubber as has been recently demonstrated[115]. These workers blended lightly doped poly(p-phenylene) (PPP) of conductance in the range $10^{-8}$ - $10^{-5}$ $Sm^{-1}$ into a silicone elastomer and then measured the shear moduli of the blends with various loadings of PPP under various levels of field. At volume fractions > 10% both storage (G') and loss (G'') moduli showed field induced enhancements with more than a doubling of G' for a 10.8 vol% dispersion under 5 $kVmm^{-1}$. Furthermore, if the blending process was done under a 1 $kVmm^{-1}$ aligning field, then those aligned blends showed an even larger effect with G', increasing from ≈0.2 to ≈1.2MPa when a 5 $kVmm^{-1}$ field was applied. These results are very interesting, not only in showing that the ER effect can operate in the solid state, but also demonstrating that an ER effect exists even when the particles are not structured. As expected, the ER phenomenon involves a force component acting over space. Experiments of the type undertaken by Shiga *et al*[115] could help in our quantitative understanding of the ER phenomenon by simplifying the normal situation in that the distribution of particles can be locked into either a random or columnar structure before any study of further field effects.

## ACKNOWLEDGEMENTS

The authors wish to thank the Engineering and Physical Sciences Research Council for support in making their ER research possible.

## REFERENCES


1 H. Block and J.P. Kelly, *Proc. IEE Colloq.* 14: 1 (1985).

2. H. Block and J.P. Kelly, *US. Patent* 4687589 (1987); *UK. Patent* 217051B (1986).

3. F.E. Filisko and W.E. Armstrong, *US Patent* 4744914 (1988).

4. H. Block, J.P.Kelly, A. Qin and T. Watson, *Langmuir* 6: 6 (1990).

5. H.Block, J.P. Kelly and T. Watson, *in*: "High Value Polymers," A.H.Fawcett, ed., RSC Special Pub. No 87, (1991).

6. H. Block, *in*: "Actuator 92 Conference Proceedings, Bremen," K. Lenz and H. Borgmann, ed., VDI/VDU-Technologiezentrum Informationstechnik Gmbh, Berlin (1992).

7. D.R. Gamota and F.E. Filisko, *J. Rheol* 35: 399 (1991).

8. Z.P. Shulman, R.G. Gorodkin, E.V. Korobko and V.K. Gleb, *J. Non-Newtonian Fluid Mechanics* 8: 29 (1981).

9. H. Block and J.P. Kelly, *J. Phys. D Appl. Physics* 21: 1661 (1988).

10. D. Brooks, *Physics World* 35 (1989).

11. M. Whittle, *Physics World* 39 (1989).

12. T.C. Jordan and M.T. Shaw, *IEEE Trans. Electrical Insulation* 24: 849 (1989).

13. W. A. Bullough, *Endeavour New Series* 15: 165 (1991).
14. H. Block, *Chem. Tech.* 369 (1992).
15. F.E. Filisko, *Chem. and Industry* 370 (1992).
16. K.M. Blackwood and H. Block, *Trends in Polymer Science* 1: 98 (1993).
17. T.C. Halsey, *Adv. Materials* 5: 711 (1993).
18. "First Internat. Symp. on ER Fluids," H. Conrad, J.D. Carlson and A.F. Sprecher, ed., North Carolina State Univ. Engineering Pub., (1989).
19. "Electrorheological Fluids - Mechanisms, Properties, Structure, Technology, and Applications," R. Tao, ed., World Sci. Pub. Co., Singapore - New Jersey - London - HongKong (1992).
20. "Proc. Int. Conf. Electrorheological Fluids, Feldkirch, Austria," R. Tao, ed., World Sci. Pub. Co., Singapore - New Jersey - London - HongKong, (1994)
21. Z.P. Shulman, V.I. Kordonsky, E.A. Zaltsgendler, I.V. Prokhorov, B.M. Khusid and S.A. Demchuk, *Int. J. Multiphase Flow* 12: 935 (1986).
22. G. Bossis, Y. Grasselli, E. Lemaire, L. Petit and J. Persello, *in*: "Electrorheological Fluids - Mechanisms, Properties, Structure, Technology, and Applications," R. Tao, ed., World Sci. Pub. Co., Singapore - New Jersey - London - HongKong (1992)
23. C.Kormann, M. Laun and G. Klett, "4th Int. Conf. New Actuators," K. Lenz and H. Borgmann, ed., Axon Tech. Consult. Gmbh, Berlin, (1994).
24. C.E. Chaffey and S.G. Mason, *J. Colloid Interface Sci.* 27: 115 (1968).
25. A. Okagawa, R.G. Cox and S.G. Mason, *J. Colloid Interface Sci.*, 47: 536 (1974).
26. P.A. Arp and S.G. Mason, *Colloid Polym. Sci.* 255: 1165 (1977).
27. A. Okagawa, R.G. Cox and S.G. Mason, *J Chem. Soc. Faraday Trans. I.*, 74: 1242 (1978).
28. D.M. Heyes and J.R. Melrose, *Molecular Simulation* 5: 293 (1990).
29. H.Block and J.P. Kelly, *in*: "First Internat. Symp. on ER Fluids," H. Conrad, J..D. Carlson and A.F. Sprecher, ed., North Carolina State Univ. Engineering Pub.,(1989).
30. R.A.Anderson, *in*: "Electrorheological Fluids - Mechanisms, Properties, Structure, Technology, and Applications," R. Tao, ed., World Sci. Pub. Co., Singapore - New Jersey - London - HongKong (1992).
31. L.C. Davis, *Appl. Phys. Lett.* 60: 319 (1992).
32. L.C. Davis, *J. Appl. Phys.* 72: 1334 (1992).
33. L.C. Davis, *J. Appl. Phys.* 73: 680 (1993).
34. J.C. Maxwell, "Electricity and Magnetism", Clarendon Press, Oxford, (1892).
35. K.W. Wagner, *Arch. Elektrotech.* 2: 371 (1914).
36. R.W. Sillars, *J. Inst. Elec. Eng.* 80: 378 (1937).
37. J.N. Foulc, N. Felici and P. Atten, *C. R. Acad. Sci. Paris, serie II* 314: 1279 (1992).
38. J.N. Foulc, P. Atten and N. Felici, *C. R. Acad. Sci. Paris, serie II* 317: 5 (1993).
39. J. Hemp, *Proc. Royal Soc. London A* 434: 297 (1991).
40. H. Block, K.M. Goodwin, E.M. Gregson and S.M. Walker, *Nature*: 275, 632 (1978).
41. H. Block, E. Kluk, J. McConnell and B.K.P. Scaiffe, *J. Colloid Interface Sci.*: 101, 320 (1984).
42. H. Block, *in*: "Polymers in Solution," W.C. Forsman., Plenum Press, New York and London (1986).
43. H. Block, P. Rattray and T. Watson *in*: "Electrorheological Fluids - Mechanisms, Properties, Structure, Technology, and Applications," R. Tao, ed., World Sci. Pub. Co., Singapore - New Jersey - London - HongKong (1992).
44. H. See, H. Tamura and M. Doi, *J. Phys. D App. Phys.* 26: 746 (1993).
45. H. Tamura, H. See and M. Doi, *J. Phys. D App. Phys.* 26: 1181 (1993).
46. H. Tamura, H. See and M. Doi, *Int. J. Mod. Phys B.* 7: 2057 (1993).
47. J.E. Stangroom, *Phys. Technol.* 14: 290 (1983).
48. H. Uejima, *Japan. J. Appl. Phys.* 11: 319 (1972).
49. T.C. Halsey and W. Toor, *Phys. Rev. Lett.* 65: 2820 (1990).
50. R. Tao and J.M. Sun, *Phys. Rev. Lett.* 67: 398 (1991).
51. R. Tao, *in*: "Electrorheological Fluids - Mechanisms, Properties, Structure, Technology, and Applications," R. Tao, ed., World Sci. Pub. Co., Singapore - New Jersey -London - HongKong (1992).
52. T.C. Halsey, *in*: "Electrorheological Fluids - Mechanisms, Properties, Structure, Technology, and Applications," R. Tao, ed., World Sci. Pub. Co., Singapore - New Jersey - London - HongKong (1992).
53. J.E. Martin, J. Odinek and T.C. Halsey, *Phys. Rev. Lett.* 69: 1524 (1992).
54. W. Toor, *J. Colloid Interface Sci.* 156: 335 (1992).
55. T-j. Chen, X. Zhang, R.N. Zitter and R. Tao, *J. Appl. Phys.* 74: 942 (1993).
56. L.C. Davis, *Phys. Rev. A* 46: R719 (1992).
57. D.J. Klingenberg, C.F. Zukoski and J.C. Hill, *J. App. Phys.* 73: 4644 (1992).

58. H. See and M. Doi, *J. Phys. Soc. Japan* 60: 2778 (1991).
59. D.J. Klingenberg, F. van Swol and C.F. Zukoski, *J. Chem. Phys.* 91: 7888 (1989).
60. M. Whittle, *J. Non-Newtonian Fluid Mechanics* 37: 233 (1990).
61. J.R. Melrose, *Phys. Rev. A* 44: (1991).
62. T.C.B. McLeish, T. Jordan and M. T. Shaw, *J. Rheol.* 35: 427 (1991).
63. D.J. Klingenberg, F. van Swol and C.F. Zukoski, *J. Chem. Phys.* 94: 6160, (1991).
64. D.J. Klingenberg, F. van Swol and C.F. Zukoski, *J. Chem. Phys.* 94: 6170 (1991).
65. D.J. Klingenberg, D. Dierking and C.F. Zukoski, *J. Chem. Soc. Faraday Trans.* 87: 425 (1991).
66. H. See and M. Doi, *J. Rheol.* 36: 1143 (1992).
67. K.C. Hass, *Phys. Rev. E* 47: 3362 (1993).
68. D.J. Klingenberg, *J. Rheol.* 37: 199 (1993).
69. W.S. Yen and P.J. Achorn, *J. Rheol.* 35: 1375 (1991).
70. J.A. Powell, *Smart Mater. Struct.* 2: 217 (1993).
71. K.M. Blackwood, H. Block, P. Rattray, G. Tsangaris and D. N. Vorobiev, *in*: "Proc. Int. Conf. Electrorheological Fluids, Feldkirch, Austria," R. Tao, ed., World Sci. Pub. Co., Singapore - New Jersey - London - HongKong, in press.
72. W.M. Winslow, *J. Appl. Phys.* 20: 1137 (1949).
73. D.J. Klingenberg and C.F. Zukoski, *Langmuir* 6: 15 (1990).
74. T-j. Chen, R.N. Zitter and R. Tao, *Phys. Rev. Lett.* 68: 2555 (1992).
75. T-j. Chen, R.N. Zitter and R. Tao, *in*: "Electrorheological Fluids - Mechanisms, Properties, Structure, Technology, and Applications," R. Tao, ed., World Sci. Pub. Co., Singapore - New Jersey - London - HongKong (1992).
76. J.M. Ginder and L.D. Elie, *in*: "Electrorheological Fluids - Mechanisms, Properties, Structure, Technology, and Applications," R. Tao, ed., World Sci. Pub. Co., Singapore - New Jersey - London - HongKong (1992).
77. P. Rattray, PhD. Thesis, University of Cranfield (1994).
78. K.D. Weiss and J.D. Carlson, *Int. J. Modern Physics* B 6: 2609 (1992).
79. E.V. Korobko and A.I. Chernobai, *Trans. Inhenerno Fiz. Zh.* 48: 219 (1985).
80. R. Roy and J. Richardson, *J. Acoust. Soc. Amer. Suppl. 1* 87: S85 (1990).
81. E.V. Korobko and S.T. Zavtrak, *in*: "Electrorheological Fluids - Mechanisms, Properties, Structure, Technology, and Applications," R. Tao, ed., World Sci. Pub. Co., Singapore - New Jersey - London - HongKong (1992).
82. J.L. Bellin, R. Holsinger and A.W. Chow, *J. Acoust Soc. Amer.* 92: 1174 (1992).
83. P. Bailey, D.G. Gillies and L.H. Sutcliffe, *J. Chem. Soc. Faraday Trans.* 86: 3309 (1990).
84. J.A. Constance and K.A. Kovaly, *Future Tech.* 54 (1988).
85. J. Dunn, *The Engineer* 37 (1989).
86. H. Aldersey-Williams, *New Scientist* 37: 40 (1990).
87. R. Pool, *Science* 247: 1180 (1990).
88. N. Webb, *Chem. Britain* 338 (1990).
89. M. Edwards, *Eureka Transfer Technology* 50, (1990).
90. "Electrorheological (ER) Fluids", US Dept. Energy DOE/ER/30172 (1993).
91. J.D. Carlson, "4th Int. Conf. on New Actuators", K. Lenz and H. Borgmann, ed., Axon Tech. Consult. Gmbh, Bremen (1994).
92. G. Parkinson, *Chem. Eng.* 35 (1992).
93. H. Block, J. Chapples and T. Watson, *UK Patent App.* 8900825.6 (1989).
94. C.J. Gow and C.F. Zukoski, *J. Colloid. Interf. Sci.* 136: 175 (1990).
95. K.M. Blackwood, G. Tsangaris, D.N. Vorobiev, H. Block and J. Akhavan, this volume.
96. H. Block and K.M. Blackwood, *UK Patent App.* 9313408.8 (1993).
97. E. Wendt and K.W. Busing, *in*: "4th Int. Conf. on New Actuators", K Lenz and H. Borgmann, ed., Axon Tech. Consult. Gmbh, Bremen (1994).
98. Bridgestone Corp., *Eur. Patent App.* 89115820.6 (1989).
99. *JETRO* 16, (1993).
100. *Japan Autotech Report* 179: 20 (1993).
101. J.M. Ginder and L.C. Davis, *in*: "Proc. Int. Conf. Electrorheological Fluids, Feldkirch, Austria," R. Tao, ed., World Sci. Pub. Co., Singapore - New Jersey - London - HongKong, in press.
102. Y. Otsubo, *Colloids and Surfaces* 58: 73 (1991).
103. T.J. Garino, D. Adolf and B. Hance, *in*: "Electrorheological Fluids - Mechanisms, Properties, Structure, Technology, and Applications," R. Tao, ed., World Sci. Pub. Co., Singapore - New Jersey - London - HongKong (1992).
104. T. Fujita, J. Mochizuki and I.J. Lin, *J Magnetism and Magnetic Mater.* 122: 29 (1993).

105. D.N. Vorobiev and V.N. Makatun, "4th Int. Conf. on New Actuators", K. Lenz and H. Borgmann, ed., Axon Tech. Consult. Gmbh, Bremen (1994).
106. S. Holmstrom and S.T. Lagerwall, *Mol. Cryst. Liq. Cryst.* 38: 141 (1977).
107. T. Honda, T. Sasada and K. Kurosawa, *Jap. J. Appl. Phys.* 17: 1525 (1978).
108. T. Sasada and K. Uchida, *Kinkichem. Soc. Jap. Chem. Express.* 5: 449 (1990).
109. I.K. Yang and A.D. Shine, *J. Rheol.* 36: 1079 (1992).
110. A.J. Bur and L.J. Fetters, *Chem. Rev.* 76: 728 (1976).
111. H. Block, J.J. Gosling and S.M. Walker, *Polymer* 25: 1465 (1984).
112. W.A. Bullough, *Advanced Materials for Optics and Electronics* 1: 159 (1992).
113. H.E. Hollmann, *J. App. Phys.* 21: 402 (1950).
114. C.A. Randall, D.V. Miller, J.H. Adair and A.S. Bhalla, *J. Mater. Res.* 8: 899 (1993).
115. T. Shiga, A. Okada and T. Kurauchi, *Macromol.* 26: 6958 (1993).

# NOVEL MATERIALS FOR ELECTRORHEOLOGICAL FLUIDS

Kathleen O'Leary Havelka

The Lubrizol Corporation
29400 Lakeland Boulevard
Wickliffe, OH 44092-2298, USA

## INTRODUCTION

Since Willis Winslow's discovery in the late 1930s, the field of electrorheology has emerged as a multidisciplinary technology that has captured the imagination of scientists and engineers world-wide.[1, 2] An electrorheological (ER) fluid typically consists of a dispersion of particles in an insulating oil. The rheological properties of an ER fluid are electric field dependent. ER fluids demonstrate orders of magnitude changes in apparent viscosity in milliseconds, for a few watts of electrical power. This fast, strong, and reversible gelation provides a novel and efficient way to control motion.

The recent increase of interest in this area stems from the prospect of utilizing the ER effect to develop novel approaches to the problem of mechanical energy transfer and control. The sensitivity of ER fluids to electric field and their fast switching capabilities allow ER devices to be directly controlled by computers. Electrorheology has many distinct merits over competitive technologies, including: the ability to achieve a continuously variable wide dynamic range that is energy efficient, fast, reliable, and compact at a reasonable cost. Specifically, electrorheology allows for an extremely simple and efficient interface between electronic controls and mechanical devices without the need for moving parts as in conventional systems.

In spite of the great potential of ER technology, no commercial ER devices have reached the marketplace. A recent Department of Energy funded study concluded that the failure to commercialize this promising technology indicates serious deficiencies in understanding the materials and mechanisms responsible for the ER effect.[3] The objective of this paper is to discuss chemical and physical properties of ER fluids in order to develop fundamental knowledge of how material properties affect ER performance. This paper

*Progress in Electrorheology*, Edited by K.O. Havelka
and F.E. Filisko, Plenum Press, New York, 1995

presents a review of recent progress in materials development conducted in both Lubrizol's laboratory and by other researchers in the field of electrorheology.

## APPLICATIONS OF ER TECHNOLOGY

Potential applications of ER fluids are numerous and diverse. An applied electric field affects a variety of ER fluid properties, including: changes in the rheological, electrical, optical, thermal, volumetric, and acoustic properties. ER applications based on each of these changes have been proposed.[4, 5] However, most applications that have been studied in detail are based on utilizing the controllable rheological properties of ER fluids to engineer systems where variable performance is desired. Analysis of the literature and patents indicates that ER controllable devices such as valves, mounts, clutches, brakes, and dampers dominate the ER applications research and development.[5]

In the past decade, considerable effort in the ER community has been focused on developing ER fluids and devices for automotive applications, where the market is expected to be substantial. An example of a successful demonstration of the viability of ER technology for automotive applications is a program between Lubrizol, RheoActive, and Monroe Auto Equipment Co. where a prototype ER semi-active suspension system was installed on a vehicle, as shown in Fig. 1.[6] This semi-active system can sense road events, process the information, and affects appropriate changes in the damping force. Improved ride and handling was observed from this system. This indicates that ER fluids are an electromechanical interface that can be directly controlled by computers to provide desirable performance characteristics.

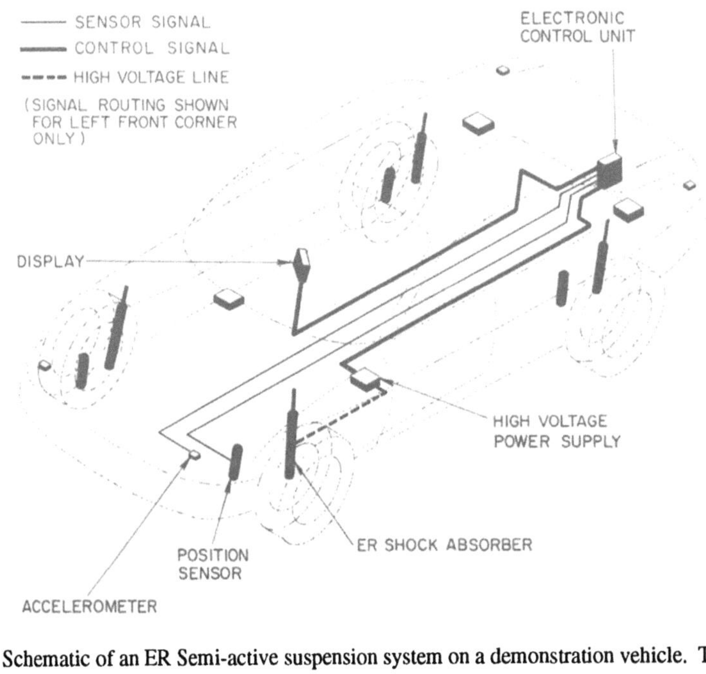


**Figure 1.** Schematic of an ER Semi-active suspension system on a demonstration vehicle. The semi-active system includes shock absorbers with an ER fluid, sensors that detect road conditions, a controller that processes the information and determines the appropriate damping, and high voltage power supplies for each wheel that facilitate independent control.

## TESTING ER FLUIDS

A widely recognized fact is that the shear stress versus shear rate properties of an ER fluid vary as a function of applied electric field. In the absence of electric field, an ER fluid demonstrates approximate Newtonian behavior. Upon application of an electric field, the fluid changes from a liquid to a semi-solid and is described as a Bingham plastic. The shear stress $\tau$ for the Bingham plastic model is given by

$$\tau = \dot{\gamma}\eta + \tau_{ER} \qquad (1)$$

where $\dot{\gamma}$ is the shear rate, $\eta$ is the viscosity, and $\tau_{ER}$ is the electric field induced dynamic yield stress.

Actual ER fluids are more complicated than this description indicates. The shear stress of an ER fluid has been shown to depend on a number of parameters, including: material properties, applied electric field, shear rate, temperature, and flow-geometry. At present, there are no standards for testing ER fluids. Often researchers report data obtained on instruments of different flow fields, gap sizes, geometries, frequencies, and shear rates. This inconsistency in test measurements makes it difficult to compare ER performance from the literature. In order to evaluate the effectiveness of ER fluids for specific applications, reporting of experimental data needs to be consistent in test methodology and terminology for each composition.

The most commonly studied applications are based on two fundamental flow geometries, Duct and Couette-flow, as shown in Table 1. Devices based on Duct-flow are characterized by shear due to fluid moving between two parallel electrodes that are stationary relative to each other, and devices based on Couette-flow are characterized by fluid being sheared between two parallel electrodes moving relative to each other.

**Table 1.** The Most Common Geometry For A Given Application

| **Duct Flow or Pressure Flow** | **Couette Flow or Shear Flow** |
|---|---|
| 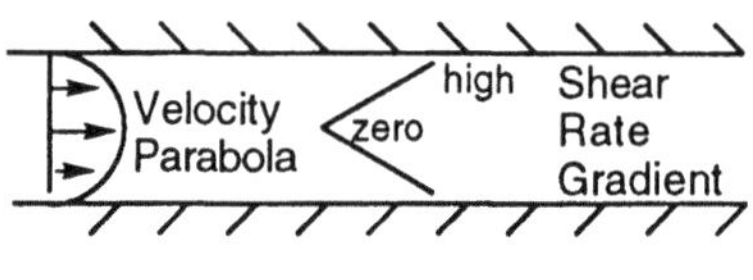  | 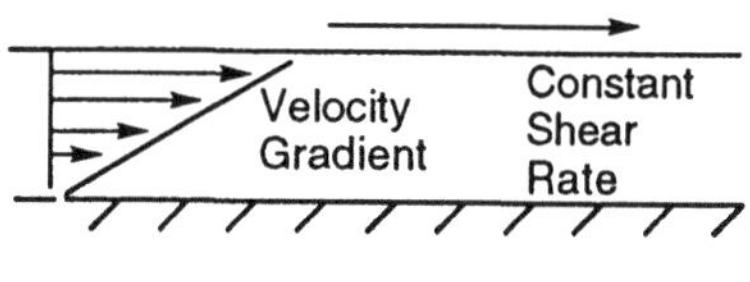  |
| Adaptive and semi-active dampers | Torque Converter Lock-up |
| Engine Mount | Cont. Variable Transmission |
| Cab Mount | Four Wheel Drive Coupling |
| Machinery Mount | Auxiliary Drive Clutch |
| Hydraulic valves | Positioning Equipment |

Increased understanding on the effects of flow geometry on ER fluids can lead to improved approximations of ER performance for a given application.

At this point a quantitative relationship that correlates ER fluid properties as a function of flow geometry does not exist. An early study provided some experimental evidence of the relationship between flow geometry and ER properties. Experimental measurements were carried out on a single ER fluid in Couette and Duct-flow geometries. The zero-field viscosity and current density measured in both geometries are approximately the same. However, the ER yield stress values varied considerably as shown in Fig. 2.[7] The effect of flow geometry is easily observed from this study. The difference in ER yield stress values from the two flow geometries indicates that an integrated systems approach is necessary to understand and predict performance of ER materials.

Lubrizol and other researchers have taken an integrated systems approach to develop and understand ER fluids. An integrated approach involves evaluating ER fluids in test cells with flow configurations that are similar to the intended application and electrode separations consistent with that in typical devices.[7] Development of novel materials in parallel with innovative device engineering allows researchers to effectively evaluate and improve ER performance.[8]

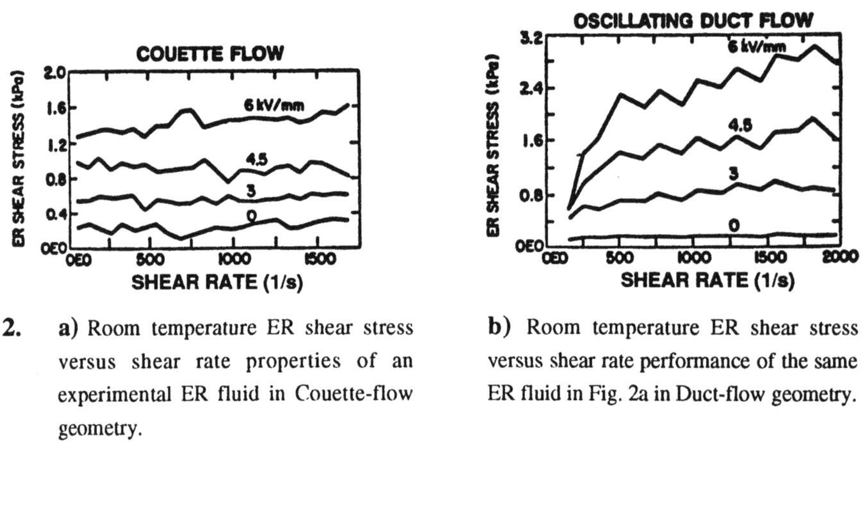


**Figure 2.** **a)** Room temperature ER shear stress versus shear rate properties of an experimental ER fluid in Couette-flow geometry. **b)** Room temperature ER shear stress versus shear rate performance of the same ER fluid in Fig. 2a in Duct-flow geometry.

## THEORIES OF THE ER PHENOMENON

ER fluids are typically composed of electrically polarizable particles dispersed in a low permittivity and low conductivity oil. Upon application of an electric field, particles become polarized and the local electric field is distorted. The migration of mobile charges to areas with greatest field intensification increases the polarizability of the particle and results in a larger dipole moment. These field induced dipoles attract one another and cause the particles to form chains or fibrillated structures in the direction of the field. These chains are held together by interparticle forces that have sufficient strength to inhibit fluid flow. Subjecting these fibers to a shearing force pulls particles apart, while charges on the particles still attract. An equilibrium is established between the formation and breaking of the chains. The increase in apparent viscosity from particle chain interactions under shear corresponds to the yield stress defined in the Bingham plastic model.[9, 10, 11] When the electric field is removed, the particles return to a random distribution and fluid flow resumes.

A generally accepted belief is that the ER phenomenon originates from electric field induced particle polarization and structure formation. Although many models for the interaction of an electric field with a dispersion of particles under shear have been proposed,[9,12,13] general consensus regarding the mechanism for the observed ER phenomenon has not been reached. The proposed mechanisms qualitatively account for a number of experimental observations.

While many factors influence ER performance, polarization and structure formation play a leading role in investigating the ER mechanism. The most widely accepted polarization mechanisms involve bulk or interfacial processes.[12, 14-17] Bulk polarization describes dipolar interactions or other forms of polarization within the confines of the particulate or continuous phase. Interfacial polarization can involve migration of charge along the surface of the particles, or within the double layer region of the dispersion. The above mechanisms may occur at varying magnitudes and rates in ER fluids and will lead to different dynamic particle interactions. The mechanisms are consistent with rheological and electrostatic models for ER activity. Increased understanding of the ER mechanism will facilitate improvements in ER material properties and fluid characteristics.

**The Electrostatic Polarization Model**

The electrostatic polarization model is a commonly employed theory to describe ER response.[12] The model assumes that ER fluids are dispersions of dielectric, non-conducting particles in a low dielectric medium and that free charges and charge transfer can be neglected. This model is based on the fact that due to the dielectric constant mismatch between the particles and the continuous phase, particles polarize and apply a force to neighboring particles.

The simplest approximation of this model is the point-dipole approximation. The point-dipole approximation treats polarization interactions between two spheres and ignores multibody interactions. If the centers of two spheres are distance r apart and orientation of the dipoles are at an angle $\theta$ from the direction of the applied electric field, each feels a force from the polarization of the other sphere. The resulting electrostatic polarization force $\mathbf{F_{el}}$ is expressed as

$$\mathbf{F_{el}} = 12\pi\varepsilon_o\varepsilon_c a^2\beta^2 E_o^2 (a/r)^4 \left[(2f_{\|}\cos^2\theta - f_{\perp}\sin^2\theta)\mathbf{e_r} + f_{\Gamma}\sin 2\theta\,\mathbf{e_\theta}\right] \quad (2)$$

where the magnitude of electrostatic polarization force depends on the, dielectric constant mismatch of the particle and base fluid $\beta$, particle size a, relative permittivity of free space $\varepsilon_o$, and relative permittivity of the continuous phase $\varepsilon_c$ and the applied electric field **E**. The applied electric field has the form $\mathbf{E} = E_0\mathbf{ez}$ where $\mathbf{e_z}$, $\mathbf{e_r}$ and $\mathbf{e_\theta}$ are unit vectors pointed in the z, r, and θ directions, respectively. The $f_{\parallel}$ and $f_{\perp}$ are dimensionless functions of a/r and $\varepsilon_o/\varepsilon_c$.

In practice, ER fluids are more complex than suggested by this model. ER fluids have finite conductivity and charge often is transferred across electrodes. A number of studies have expanded this model to include conductivity and other parameters.[18, 19] The polarization model and expanded models qualitatively account for a number of phenomenona such as suspension structure. However, the models cannot quantitatively predict ER performance based on material properties.

## MATERIAL PROPERTIES

A vast number of materials of very different chemical nature demonstrate ER activity. An ER fluid typically is a dispersion of particles, 1 to 100 μm in size, in an insulating oil. Most researchers find a volume fraction between 10 and 40 percent to be the optimum particle loading for a practical balance between zero field viscosity and dynamic yield stress.[15] When formulating an ER fluid using a particular material a number of properties need to be optimized, as shown in Table 2.

In order to accurately compare the effectiveness of a particular material in an ER fluid, materials must be tested and reported in a consistent manner. A number of interdependent performance properties have been identified that need to be evaluated to determine commercial viability of a particular ER fluid. Although it is likely that ER fluid properties will need to be tailored for specific applications, the values shown in Table 2 are consistent with performance requirements typically found in the literature for a variety of applications.[7]

**Table 2.** ER Fluid Requirements for a Variety of Applications

| ER Fluid Performance | Value |
|---|---|
| Dynamic yield stress at 6.0 kV/mm | > 4.0 kPa |
| zero-field viscosity | < 100 cP |
| Current density at 6.0 kV/mm (DC) | < 300 $mA/m^2$ |
| Operating temperature range | -20 to 140°C |
| Response Time | milliseconds |
| Dispersion Properties | stable |
| Tribology properties | non-abrasive |
| Environmental properties | non-toxic |

ER fluids consist of dielectric particles surrounded by an insulating medium, and in a device they essentially function as leaky capacitors. The transfer of charge between particles results in the measurement of an electric current through the fluid. The current density associated with a particular ER fluid is useful for estimating the power consumption of devices using this fluid. Reducing the increase in current density as a function of temperature without adversely affecting the other ER properties has been the focus of much of the materials research conducted at Lubrizol and in other laboratories.

For simplicity, research in ER particulate materials can be classified into two general categories: (i) Extrinsically polarizable materials that require an activator and most likely involve ionic polarization and (ii) Intrinsically polarizable materials that do not require an activator and involve electronic polarization. These two material types are discussed in more detail in the following sections.

**Extrinsically Polarizable Materials**

It was originally felt that water was necessary for the ER effect.[20] Extrinsically polarizable ER fluids are composed of low dielectric particles that require water or a polar-activator such as alcohols to obtain measurable ER activity. Particles such as silica, polysaccharides, polyelectrolytes, and other porous supports when activated demonstrate an ER effect.[15] Although suitable to demonstrate the ER phenomenon, extrinsically polarizable ER fluids have significant limitations at low and high temperatures. At both extremes the polar activator frequently is not available to activate the ER fluid due to the activators inherent thermal properties. However, an even more serious concern is the very high thermal coefficient of conductance that extrinsically polarizable ER fluids show. Frequently, the fluid conductivity doubles every 5 degrees.[21]

Understanding the role of the polar activator is necessary to develop improved fluids. The most commonly accepted mechanism for polar activated systems involves interfacial polarization.[15] Polar activators are speculated to enhance ionic migration or alignment of dipoles. One theory considers the capillary forces of water in ER fluids. It suggests that the surface tension between the water and the solvent leads to a cohesive force that holds the particles together.[20] More studies are necessary to explain particle-activator interactions.

Information on the complex dielectric of polar activated particles would aid in understanding the interactions between the particle and the polar activator. The conductivities of particles associated with an activator are not known and the measurements present a considerable challenge. Accurate dielectric measurement of activated particles and of extrinsically polarizable ER fluids will increase the understanding of how these materials polarize in an electric field.

**Intrinsically Polarizable Materials**

The discovery that water is not necessary to obtain ER activity catalyzed a resurgence of interest in electrorheology.[17, 22] Intrinsically polarizable ER fluids offer the possibility of expanding the temperature range for ER activity. Intrinsically polarizable materials that have been investigated include ferroelectrics, e.g., $BaTiO_3$, other high dielectric inorganics, e.g., zeolites,[22] and electronic semiconductors.[17, 23, 24]

Electronic semiconductors, such as polyacenequinones and polyaniline, have a much lower thermal coefficient of conductance than extrinsically polarizable systems and demonstrate a measurable ER response. The drawbacks cited for semiconducting polymers in the electronics industry, such as low conductivity and insolubility, are attractive for electrorheological materials. The synthetic versatility of semiconducting polymers, such as polyaniline, enables researchers to alter its permittivity and conductivity. This ability to tailor electrical properties may allow researchers to obtain the desired rate and magnitude of polarization for ER performance.

A number of studies have been conducted at Lubrizol and by other researchers on polyaniline to determine how the complex dielectric affects ER performance.[17, 25-27] Polyaniline's conductivity was varied by adjusting dopant level and type. Electrical properties were shown to affect the rate and magnitude of polarization.[27] The ER results are consistent with the polarization model. However, more research is needed to quantify the effect of polarization and material properties on ER performance.

**Recent Advances in ER Materials**

Lowering current density continues to be a challenge in both extrinsically and intrinsically polarizable ER fluids. In an attempt to reduce the current density of ER fluids, insulating coatings and surfactants have been investigated. Since particle-particle separation plays an important role in the polarization force, a strong dependence $(a/r)^4$ is indicated in the electrostatic polarization model,[12] it is expected that ER shear stress will also decrease from this approach. However, in a number of systems this compromise is not observed.

Polysaccharide coatings on polyaniline dispersions demonstrate desirable ER properties, as shown in Fig. 3. These properties include reduced current density, increased steady-state ER shear stress values, and reduced zero-field viscosity.[28] $20\%_w$ blends of polyaniline coated with a polysaccharide were tested in a duct-flow device with a 1 mm electrode separation. The test was run over a 20 to 80°C temperature range. Data were collected over a shear rate range from 1,200 $s^{-1}$ to 37,000 $s^{-1}$ at 6 kV/mm. The current density and dynamic shear stress values were taken at 25,000 $s^{-1}$. The results presented in Fig. 3 indicate that appropriate choice of particle and coating leads to improved ER performance.

On the basis of polarization alone, highly conducting particles such as metals dispersed in an insulating base fluid should provide optimum ER shear stresses. In practice, dispersions of metal particles are very conductive and do not provide fluids of significant stress values.[29] Researchers at Lubrizol felt that a very thin coating or oxide layer on the metal was needed to investigate bulk and surface polarization. The investigations using dc electric fields indicated that uncoated metal particles were too conductive to test. Therefore a coating was put on the metal particles to reduce the current density. Even with the insulating coating, the metal particles demonstrated weak ER shear stresses. The failure of good conductors to make good ER substrates implies that there may be an optimum level of conductivity in the semi-conducting region or that the coatings were not adequate.[21] More work is needed to understand particle-coating interactions, and its affect on ER performance.

Research conducted at Lubrizol indicates that one can dramatically increase ER shear stress by preparing novel materials and innovative formulations. ER shear stresses of 50 kPa have been predicted,[30] and dynamic shear stress values ranging from 15 to 20 kPa have been observed at Lubrizol using high electric fields.[31, 32] However, extremely high current densities have been associated with the high shear stress values. An example of an experimental fluid that had 20 kPa shear stress is shown in Fig. 4.[31] The experimental fluid was tested at 20°C in the duct-flow device over a shear rate range of 1,200 $s^{-1}$ to 37,000 $s^{-1}$

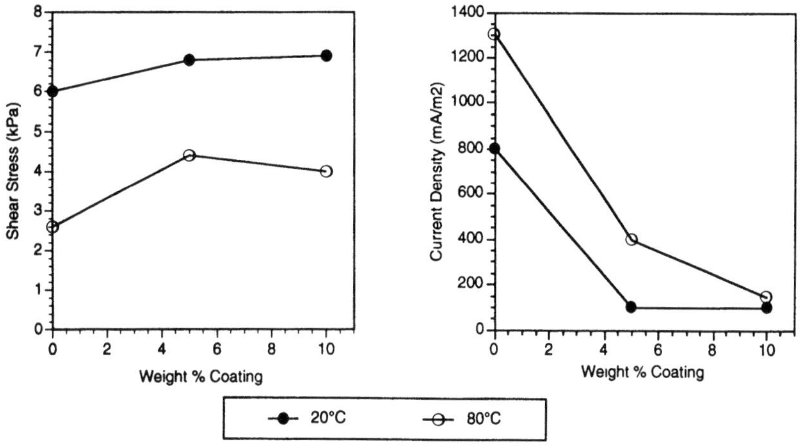


**Figure 3.** **a)** ER shear stress versus weight percent coating on polyaniline tested at 20 and 80°C.

**b)** Current density versus weight percent coating on polyaniline tested at 20 and 80°C.

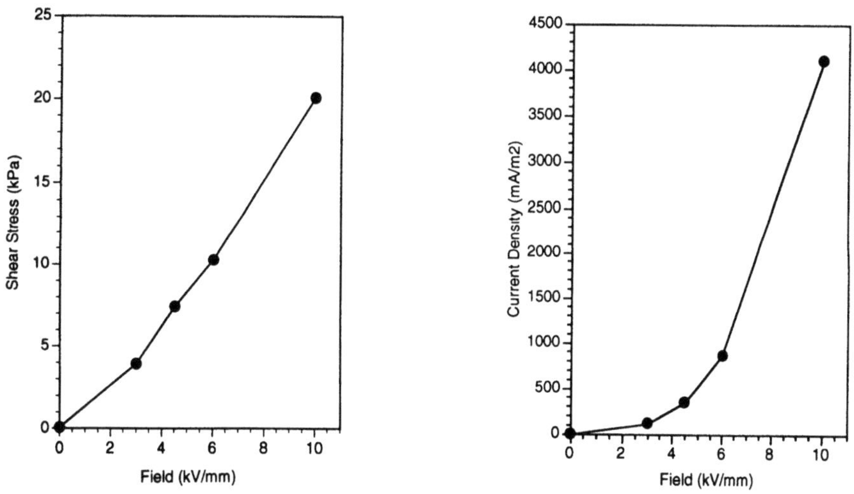


**Figure 4.** **a)** ER yield stress versus electric field on an experimental ER fluid tested at 20°C.

**b)** Current density versus electric field on an experimental ER fluid tested at 20°C.

at electric fields of 3, 4.5, 6, and 10 kV/mm. While the high shear stress value of 20 kPa is exciting from an intellectual standpoint, the high current density associated with the fluid is an order of magnitude greater than acceptable for most applications. This indicates that all of the performance properties of an ER fluid must be evaluated and reported to determine commercial viability.

Dispersion and ER properties need to be optimized together. Since micron size particles tend to settle and many of the applications may promote settling through vibration or centrifugation of the fluid, dispersion stability is a challenging requirement. A number of approaches have been investigated to reduce settling and possibly improve redispersibility, including: all liquid systems, density matching of base fluid and particle, steric stabilization, electrostatic stabilization, and additives that are shear thinning.[3, 15, 33] Each of these approaches will most likely affect both the dispersion properties and the ER properties, including: ER yield stress, current density, and zero field viscosity. More research is needed to understand how dispersion and ER properties can be optimized together.[34]

### All Liquid ER Fluids

Some encouraging advances have been made in all liquid ER fluids based on liquid crystalline polymer (LCP) solutions.[33] This one-phase system eliminates particle settling problems inherent to conventional ER fluids. The ER effect in LCPs is attributed to electric field-induced orientation of the molecular axis[33] or orientation of the mesogen groups.[35] Shine et. al., reported that a solution of $26\%_w$ poly(n-hexyl isocyanate) in p-xylene showed a viscosity enhancement of 35 times. This is exciting from both a performance and phenomenological standpoint. Increases like this are relatively small for ER dispersions but are novel for one-phase systems. Based on these results, further investigations should be conducted on LCP with a rigid backbone structure, a large permanent axial dipole moment, and solubility in non-polar solvents to develop a better understanding of this system.

## CONCLUSION

ER fluids provide a unique electromechanical interface for which exciting new classes of devices are being engineered. Success in developing a commercially viable ER systems depends on both novel fluid formulation and innovative device engineering, hence the importance of the integrated systems approach that is being pursued by Lubrizol and others in the field. Developing an appropriate ER fluid for a specific application means balancing often conflicting requirements, such as high yield stress and low power consumption. The key to improved ER fluid performance lies in developing a fundamental understanding of the underlying ER mechanism. Improved understanding of the mechanism will enable researchers to design, synthesize, and formulate ER fluids with desired performance properties.

## ACKNOWLEDGMENTS

I would like to thank the members of the ER Fluids Team at Lubrizol and RheoActive, and Chip Zukoski for their valuable research contributions.

## REFERENCES


1. Winslow, W. M. U.S. Patent 2 417 850, 1947.
2. Winslow, W. M. *J. Appl. Phys.* **1949**, *20* 1137.
3. *Electrorheological (ER) Fluids:* 1993; U.S. Department of Energy. Office of Energy Research. Office of Program Analysis. U.S. Government Printing Office: Washington DC, 1993; DE-AC02-91ER30172.
4. Shulman, Z. P.; Gorodkin, R. G.; Korobko, E.V.; Gleb, V. K. *J. of Non-Newtonian Fluid Mechanics* **1981**, *8*, 29.
5. Weiss, K. D.; Carlson, J. D.; Coulter, J. P. *J. of Intelligent Materials Systems and Structures* **1993** *4*, 13-34 and 248-259.
6. Petek, N. K.; Romstadt, D. J.; Lizell, M. B.; Weyenberg, T. R. *SAE Paper 950586* (1995).
7. Boyle, F. P. In *Proc. Int. Conf. Electrorheological Fluids*; Tao, R., Ed.; World Scientific: Singapore, 1992; p. 236.
8. Petek, N. K. U.S. Patent 5259487, 1993.
9. Klingenberg, D. J. and Zukoski, C. F. *Langmuir*, **1990**, *6*, 15.
10. Deinega, Y. F. and Vinagradov, G. V. *Rheo. Acta* **1984**, 23, 636.
11. Shulman, Z. P.; Matsepuro, A. D.; Novichenok, L. N.; Demchuk, S. A. and Svirnovskaya, I. L. *J. Eng. Phys.*, **1974**, *27(6)*, 1569.
12. Gast, A. P. and Zukoski, C. F. *Adv. Coll. Int. Sci.* **1989**, *30,* 153.
13. Brooks, D.; Goodwin, J.; Hjelm, C.; Marshall, L. and Zukoski, C. *Colloids and Surfaces*, **1986**. *18*. 293.
14. Klass, D.L. and Martinek, T.W. *J. Appl.* Phys. **1967a**, *38*, 67.
15. Block, H.; Kelly, J. P. *J Phys. D.* **1988**, *21*, 1661.
16. Klass, D.L. and Martinek, T.W. *J. Appl.* Phys. **1967b**, *38*, 75.
17. Block, H.; Kelly, J.P.; Qin, A.; Watson, T. *Langmuir* **1990**, *6*, 6.
18. Anderson, R. A. In *Proc. Int. Conf. Electrorheological Fluids*; Tao, R., Ed.; World Scientific: Singapore, 1992; p. 81.
19. Zukoski, C. F. *Annu. Rev. Mater. Sci.* **1993**, *23*, 45.
20. Stangroom, *J. E. Phys. Technol.* **1983**, *14*, 290; See, H.; Tamura, H.; Doi, M. *J. Phys. D: Appl. Phys.*, **1993**, *26*, 746.
21. Blackwood, K. M. and Block, H. *TRIP* **1993**, 1(4), 98.
22. Filisko, F. E. and Radzilowski, L. H. *J. of Rheology*, **1990**, *34(4)*, 539.
23. Block, H.; Kelly, J. P. U.S. Patent 4 687 589, 1987.
24. Pialet, J. W.; Bryant, C. P.; Lal, K. WO 9307244.
25. Gow, C. J. Zukoski, C. F. *J. Colloid Interface Sci.*, **1990**, *136*, 175.
26. Block, H.; Rattray, P.; Watson, T. In *Proc. Int. Conf Electrorheological Fluids*; Tao, R., Ed.; World Scientific: Singapore, 1992; p. 93.
27. Webber, R. M. *Polymer Preprints*, **1994**, *35(2)*, 387.
28. Pollack R. A. U.S. Patent applied for.
29. Kelly, J. P. MSc Thesis, University of Liverpool, 1986.
30. Conrad, H.; Chen, Y.; and Sprecher, A.F. In *Proc. Int. Conf. Electrorheological Fluids*; Tao, R., Ed.; World Scientific: Singapore, 1992; p. 195.
31. Schober, B. J.; Pialet, J. W.; Piedrahita, C. A. U.S. Patent applied for.
32. Havelka, K. O.; Collins, E. A. U.S. Patent applied for.
33. Shine, A. D.; Yang, I. K. ; Tse, K. L. *Electrorheological Flow-1993, ASME Fluids Engineering Division,* **1993**, *164*, 49.

34. Webber, R. M.; Kornbrekke, R. E.; Schober, B. J.; Soukup, B. L.; Kirchner, T. M. *68th ACS Colloid & Surface Science Symposium*, Stanford University, Stanford, CA, June 1994.
35. Minegawa, K; Kimura, H.; and Koyama, K *Polymer Preprints,* **1994**, *35(2)*, 381.

# ELECTRICAL PROPERTIES AND THE STRENGTH OF ELECTRORHEOLOGICAL (ER) FLUIDS

Hans Conrad and Ying Chen

Materials Science and Engineering Department
North Carolina State University
Raleigh, N. C. 27695–7907

## 1. INTRODUCTION

The shear resistance or flow stress $\tau$ of electrorheological (ER) fluid suspensions can be considered to consist of two components

$$\tau = \tau_E + \tau_v \tag{1}$$

where $\tau_E$ is the component which contains all effects of the electric field. $\tau_v$ is the flow stress at zero electric field and is usually termed the hydrodynamic or viscous component:

$$\tau_v = \eta\dot{\gamma} \tag{2}$$

where $\eta$ is the viscosity of the fluid and $\gamma$ the shear rate. Assuming that $\tau_E$ reflects a polarization mechanism, theoretical treatments give [1–3]

$$\tau_E \propto \phi K_f E^2 f(\beta) \tag{3}$$

where $\phi$ is the volume fraction of particles, $K_f$ the dielectric permittivity of the host liquid, E the electric field and $\beta$ the dielectric mismatch parameter given by

$$\beta = (K_p - K_f)/(K_p + 2K_f) \tag{4}$$

$K_p$ is the dielectric permittivity of the particles.

In many cases ER behavior is associated with significant conductivity of the suspension, resulting either from the bulk conductivity of the particles or from interfacial conductivity at the particle-host liquid interface. Such conductivities require that $\beta$ be given in terms of the complex permittivities of both the host liquid and the particles, i.e. in terms of

*Progress in Electrorheology*, Edited by K.O. Havelka and F.E. Filisko, Plenum Press, New York, 1995

$$K^* = K' - iK'' = K' - i\sigma / 2\pi f\varepsilon_o \quad (5)$$

where K' and K" are the real and loss components, $\sigma$ the conductivity, f the frequency of the alternating electric field, $\varepsilon_0$ the permittivity of free space and $i = \sqrt{-1}$. The conductivity usually leads to an electric current between the electrodes, which increases continuously with electric field and temperature. In contrast, the strength component $\tau_E$ generally goes through a maximum with increase in temperature or other factors which give rise to an increase in conductivity [4–6].

Additional factors which may affect $\tau_E$ but are not directly included in Eqns. 3–5 are listed in Table 1. The references given therewith provide only a few examples and are not intended to be all-inclusive. It is generally assumed that the influence of these additional factors on $\tau_E$ is through the electrical properties in the complex mismatch parameter

$$\beta^* = \left(K_p^* - K_f^*\right) / \left(K_p^* + 2K_f^*\right) \quad (6)$$

A major focus of our research on ER fluids has been to characterize these additional effects, with the objective of identifying the operative mechanism. Special attention has been directed to providing information which contributes to resolving the following questions:

(a) What relationship exists between the permittivity measured with a dc electric field compared to an a.c. field?
(b) Does the permittivity of an aligned or fibril arrangement of the particles differ from that of a random dispersion?
(c) Although $K_f'$, $K_f''$ and $\sigma_f$ can be measured directly on the host liquid, the corresponding properties for the particles must generally be deduced from those of the ER suspension. What are the rigorously correct relationships which give $K_p'$, $K_p''$ and $\sigma_p$ of the particles from $\overline{K}'$, $\overline{K}''$ and $\overline{\sigma}$ of the suspension?
(d) What relationship exists between $\tau_E$ and the current density $\overline{j}$ or conductivity $\overline{\sigma}$ of the suspension?

To be noted is that a bar over the electrical property will be used to refer to the suspension throughout the paper.

**Table 1.** Additional factors which may influence $\tau_E$.

| | | |
|---|---|---|
| Str. | = | Structure (geometrical arrangement of particles, e.g. single row chains or clusters) [7–9] |
| C | = | Composition of the particles, including $H_2O$ content and other additives [4–12] |
| $\dot{\gamma}$ | = | Shear rate [8] |
| T | = | Temperature [5,11] |
| f | = | Frequency of applied electric field [12] |
| j | = | Current density in the ER fluid [4–6] |

The present paper addresses the above-listed questions referred mainly to suspensions of zeolite particles in various host oils. These particular suspensions were chosen since the authors and their coworkers had previously obtained considerable data on their ER response [4,5,7,8,13]. To reduce hydrodynamic effects, only $\tau_E$ at low shear rates are considered, i.e., in the quasi-static regime with $\dot{\gamma} < 10\ s^{-1}$ [8], where $\tau_E >> \tau_v$ and the angular rotation of the particles during shear is small. Further, $\tau_E$ in the post-yield region of shear stress vs shear strain curves at constant shear rate will mainly be considered, where a steady-state condition is approximated [8].

## 2. EXPERIMENTAL

### 2.1 Materials

The host oils employed in the present studies are presented in Table 2; included are their viscosity and d.c. electrical properties determined in the manner described below. The listed values are in agreement with those provided by the suppliers. Regarding the conductivity $\sigma_f$, it is important to note that the oils generally exhibited non-ohmic behavior in which case at high electric fields the value given by j/E is usually less than the instantaneous value dj/dE. To reduce and standardize the water content of the oils, they were generally heated 1.5–2 hr at 100°C prior to mixing with the zeolite particles.

**Table 2.** The host oils used to prepare the zeolite/oil suspensions, their viscosity and their d.c. electric properties at 25°C.

| Oil | Source | $\eta_f$ (Pa-s) | $K_f$ (E≈2 V/mm) | $\sigma_f$ (E=2kV/mm) (S/m) j/E | dj/dE |
|---|---|---|---|---|---|
| Mineral | Ward Transformer Co. | 0.035 | 2.19 | $0.7\text{x}10^{-12}$ | $1.6\text{x}10^{-12}$ |
| Silicone | Dow Corning | 0.053 | 2.61 | $2.5\text{x}10^{-12}$ | $3.6\text{x}10^{-12}$ |
| Corn | CPC International Inc. | 0.080 | 3.28 | $4.1\text{x}10^{-11}$ | $4.7\text{x}10^{-11}$ |
| Caster | High Energy Corp. | 0.980 | 4.24 | $2.4\text{x}10^{-10}$ | $2.6\text{x}10^{-10}$ |
| DOP | High Energy Corp. | 0.071 | 5.35 | $2.1\text{x}10^{-10}$ | $2.9\text{x}10^{-10}$ |
| Chlorinated Hydorcarbon | Dover Chemical Co. | 0.055 | 7.20 | $2.3\text{x}10^{-8}$ | $2.3\text{x}10^{-8}$ |

$\eta_f$ = viscosity; $K_f$ = dielectric permittivity; $\sigma_f$ = electrical conductivity; j = current density; E = electric field.

The zeolite particles were purchased from the Sigma Chemical Company and are designated Z3125. Information provided by the supplier indicated that their chemical composition was $Na_{12}[(Al_2O_3)_{12}(SiO_2)_{12}]$ x $H_2O$ with a maximum of 25 wt.% $H_2O$. This composition is in keeping with $Na_{12}$–A zeolite whose crystal structure consists of a framework composed of truncated octahedra joined in a cubic array [14–16]; see for example Fig. 1. The truncated cube octahedron has an internal cavity of 11 Å in dia. ($\alpha$ cage) which is entered through 6 apertures (4.2 Å dia.) formed by a nearly regular ring of 8 oxygen atoms. This gives a 3-dimensional pattern of unduloid-like channels with a maximum dia. of 11 Å and a minimum of 4.2 Å. The framework density of the zeolite is 1.27 g/cm$^3$, the dehydrated crystal density is 1.99 g/cm$^3$, the void volume is 0.30 cm$^3$/g and the void fraction 0.47. Water is mainly absorbed in the channels of zeolite particles. In our experiments the water content of the particles was varied by heating in air the as-received zeolite particles for 6 hours at various temperatures and comparing the weight loss at a given temperature with that at 750°C [17], where it was assumed that the retained water was ~ 0 wt.% [15].

The appearance of the Z3125 zeolite particles observed by SEM is shown in Fig. 2. They are generally cubic in shape with some rounding of the edges. Their mean linear intercept size is 2.1 μm with a relatively uniform distribution of sizes about this average, i.e., there was no indication of a bimodal distribution. The typical chain or fibril structure which formed with application of an electric field is illustrated in Fig. 3.

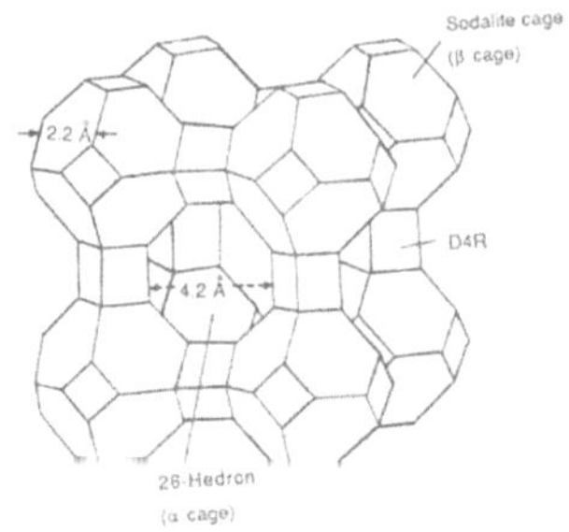


Fig. 1 Structure of zeolite A. From Dyer [16].

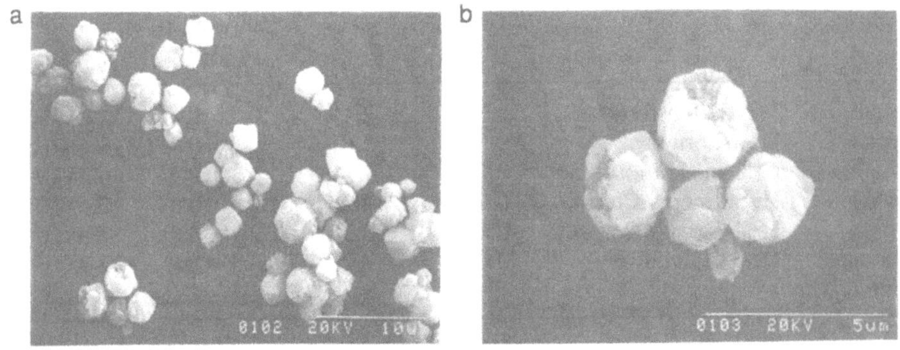


Fig. 2 SEM micrographs of the Z3125 zeolite at two magnifications.

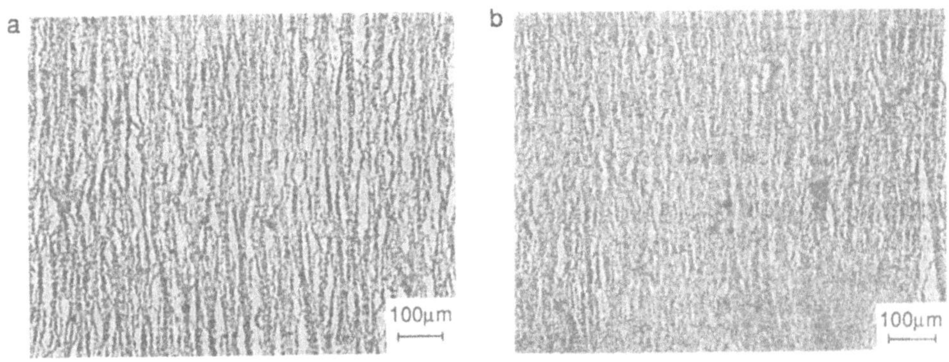


Fig. 3 Optical micrographs of the fibril structure of zeolite/silicone oil suspensions with E = 2 kV/mm at 25°C: (a) $\phi = 0.045$ and (b) $\phi = 0.084$.

The ER fluid suspensions were prepared by thoroughly mixing the previously heated zeolite particles in the dried oils and storing the suspensions in sealed glass jars for testing, which usually took place within 24–48 hr following mixing. It was established that the ER properties of the suspensions did not change noticeably during the time between mixing and testing. The test variables considered were: (a) the host oil, (b) the water content of the zeolite particles, (c) the volume fraction of particles, (d) the test temperature, (e) the magnitude and frequency of the electric field and (f) the influence of heating the prepared suspensions prior to testing at 25°C.

## 2.2 Electrical Properties

**2.2.1 Dielectric Permittivity.** Three methods were employed to measure the dielectric constants of the host oils and the various suspensions: (a) CIE 7705 capacitance meter at a low voltage (~ 2V/mm) dc [5], (b) Hewlett-Packard Model 4192A LF impedance analyzer at low voltage (~ 1V/mm) ac and (c) measurement of resistance and reactive components of the time-varying current and voltage at high voltage (0.5–3 kV/mm) ac [5]. Both random and fibril arrangements of the particles were measured by Method a. The random arrangement was obtained by vigorously stirring the suspension prior to the measurement, whereas the aligned or fibril arrangement was obtained by subjecting the ER fluid to a high electric field immediately prior to the measurement at the low field. The high electric field was turned off both prior to shearing the suspension and following straining. Only the random arrangement was measured with the impedance analyzer, while only the fibril structure existed with the high ac field used for the time-varying current and voltage measurements.

Fig. 4 shows the results obtained with the impedance analyzer (Method b) for the host silicone oil and for suspensions of zeolite particles with three water contents in the oil. Also included are the values for the dc measurements (Method a) on the random arrangement of the particles. The following are noted from Fig. 4: (a) the values of $K_f'$ and $K_f''$ ($\approx 0$) obtained for the silicone oil are in accord with those given by the supplier, providing support for the accuracy of our measurements and (b) the variation of $\bar{K}'$ and $\bar{K}''$ with frequency for the suspensions are in keeping with a Maxwell-Wagner interfacial polarization mechanism [18–20]. The values of the relaxation time $\bar{t}_R = 1/2\pi f_m$ given by the frequency $f_m$ at which $\bar{K}''$ is a maximum ranges from $1.0 \times 10^{-4}$ to $2.5 \times 10^{-7}$ $s^{-1}$ as the water content of the zeolite particles is increased from 3 to 9 wt.%.

Further support that the dielectric dispersion in Fig. 4 is in accord with Maxwell-Wagner interfacial polarization is provided by the Cole-Cole plots [21] of Fig. 5. The data points for the three water contents fall on segments of circular arcs whose centers lie below the abscissa. In Fig. 5 the small circles are the actual data points, while the triangles give the corrected data points assuming that the value of $\bar{K}''$ at low frequency includes a conduction contribution $\bar{K}_c'' = \bar{\sigma}/2\pi f\varepsilon_o$. The method by which the conduction of the suspension $\bar{\sigma}$ was determined is given in the next section. By substracting this conduction contribution from the measured $\bar{K}''$, a better fit is obtained to the circular arc for all three water contents, with a rather good fit at 6.7 and 9 wt.% $H_2O$.

The angle between the radius of the circular arc and the abscissa in Fig. 5 defines the quantity $\alpha\pi/2$, where $\alpha$ is a measure of the range associated with the symmetrical distribution of relaxation times. It is the exponent in the expression for the complex permittivity [21]

$$K^* = \frac{K_s - K_\infty}{1 + (i2\pi f t_R)^{1-\alpha}} \qquad 0 < \alpha < 1 \tag{7}$$

where $K_s$ is the dielectric permittivity at very low ac electric field frequency and $K_\infty$ that at high frequency. The $\alpha$ value decreases with water content of the zeolite particles being 0.52, 0.47 and 0.43 for 3, 6.7 and 9 wt.% $H_2O$, respectively.

A further check on the validity of the Cole-Cole relationship is to measure the lengths of the chords U and V for the semicircular arc diagram in Fig. 5 and plot log (V/U) vs log ($\omega = 2\pi f$). This should yield a straight line with slope ($1-\alpha$) and intercept ($1-\alpha$) log $t_R$ [22]. Such plots were made and yielded straight lines, giving the following values of $\alpha$ and $t_R$ for the three increasing water contents respectively: $\alpha$ = 0.49, 0.45, 0.40 and $t_R$ = 5.5 x $10^{-5}$, 2.3 x $10^{-6}$ and 2.1 x $10^{-7}$ s. These values are consistent with those taken directly from the Cole-Cole plots of Fig. 5 and the peaks in $\overline{K}''$ in Fig. 4.

Figs. 4 and 5 thus indicate that the dielectric behavior of random zeolite/silicone oil suspensions at low electric fields is in accord with Maxwell-Wagner interfacial polarization and moreover that $\overline{K}_{dc} \approx \overline{K}_s'$ for low frequency ac. However, pronounced ER behavior in these suspensions only occurs at E > 0.5 kV/mm, where the particles are arranged in chains or fibrils extending between the electrodes. Fig. 6 shows that $\overline{K}_{dc}$ and $\overline{K}_s'$ are larger for the fibril structure compared to the random arrangement. Evident in Fig. 7 is that these dielectric constants increase with water content of the zeolite particles, but for both arrangements it appears that they attain a maximum value at ~ 8–9 wt.% $H_2O$. Also seen is that the difference in dielectric permittivity between the fibril and random arrangements increases with water content of the particles.

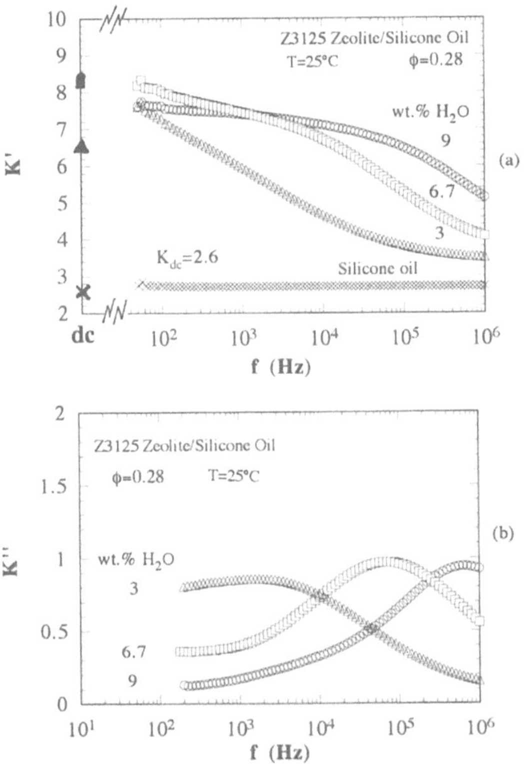


Fig. 4 The dielectric constants, $\overline{K}'$ and $\overline{K}''$ for the silicone oil and suspensions ($\phi$ = 0.28) of zeolite particles with three water contents and randomly arranged in silicone oil measured by Method b at 25°C. Also included are the values of the permittivity $\overline{K}_{dc}$ determined with low voltage dc (Method a).

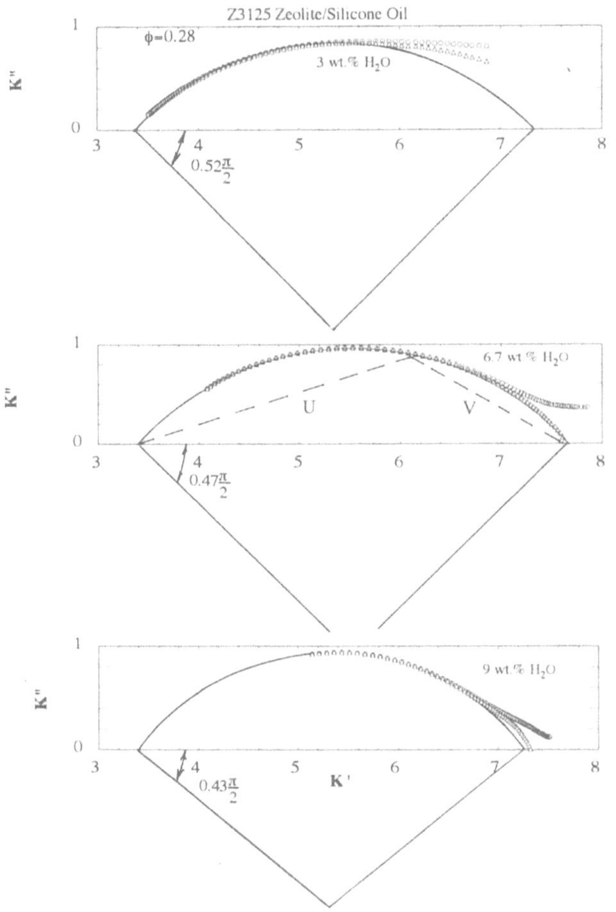


Fig. 5 Cole-Cole plots of the dielectric dispersions shown in Fig. 4. O the actual data points, Δ the corrected values of $\overline{K}''$ obtained by subtracting $\overline{K}_c'' = \overline{\sigma}/2\pi f\varepsilon_o$ from the measured values.

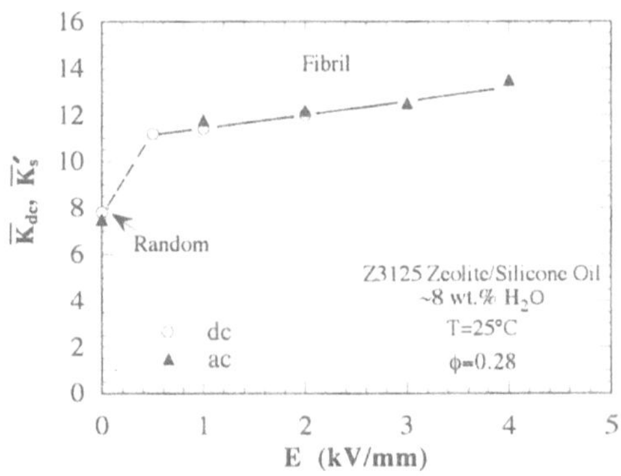


Fig. 6 Effect of electric field, and in turn the arrangement of particles, on $\overline{K}_{dc}$ and $\overline{K}_s'$ for a suspension ($\phi$ = 0.28) of zeolite (~ 8 wt.% $H_2O$) in silicone oil at 25°C. All dc measurements were by Method a; the ac measurements were by Method b for the random arrangement and Method c for the fibril structure.

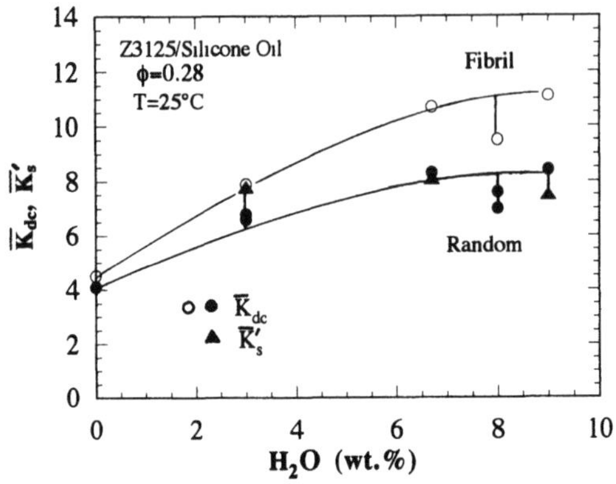


Fig. 7 Effect of water content of the zeolite particles on $\overline{K}_{dc}$ and $\overline{K}_s'$ at 25°C of suspensions in silicone oil. Measurements on the random arrangement of the particles were by Methods a and c; fibril arrangement by Method a.

The effects of volume fraction $\phi$ of particles and dielectric constant $K_f$ of the host oil on $\overline{K}_{dc}$ for the random arrangement using Method a is presented in Fig. 8. The data in Fig. 8a suggest that $\overline{K}_{dc}$ increases in a linear fashion with $\phi$ for the various oils considered, whereas those in Fig. 8b indicate that the concentration effect is approximately proportional to the dielectric constant of the oil, giving for zeolite particles with ~ 4 wt.% $H_2O$ tested at 25°C

$$\overline{K}_{dc} = K_f(1+6\phi) \tag{8}$$

The effect of temperature on the dielectric constant of the silicone oil $K_f$ and on the dielectric constants $\overline{K}_{dc}$ and $\overline{K}_s'$ for the suspension ($\phi$ = 0.28) of zeolite particles (3 wt.% $H_2O$) in silicone oil is shown in Fig. 9. Evident is that $K_f$ decreases slightly with temperature, whereas $\overline{K}_s'$ and $\overline{K}_{dc}$ increase with temperature to ~ 100°C and then remain constant to 160°C. Similar effect of temperature occurs for the fibril structure as for the random arrangement of particles. Again, at all temperatures the permittivity of the fibril structure is larger than that of the random, the difference in this case increasing with temperature. Also, again $\overline{K}_{dc} \approx \overline{K}_s'$ (ac).

Fig. 10 shows the effect of frequency f of the electric field in the low frequency regime on $K_f$ of several oils and on $\overline{K}_s'$ of suspensions ($\phi$ = 0.28) of zeolite particles (8 wt.% $H_2O$) in silicone and caster oils. It is here seen that the dielectric permittivities of the oils and the suspensions are relatively independent of frequency over the range considered. Again, $\overline{K}_{dc}$ is approximately equal to $\overline{K}_s'$.

The following is a summary of the results obtained regarding the dielectric constants of suspensions consisting of zeolite particles in various oils:

(a) The dielectric dispersion is in accord with a Maxwell-Wagner interfacial polarization mechanism.

(b) The dielectric constant obtained with dc is in accord with that obtained with low frequency (~ 50 Hz) ac, i.e. $\overline{K}_{dc} \approx \overline{K}_s'$ (a.c.)

(c) The dielectric constant for a fibril arrangement of the particles is up to 50% greater than that for a random arrangement.

(d) The dc and low frequency ac dielectric constants increase with $\phi$, $H_2O$ content of the particles, temperature and $K_f$, with maximum values for a given $\phi$ and $K_f$ occurring at ~ 9 wt.% $H_2O$ and T ≈ 100°C

(e) $\overline{K}_{dc} \equiv \overline{K}_s' \approx K_f(1 + 6\phi)$ at 25°C and with $H_2O$ content of the particles being ~ 4 wt.%.

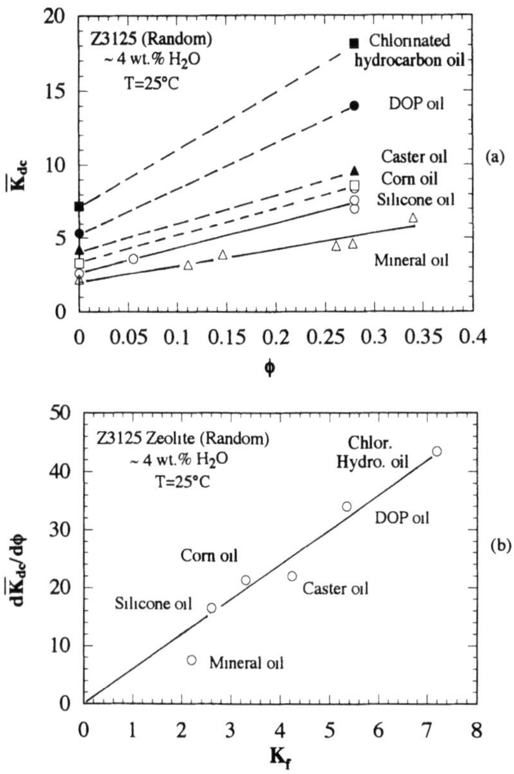


Fig. 8 (a) Effect of volume fraction of zeolite particles in various oils on $\overline{K}_{dc}$ for a random arrangement of the particles using Method a. (b) The slopes of the straight lines in Fig. 8a vs the dielectric constant $K_f$ of the oils.

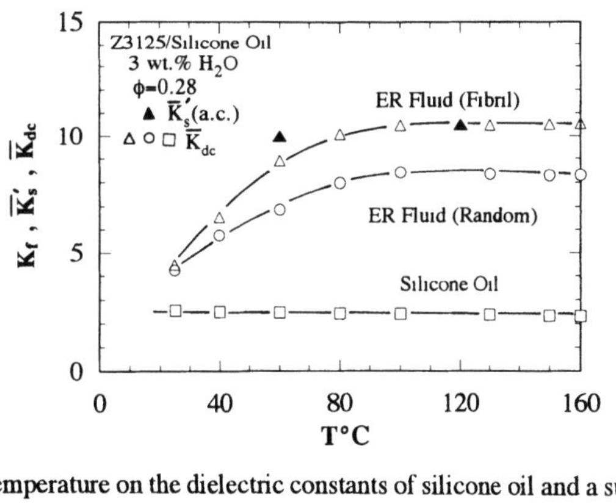


Fig. 9 Effect of temperature on the dielectric constants of silicone oil and a suspension ($\phi = 0.28$) of zeolite particles (3 wt.% $H_2O$) in silicone oil with both random and fibril arrangements of the particles. The dielectric constants were obtained using Method a (dc) and Method c (ac).

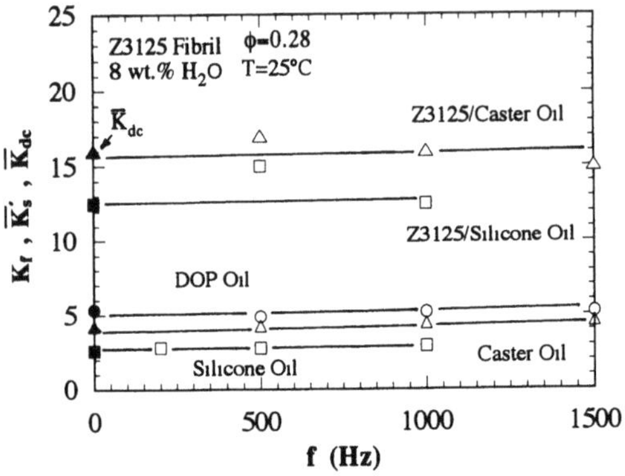


Fig. 10 Effect of frequency of the electric field on the dielectric permittivity of several oils and suspensions ($\phi = 0.28$) of zeolite particles (8 wt.% $H_2O$) at 25°C and E = 2 kV/mm.

**2.2.2 Conductivity.** The dc conductivity of the host oils and the ER suspensions was determined from the slopes of plots of the current density j vs dc electric field E; see, for example, Figs. 11 and 12. The dc conductivity $\sigma = dj/dE$ in both cases increases with electric field, the value for the suspension $\bar{\sigma}$ being orders of magnitude larger than that of the

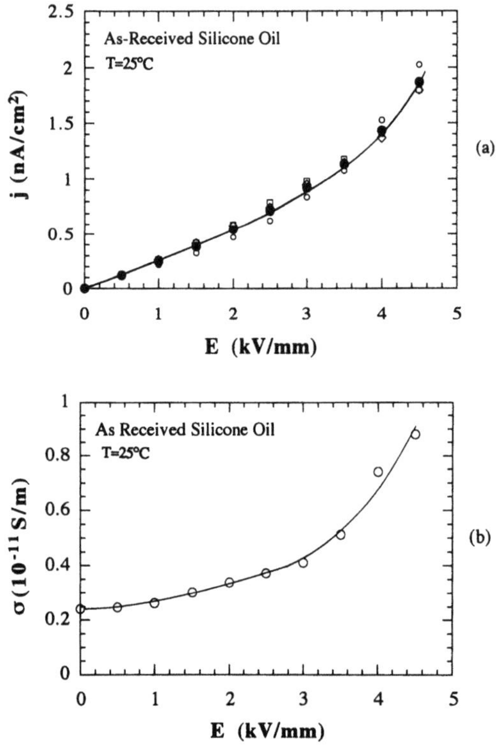


Fig. 11 (a) DC current density vs electric field E for as-received silicone oil at 25°C. (b) DC conductivity $\sigma_f = dj/dE$ vs E of as-received silicone oil given by the slope of the curve in Fig. 11a. From Wu et al [23].

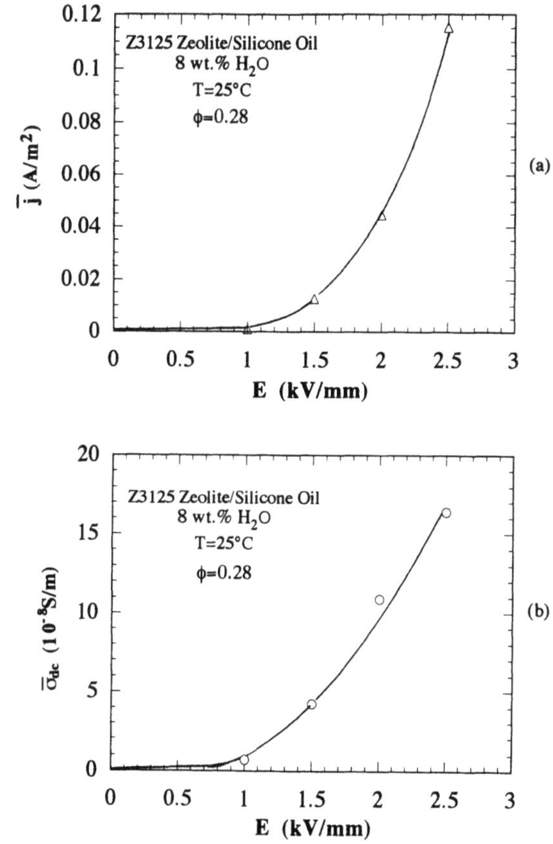


Fig. 12 (a) DC current density $\bar{j}$ vs electric field and (b) dc conductivity $\bar{\sigma} = d\bar{j}/dE$ of the suspension ($\phi$ = 0.28) of zeolite particles (8 wt.% $H_2O$) vs electric field at 25°C.

host oil $\sigma_f$. This is further illustrated in the semilog plot in Fig. 13, which shows the effects of volume fraction and water content of the zeolite particles on the conductivity of the suspensions, and for comparison includes the conductivity of the host mineral and silicone oils. It is here seen that the conductivity of the suspensions with zeolite particles of either the lowest volume fraction or water content is 2.5 to 3 orders of magnitude greater than that of the host oil. This difference increases an additional 2 orders of magnitude as the water content of the particles is increased to ~ 8 wt.% or the volume fraction to ~ 0.35. Similar effects of water content of the particles on $\bar{\sigma}$ were noted by Filisko and Radzilowski [24] for Linde 3A zeolite particles of composition $K_9Na_3(AlO_2)_{12}(SiO_2)_{12}]xH_2O$ suspended in light paraffin oil. Of interest is that heating suspensions ($\phi$ = 0.28) of Z3125 zeolite particles (0 and 8 wt.% $H_2O$) at 80° to 160°C produced a decrease in their conductivity determined subsequently at 25°C [13].

Arrhenius-type plots of $\bar{\sigma}$ vs 1/T for a suspension ($\phi$ = 0.28) of zeolite particles (3 wt.% $H_2O$) in silicone oil at two electric fields are presented in Fig. 14. Reasonably good straight lines are obtained whose slopes are relatively independent of electric field, yielding an activation energy $\Delta H_{\bar{\sigma}}$ = 16.1 kcal/mole. Similar plots for the effect of temperature on the conductivity of a suspension ($\phi$ = 0.28) of zeolite particles with 6 wt.% $H_2O$ in silicone oil yielded $\Delta H_{\bar{\sigma}}$ = 16.8 kcal/mole [4]. Considering both sets of results on the effect of T on $\bar{\sigma}$ for zeolite particles in silicone oil and all data on the effects of electric field, water content and volume fraction of particles, we obtain

$$\bar{\sigma} \approx 4.5 \times 10^{-13} \exp(0.5W + 13.4\phi) E^{2.2} \exp\left(\frac{-16.5 \pm 0.4 \text{ kcal/mole}}{RT}\right) \quad S/m \qquad (9)$$

where W is the water content of the particles in wt.% and E the applied electric field in V/m.

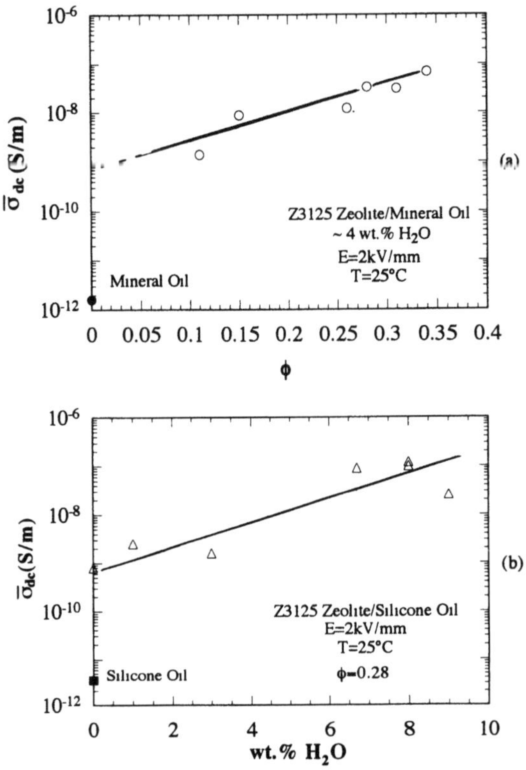


Fig. 13 Semilog plots of: (a) Effect of volume fraction of zeolite particles containing 8 wt.% $H_2O$ in mineral oil on the dc conductivity of the suspension and (b) effect of water content of the zeolite particles in silicone oil on the dc conductivity of the suspension ($\phi$ = 0.28).

The effect of temperature on the conductivity of the silicone oil $\sigma_f = dj/dE$ is presented as Arrhenius plots in Fig. 15. The activation energy $\Delta H_{\sigma_f}$ derived from the limited data in the plot is ~ 6.4 kcal/mole, which is substantially smaller than that for the suspensions of zeolite particles in silicone oil. It is however in accord with that reported by Yasufuku and Inuishi [25] for silicone oil. The present data give for the conductivity of silicone oil

$$\sigma_f \approx 1.1 \times 10^{-7} \exp\left(\frac{-6.4 \text{kcal/mole}}{RT}\right)\left[1 + 5 \times 10^{-3} \exp\left(\frac{E}{E_c}\right)^{1/2}\right] \quad S/m \qquad (10)$$

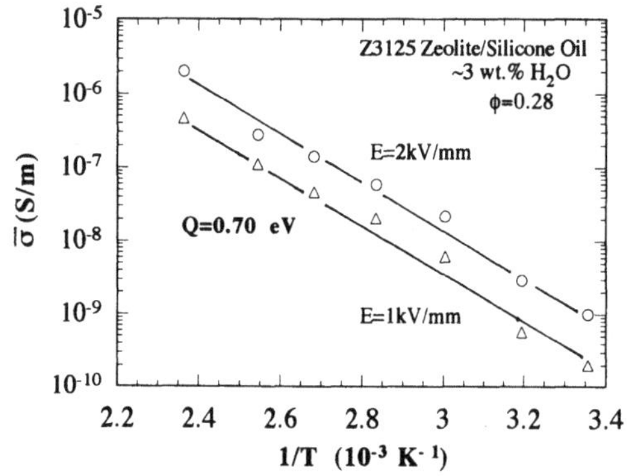


Fig. 14 Arrhenius-type plots of dc conductivity of a suspension ($\phi$ = 0.28) of zeolite particles (3 wt.% $H_2O$) in silicone oil for E = 1 kV/mm and 2 kV/mm.

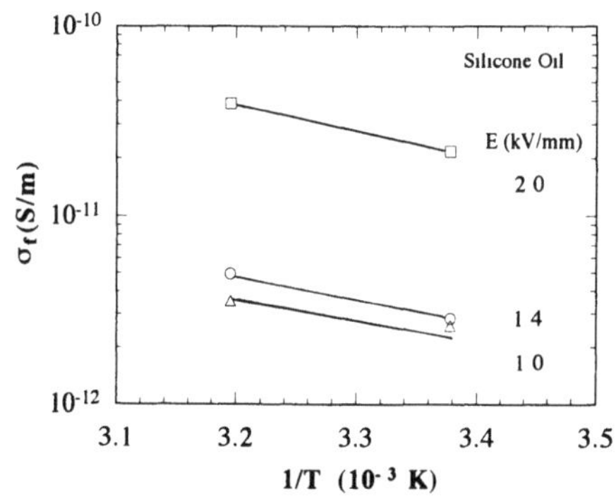


Fig. 15 Arrhenius-type plots of dc conductivity of the silicone oil for three electric fields.

which is similar in form to the simplified expression derived by Atten et al [26] for the conductivity of non-polar liquids based on Onsager theory [27]. E is in units of MV/m and $E_c$ = 0.11 MV/m.

It was noted that the permittivity of the suspensions $\overline{K}_{dc}$ increased with the same factors which gave an increase in their conductivity. This suggests that the two electrical properties may be related. The degree of correlation is shown in the semilog plot of $\overline{K}_{dc}$ vs $\bar{\sigma}$ for E = 2 kV/mm in Fig. 16. It is here seen that the data points for various host oils, water contents and volume fractions of particles scatter about a single, sigmoidal-shaped curve for about seven orders of magnitude in conductivity.

The following is a summary of the results pertaining to the conductivity measurements:

(a) The conductivity of the suspensions was in all cases orders of magnitude greater than that of the host oil.

(b) The conductivity of the suspensions increased with water content and volume fraction of the particles and with electric field and temperature.

(c) The activation energy for the conductivity of suspensions of zeolite particles in silicone oil was 16.5 ± 0.4 kcal/mole; that for the host oil was 6.4 kcal/mole.

(d) The conductivity at 25°C of suspensions of zeolite particles (0 and 8 wt.% $H_2O$) in silicone oil decreased as a result of prior heating at 80° to 160°C.

(e) A strong correlation existed between the dielectric permittivity of the suspensions of zeolite particles in various oils and their conductivity over seven orders of magnitude in conductivity.

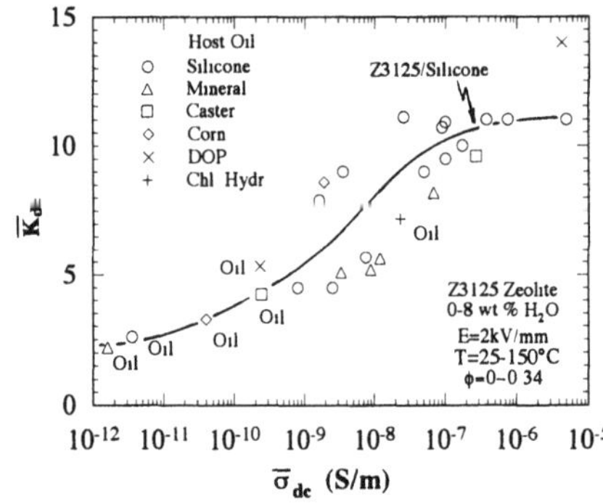


Fig. 16 Semilog plot of $\bar{K}_{dc}$ vs $\bar{\sigma}_{dc}$ for suspensions of zeolite particles with various water contents and volume fractions in several host oils. The curve drawn is for zeolite particles in silicone oil.

## 2.3 RHEOLOGY

As mentioned in the Introduction section, the rheological property we will consider is the electric field component $\tau_E$ pertaining to the post-yield, approximately steady-state region of the shear stress-strain curve obtained at a constant shear rate $\dot{\gamma} = 0.085\ s^{-1}$ in the quasi-static strain rate regime. An example of the shear stress-strain curves obtained at 25°C for these conditions is presented in Fig. 17. To be noted is that the stress increases rapidly with strain in the pre-yield region and then changes only slightly in the post-yield region. By taking the flow stress $\tau$ at a fixed strain ($\gamma = 5$), strain rate $\dot{\gamma}$, temperature T and electric field E, we rigorously define the rheological parameter under consideration [8]. The importance of defining the shear rate regime is illustrated in Fig. 18, which shows that at shear rates $\dot{\gamma} < 10\ s^{-1}$ (the quasi-static regime) $\tau_E$ decreases with increase in $\dot{\gamma}$, whereas at $\dot{\gamma} > 10\ s^{-1}$ (the dynamic regime) $\tau_E$ increases with $\dot{\gamma}$.

To be noted in Fig. 17 is that the flow stress increases significantly with electric field. The plots in Fig. 19 indicate that $\tau_E \propto E^2$, the proportionality constant varying with host oil and the frequency of the electric field. Log-log plots of $\tau_E$ vs E of *all data* obtained by us for the various conditions and factors investigated, including low frequency ac, gave the exponent n over E equal to 2.0 ± 0.5, the lower values occurring for the highest electric fields and lowest shear rates.

The effect of water content of the zeolite particles on $\tau_E$ for E = 2 kV/mm at 25°C is given in Fig. 20. Indicated is a parabolic effect with the exponent $q \approx 0.4$. A significant increase in $\tau_E$ with $H_2O$ content of the zeolite particles was also found by Filisko and Razilowski [24] for Linde 3A zeolite particles in paraffin oil. Thus, the strength of an ER suspension with zeolite particles is sensitively dependent on the water content of the particles, which must be taken into account in any consideration of the rheology of such suspensions.

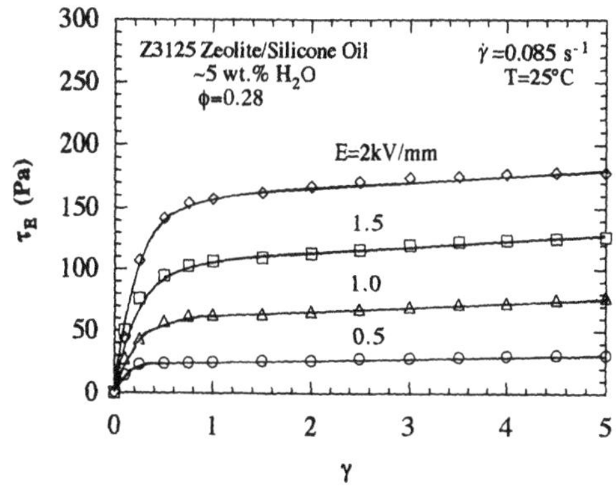


Fig. 17 Shear stress component $\tau_E$ vs shear strain curves for a zeolite (~ 5 wt.% $H_2O$)/silicone oil suspension ($\phi$ = 0.28) at 25°C and E = 0.5, 1.0, 1.5 and 2.0 kV/mm dc.

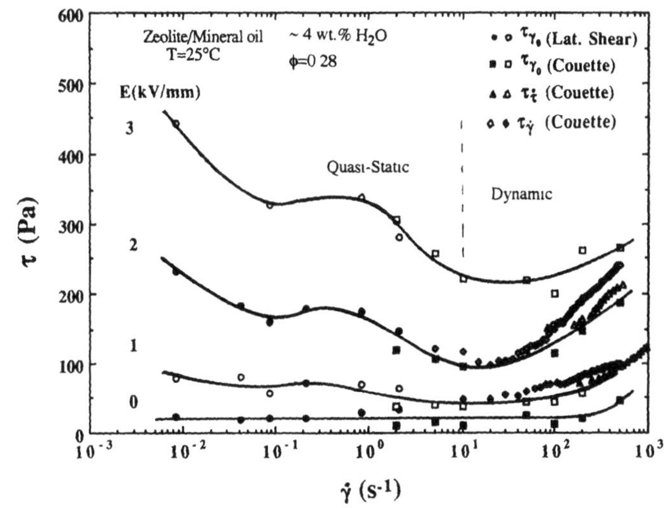


Fig. 18 The effect of shear rate on the flow stress $\tau$ for a suspension ($\phi$ = 0.28) of zeolite particles (8 wt.% $H_2O$) in mineral oil at 25°C determined by several test methods. From Conrad et al [8].

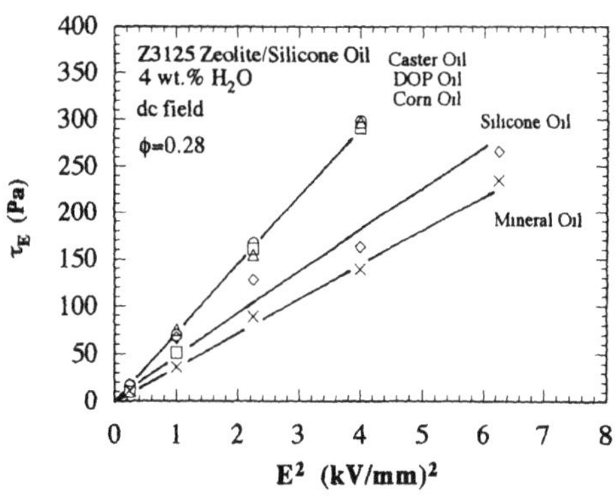


Fig. 19 $\tau_E$ vs $E^2$ dc for suspensions ($\phi$ = 0.28) of zeolite particles (4 wt.% $H_2O$) in various oils.

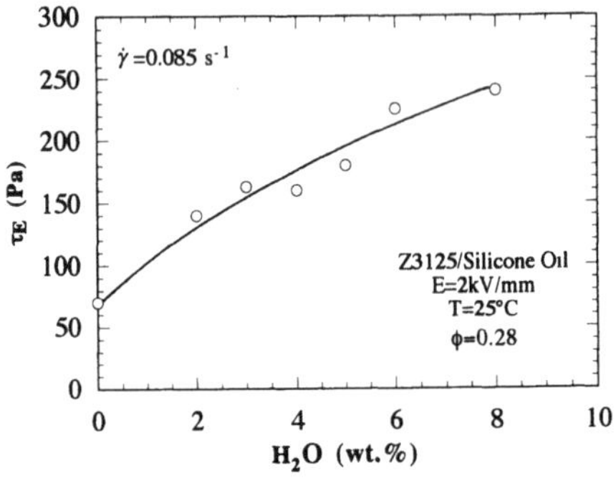


Fig. 20 $\tau_E$ vs water content of zeolite particles in silicone oil ($\phi$ = 0.28) with E = 2 kV/mm dc at 25°C.

The influence of volume fraction of zeolite particles with 4 wt.% $H_2O$ in mineral oil on $\tau_E$ at 25°C and E = 2 kV/mm is presented in Fig. 21. The plot shows a smaller effect at $\phi <$ 0.1 than at higher concentrations. Optical microscopy observations of the fibril structure revealed that for $\phi < 0.1$, the chains were mainly single rows of particles, whereas at higher concentrations the structure consisted largely of clusters or columns of particles; see for example Fig. 3. A concentration effect similar to that in Fig. 21 is also indicated for ER suspensions consisting of $BaTiO_3$ particles in dodecone oil [28] and lithium poly (methacrylate) in chlorinated hydrocarbon oil [29]. In contrast, results for corn starch/corn oil [6] and $TiO_2$/dodecane oil [28] suspensions indicate a proportionality between $\tau_E$ and $\phi$ over a similar concentration range.

An example of the influence of temperature on the flow stress of zeolite/silicone oil suspensions is given in Fig. 22. Evident is an initial increase in $\tau_E$ with temperature followed by a maximum at ~ 100°–125°C, and then a decrease at higher temperatures. Similar behavior was observed for zeolite particles with a lower water content, except the peak was not as pronounced [5]. A similar effect of temperature has also been reported for other ER suspensions [30]. It was pointed out in [4,5] that the initial rise in $\tau_E$ with temperature has the form of an Arrhenius rate equation with an apparent activation energy of ~ 3.5 kcal/mole, relatively independent of electric field and water content of the particles. Moreover, it was noted by these authors that $\tau_E$ appeared to correlate rather strongly with the current density up to the maximum in $\tau_E$.

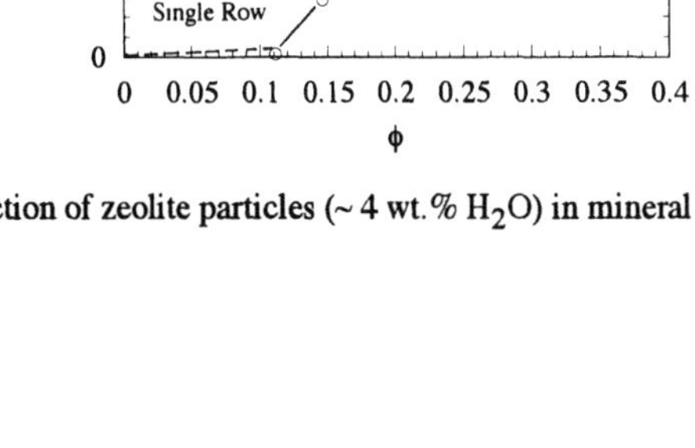


Fig. 21 $\tau_E$ vs volume fraction of zeolite particles (~ 4 wt.% $H_2O$) in mineral oil with E = 2 kV/mm dc at 25°C.

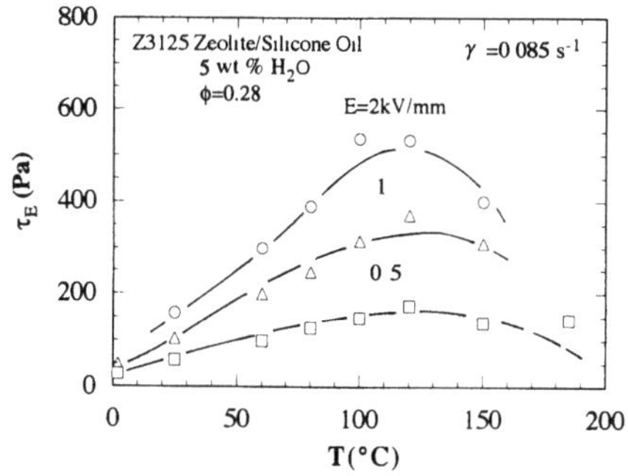


Fig. 22 Effect of temperature on $\tau_E$ of a suspension ($\phi$ = 0.28) of zeolite particles (6 wt.% $H_2O$) in silicone oil at three dc electric fields.

The influence of temperature on $\tau_E$ can conceivably consist of both reversible and irreversible components. To determine the reversible effects of temperature, suspensions ($\phi$ = 0.28) of zeolite particles (0 and 8 wt.% $H_2O$) were heated for 2.5 hr at 80°C to 160°C and then following cooling their shear resistance was measured at 25°C. The results obtained are presented in Fig. 23. It is seen in Fig. 23a that the heating had essentially no effect on $\tau_E$ for the particles with 0% water, whereas a maximum in $\tau_E$ occurred for prior heating at 80°C of suspensions with particles containing 8 wt.% $H_2O$ (Fig. 23b). Further, the stress for prior heating at 160°C was generally higher than that for the suspension without any heating. The influence of T on $\tau_E$ given here and in the literature thus include such irreversible effects.

The influence of the host oil on $\tau_E$ is shown in Fig. 24, which presents a plot of $\tau_E$ vs the dielectric constant $K_f$ of the oil. It is seen that $\tau_E$ is approximately proportional to $K_f$, giving $\tau_E \approx 80\ K_f$ for the test conditions considered.

The following is a summary of the influence of various factors ($\phi$, $H_2O$, E, host oil and T) on the quasi-static shear strength parameter $\tau_E$ at $\dot{\gamma}$ = 0.085 $s^{-1}$. As a first order approximation for temperatures up to 100°C we obtain

$$\tau_E \propto \phi^m K_f E^n W^q \exp(-Q_\tau / RT) \tag{11}$$

where $m \approx 1$, $n \approx 2$, $q \approx 0.4$ and $Q_\tau \approx 3.5$ kcal/mole. A more detailed consideration reveals the following:

(a) The influence of $\phi$ at low concentrations ($\phi < 0.1$), where the fibril structure consists mostly of single row chains, is less than at higher concentrations where the structure consists of columns or clusters of particles.

(b) The exponent over the electric field $n = 2.0 \pm 0.5$ for all conditions investigated, the lowest values tending to occur at the highest electric fields and lowest shear rates.

(c) For suspensions of zeolite particles in silicone oil $\tau_E$ initially increases with temperature to ~ 100°C and then decreases at higher temperatures. The effect of temperature includes a positive irreversible component, which is maximum at ~ 80°C.

(d) The effect of temperature on $\tau_E$ up to the maximum in $\tau_E$ closely follows that for the influence of temperature on the current density.

(e) $\tau_E$ increases parabolically with the water content of the particles up to ~ 9 wt.% with some indication that it may have reached a maximum at this water content and perhaps decreases at higher contents [4].

(f) For a given set of test conditions $\tau_E$ is approximately proportional to the dielectric constant of the host oil.

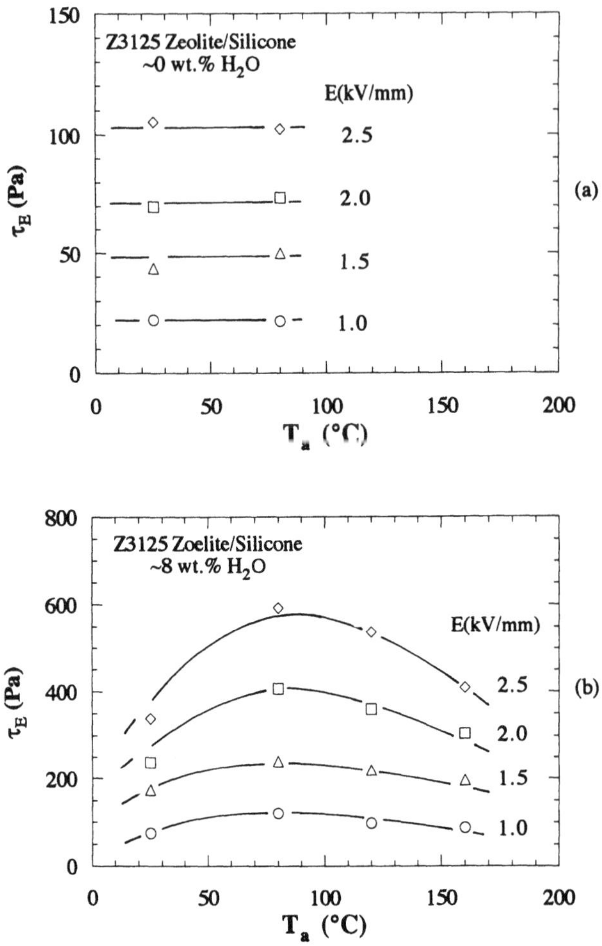


Fig. 23 Effect of prior heating temperature $T_a$ on $\tau_E$ of zeolite/silicone oil suspensions tested at 25°C as a function of dc electric field: (a) zeolite particles with ~ 0 wt.% $H_2O$ and (b) zeolite particles with 8 wt.% $H_2O$. From Conrad, Shih and Chen [13].

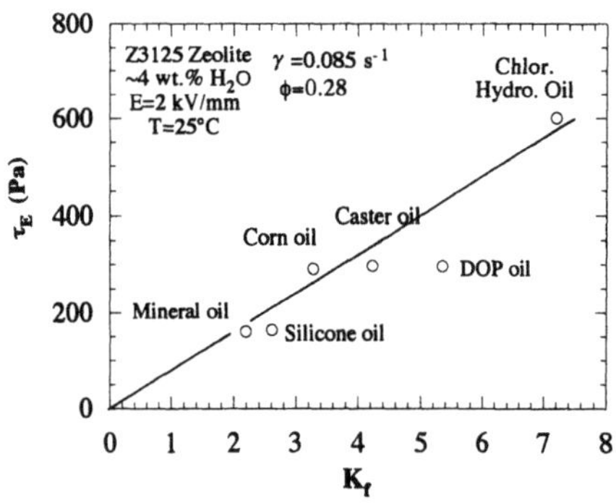


Fig. 24 $\tau_E$ vs the dielectric constant of the host oil for ER suspensions ($\phi$ = 0.28) of zeolite particles (~ 4 wt.% $H_2O$) in the various oils at E = 2 kV/mm dc and 25°C.

## 3. DISCUSSION

As mentioned in the Introduction, the polarization model for the strength of ER suspensions requires a knowledge of the complex dielectric mismatch parameter $\beta^*$, which is determined by the ratio of the complex permittivity of the particles $K_p^*$ to that of the host oil $K_f^*$. Although $K_f^*$ can be measured directly, $K_p^*$ is often derived from measurements on the suspension, since the permittivity of the particles may depend on the host oil (Fig. 8) the specific arrangement of the particles (Fig. 6 and 7) and the concentration and water content of the particles (Figs. 7 and 8). Since the form of the dielectric dispersion in Figs. 4 and 5 is in general accord with the Maxwell-Wagner interfacial polarization model [18–20, 31–34], it seems reasonable to begin our consideration of the present results in terms of this model and the relationships pertaining thereto.

The orders of magnitude greater conductivity of the zeolite/oil suspensions compared to the host oil (Fig. 13) suggests as a first consideration that the conductivity of the particles $\sigma_p$ may be considerably greater than that of the host oil $\sigma_f$. Schwarz [31] has pointed out that for suspensions containing conducting particles the Maxwell-Wagner model gives for the complex dielectric constant of the suspension

$$\overline{K}^* = K_f^*\left(\frac{1+2\beta^*\phi}{1-\beta^*\phi}\right) \tag{12}$$

Furthermore, when $\sigma_p >> \sigma_f$ the Maxwell-Wagner equations for the dielectric constant of a *dilute* heterogeneous mixture in terms of the electric properties of the components reduce to [18]

$$\overline{K}_s = K_f\frac{(1+2\phi)}{(1-\phi)} \tag{13}$$

$$\overline{K}_\infty = K_f[1+3\phi\beta] \tag{14}$$

and

$$t_R = \varepsilon_o\frac{\left(K_p^\infty + 2K_f\right)}{\sigma_p} \tag{15}$$

where $\varepsilon_o$ is the permittivity of free space and $K_p^\infty$ is the dielectric constant of the particles obtained at very high electric field frequency. To be noted is that Eqns. 13–15 do not yield directly the value of the dielectric constant of the particles $K_p^s$ from measurements on the suspension at low frequency ac or dc electric field. In the Maxwell-Wagner polarization model it is assumed that $K_p'$, $K_f'$, $\sigma_p$ and $\sigma_f$ do not vary with the frequency of the electric field for the range that is normally considered, and hence $K_p'^{\,s} = K_p^\infty$ and $\sigma_p^s = \sigma_p^\infty$.

The general solutions of the complex analog of the Maxwell-Wagner equations (including *concentrated* suspensions) were determined by Hanai [32] and are listed in [18]. Equations of interest here for $\sigma_p >> \sigma_f$ include

$$\left(\frac{\overline{K}^*-K_p^*}{K_f^*-K_p^*}\right)\left(\frac{K_f^*}{\overline{K}^*}\right)^{1/3} = 1-\phi \tag{16}$$

$$\left(\frac{\bar{K}_\infty - K_p^\infty}{K_f - K_p^\infty}\right)\left(\frac{K_f}{\bar{K}_\infty}\right)^{1/3} = 1-\phi \tag{17}$$

$$\bar{K}_s\left(\frac{3}{\bar{\sigma}_s - \sigma_p} - \frac{1}{\bar{\sigma}_s}\right) = 3\left(\frac{K_p^s - K_f}{\sigma_p - \sigma_f} + \frac{K_p^s}{\bar{\sigma}_s - \sigma_p}\right) - \frac{K_f}{\sigma_f} \tag{18}$$

$$\left(\frac{\sigma_p - \bar{\sigma}_s}{\sigma_p}\right)\left(\frac{\sigma_f}{\bar{\sigma}_s}\right)^{1/3} = 1-\phi \tag{19}$$

$$\bar{\sigma}_\infty\left(\frac{3}{\bar{K}_s - K_p} - \frac{1}{\bar{K}_\infty}\right) = 3\left(\frac{\sigma_p}{K_f - K_p} + \frac{\sigma_p}{\bar{K}_\infty - K_p}\right) - \frac{\sigma_f}{K_f} \tag{20}$$

In the above equations the subscripts and superscripts s and ∞ refer to the limiting cases of very low (s) and very high (∞) electric field frequencies.

Two additional, but less rigorous, relationships were also considered in the present analysis. One is the cubic expression proposed by Looyenga [35] which applies to *concentrated* suspensions

$$\bar{K}^{1/3} = \phi K_p^{1/3} + (1-\phi)K_f^{1/3} \tag{21}$$

The other is that originally derived by Wiener [36] for a dielectric with pores or columns of a different material connecting the electrodes, which gives a linear rule of mixtures for $\bar{K}'$ [18]

$$\bar{K}' = \phi K_p + (1-\phi)K_f \tag{22}$$

and for $\bar{K}'$ gives

$$\bar{K}'' = \sigma_p / 2\pi f \varepsilon_o \qquad \sigma_p >> \sigma_f \tag{23}$$

Miller and Jones [37] have shown that the Wiener equation is approached for columns of dielectric spheres as the spacing between the spheres becomes small.

The electrical properties of the zeolite/host oil suspensions will now be compared with the various relationships given by Eqns. 12 to 23. The effects of volume fraction of particles (Fig. 25) and of the host oil (Fig. 26) on $\bar{K}_{dc}$ indicate reasonable agreement with the simplified version of the Maxwell-Wagner equation for $\sigma_p >> \sigma_f$ given by Eqn. 13. In Fig. 25 both the intercept and slope of the line are equal to the dielectric constant of the host oil as required by Eqn. 13, while in Fig. 26 the line passes through the origin with a slope = 2.5, which is only slightly larger than the concentration parameter $(1 + 2\phi)/(1 - \phi) = 2.2$. The results of Figs. 25 and 26 thus indicate reasonable agreement with a Maxwell-Wagner interfacial polarization model with $\sigma_p >> \sigma_f$. Moreover, comparing Eqns. 12 and 13, it appears from the results in Figs. 25 and 26 that $\beta^* \approx 1$ for the suspensions and test conditions considered.

Upon comparing $K_p^\infty$ for the particles derived from $\bar{K}_\infty$ for the random arrangement in the suspensions using Eqns. 14, 17, 21 and 22, it is found that $K_p^\infty$ (random) = 6.5 ± 0.7 for zeolite particles in silicone oil ($\phi = 0.28$), relatively independent of the water content (3 to

9 wt.%) of the particles. The Wiener equation gave the lower limit, the Hanai equation the higher.

In contrast to the relative independence of $K_p^\infty$ on water content of the zeolite particles, the conductivity of the particles $\sigma_p$ obtained using the simplified Maxwell-Wagner Eqn. 15 (using $K_p^\infty$) depended sensitively on water content, increasing from 1.2 x $10^{-6}$ S/m to 9.8 x $10^{-4}$ S/m as the $H_2O$ content was increased from 3 to 9 wt.%. Similar values of $\sigma_p$ were obtained using the complete Maxwell-Wagner equation given in [18–20]. In contrast, the Hanai equation (Eqn. 19) gave negative values for $\sigma_p$ for E = 1 – 2 kV/mm and thus as such is not strictly valid for the present zeolite suspensions.

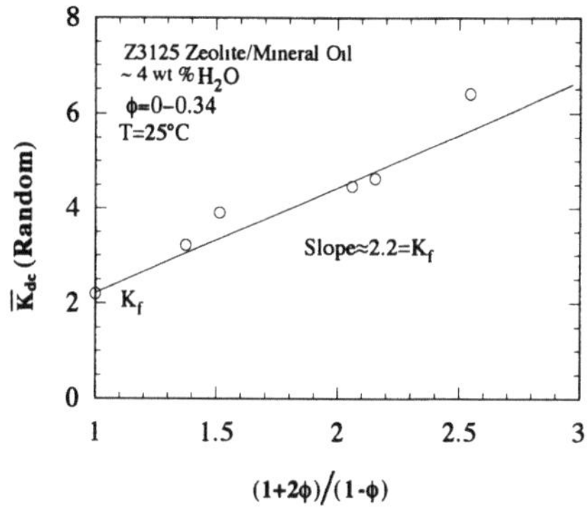


Fig. 25 DC dielectric constant of suspensions of zeolite particles (~ 4 wt.% $H_2O$) in mineral oil vs the concentration parameter $(1 + 2\phi)/(1-\phi)$.

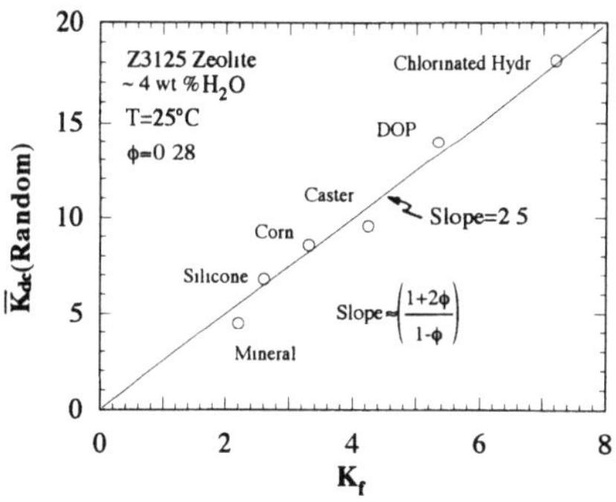


Fig. 26 DC dielectric constant of suspensions ($\phi$ = 0.28) of zeolite particles (~ 4 wt.% $H_2O$) in various oils vs the dielectric constant of the oil.

Negative values of $K'^s_p$ at low electric field frequencies were also obtained from $\overline{K}_{dc}$ or $\overline{K'}_s$ when the Hanai Eqn. 16 was employed. In contrast, the Wiener (Eqn. 22) and Looyenga (Eqn. 21) relationships gave positive values for $K'^s_p$, the Looyanga values being about twice those from the Wiener equation. The values of $K'^s_p$ obtained using the Wiener equation ranged from 7.5 to 34.3 for the factors which affected $K'^s_p$, namely the water content of the particles, the host oil, the arrangement of the particles and the test temperature. Use of the Schwartz relationship given by Eqn. 12 yielded $\beta^* > 1$, indicating that this equation is not strictly correct for obtaining $K'_p{}^s$ at low frequencies from $K'_s$, similar to what appears to be the case for the other equations given in Ref. [17–19] based on the Maxwell-Wagner polarization model. The discrepancies may be related to the requirement of the Maxwell-Wagner model that $K_p'$ is a constant independent of frequency f, whereas in fact $K_p'$ for the zeolite particles increases with f, as found for bulk specimens [38].

Upon comparing the electrical properties of the zeolite particles determined from those of the ER suspensions with the *bulk* properties, we find the following:

(a) The values of $K_p^\infty$ (3–6.5) and $K'_p{}^s$ (7–34) are similar (but not exactly equal) to those for $Na_{12}$–A reported by Ohgushi et al [38]; they are however larger than those given by Breck [15].

(b) The conductivity of the particles $\sigma_p$ ($10^{-6}$ – $10^{-5}$ S/m) is in accord with that given in [15,18] for $Na_{12}$–A.

(c) The activation energy associated with the conductivity of the suspensions ($\Delta H_\sigma$ = 16.4 kcal/mole) is in accord with that for $Na^+$ cations in zeolite 4A [15].

(d) The relaxation time $\overline{\tau}_R = 2.5 \times 10^{-7}$ s for 9 wt.% $H_2O$ is in reasonable accord with that reported for Na–A ($4.9 \times 10^{-7}$ s) with a slightly higher water content [15].

Hence, it appears that the electrical properties of the Z3125 particles in the oil suspensions are qualitatively similar to those reported in the literature for bulk specimens of this material. Any difference may be due to a difference in chemical composition, since the results in the literature on bulk specimens differ among themselves to about the same degree as the difference between particles in suspension compared to the bulk.

Pertinent to the present discussion is that the activation energy for the dielectric loss peak in bulk $Na_{12}$–A zeolite is equal to that for conduction, which in turn is equal to that for the migration of $Na^+$ cations [15]. Based on this correspondence of activation energies, Ohgushi et al [38] attribute the loss peak in $Na_{12}$–A to the jump of $Na^+$ ions between the 4-ring sites in the zeolite crystal structure. It is therefore very likely that the conduction and the dielectric loss peaks in the $Na_{12}$–A zeolite particles in the present ER suspensions are also associated with this $Na^+$ ion jump mechanism.

## 3.1 Strength

We will examine the results obtained on the zeolite/oil suspensions with the assumption that the host oil and the particles behave as ideal dielectric spheres, and the particles are aligned in chains or columns between the electrodes. With these assumptions, the theoretical analysis of Chen et al [2] gives for the polarization component of the yield or steady-state flow stress [8]

$$\tau_E = 44.1 A_s \phi \varepsilon_o K_f (\beta E)^2 \Big| \left\{ \exp\left[ 14.84 - 6.165\left(\frac{R}{a}\right)\beta^2 \right] \right\}$$

$$\times \frac{1}{(R/a)^4}\left(1 - \frac{4}{(R/a)^2}\right)^{1/2} \Big|_{max} \qquad (24)$$

Here $A_s$ is taken to be a structure factor pertaining to the alignment of the particles. It is equal to one for perfectly aligned single-row chains and may attain a value of of the order of 10 for multi-row chains or columns [7,8]. R/a is the ratio of the separation of the sphere centers to their radius. Some confidence in this model is provided by the fact that the measured forces between glass beads aligned in a single row chain and in a column of several rows were in reasonable accord with those predicted [9, 39]. Further support for this model is provided by the results in Fig. 27, where it is seen that the structure parameter $A_s = 1$, i.e., that the ratio of the measured stress to that predicted by Eqn. 24 is one for $TiO_2$ particles in dodecane. The $K_p$ value employed to obtain β for these suspensions was that determined on bulk specimens of $TiO_2$.

Since the magnitude of $K_p$ for the zeolite particles depended on the equation which was used to derived it from the dielectric permittivity of the corresponding suspension, in turn the ratio of measured-to-calculated shear stress ratio will vary with the method. This is illustrated in Fig. 28, which gives a plot of $A_s \equiv \tau_{meas}/\tau_{calc}$ for the various methods used to obtain $K_p$. $\tau_{calc}$ was derived using Eqn. 24 and taking $A_s = 1$. The "corrected Maxwell-Wagner method" consisted of empirically correcting the value of $K_p^{\infty}$ for the test conditions at low frequency and a fibril arrangement of the particles, i.e. by multiplying $K_p^{\infty}$ by the ratios $\overline{K}_s/\overline{K}_{\infty}$ and $\overline{K}_s$ (fibril)/$\overline{K}_s$ (random).

$$K_p^s(M-W) = \left[K_p^{\infty}(M-W)\right]\left(\frac{\overline{K}_s}{\overline{K}_{\infty}}\right)\left[\frac{\overline{K}_s(\text{fibril})}{\overline{K}_s(\text{random})}\right] \tag{25}$$

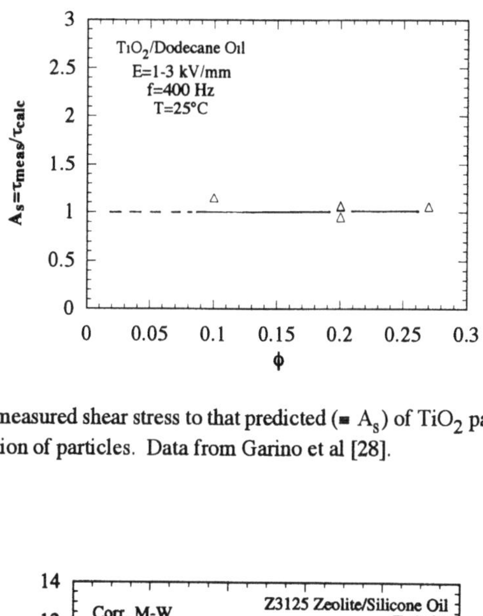


Fig. 27 The ratio of the measured shear stress to that predicted ($\equiv A_s$) of $TiO_2$ particles in dodecane oil vs the volume fraction of particles. Data from Garino et al [28].

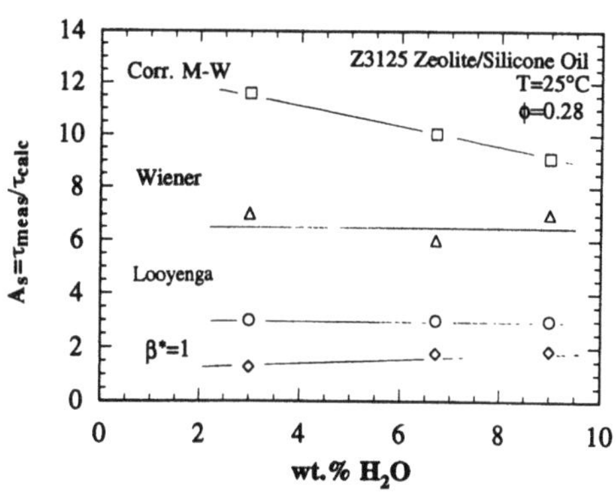


Fig. 28 The ratio of the measured shear stress to that predicted ($\equiv A_s$) for zeolite particles in silicone oil vs the water content of the particles using various relationships to obtain $K_p$.

It is seen in Fig. 28 that $\tau_{meas}/\tau_{calc}$ ranges between 1 and 12 depending on the method used to determine $K_p$, the smallest value being obtained by assuming $\beta^* = 1$ and the largest with the corrected Maxwell-Wagner equation. Unreasonable values of $\tau_{meas}/\tau_{calc}$ = 60–100 were obtained by using $K_p^{\infty}$ derived from $K_{\infty}$ at high field frequency. Further indication that $\beta^*$ may be ≈ 1 is provided by the plot in Fig. 24. It is here seen that $\tau_E$ is proportional to $K_f$, the proportionality constant yielding $A_s = 1.6$ based on Eqn. 24 and taking $\beta^* = 1$.

Values of $A_s$ within ± 50% of those in Fig. 28 for each method were obtained for all of the various factors which were investigated pertaining to the zeolite suspensions, namely the effects on $\tau_E$ at a constant shear rate of 0.085 $s^{-1}$ of: (a) host oil; (b) wt.% $H_2O$ in the particles; (c) $\phi$; (d) E; (f) T and (g) prior heating of the suspension. By taking $\beta^* = 1$ we obtain $A_s$ = 1–2 for all of the test conditions at $\gamma$ = 0.085 $s^{-1}$ except at the higher test temperatures, where $A_s$ increases from 2.8 at 80°C to 4.3 at 150°C. Also, considering the results in Fig. 18, the value of $A_s$ increases to ~ 4 as the shear rate is decreased by an order of magnitude at 25°C. $A_s > 1$ is in keeping with the increase in force between particles which occurs when the number of rows in a chain or cluster of particles increases [9]. The increase in $A_s$ with $\phi$ beyond a critical concentration for $BaTiO_3$ in dodecance oil shown in Fig. 29 may also be due to the formation of multi-row chains at the higher concentrations, as in the case of the zeolite/oil suspensions (Fig. 21). The value of $K_p$ for $BaTiO_3$ is that taken from measurements on bulk specimens [28], giving $\beta^* = 1$.

Returning our attention to the present $Na_{12}$–A zeolite/oil suspensions, the above analysis based on the polarization model indicates reasonable agreement between the measured and predicted values of $\tau_E$ for $\beta^* \approx 1$ with a structure factor $A_s$ = 1–4 for all test conditions investigated. A value of $\beta^* = 1$ would occur if $\sigma_p \gg \sigma_f$, because then

$$\beta^* = \frac{\sigma_p - \sigma_f}{\sigma_p + 2\sigma_f} = 1 \qquad (26)$$

Davis [3] has shown that Eqn. 26 applies within the Maxwell-Wagner approach in the dc limit when $\sigma_p/\sigma_f$ exceeds $K_p/K_f$. Although the values of $\sigma_p$ and $\sigma_f$ in Eqn. 26 are generally considered at the applied electric field E, the magnitude of $\sigma_f$ should be taken for the local electric field $E_{local}$ in the gap between the particles. This local electric field can be an order of magnitude larger than the applied field [2]. If we assume that even by taking the local field into account $\sigma_p \gg \sigma_f$ for all test conditions investigated, it is then difficult to explain on the basis of Eqn. 26 the observed increase in $\tau_E$ with water content of the particles and with test temperature.

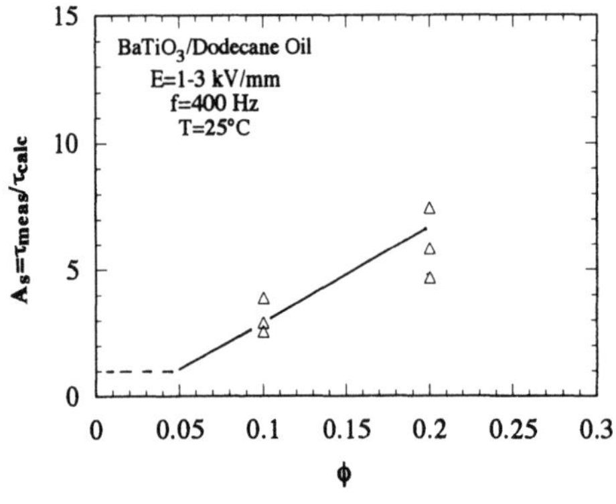


Fig. 29 The ratio of the measured shear stress to that predicted (= $A_s$) for $BaTiO_3$ in dodecane oil vs the volume fraction of particles. Data from Garino et al [28].

If instead of taking $\beta^* = 1$ we assume that $\beta^* = (K_p' - K_f')/(K_p' + 2K_f')$ with the values of $K_p'$ derived from measurements on the suspensions employing either the Wiener (Eqn. 22) or the Looyenga (Eqn. 21) relationships, we obtain that $\beta^*$ initially increases with water content of the particles and with test temperature and then approaches a maximum of near unity. In this case the structure factor takes on values in the range of 3–20, the lower values being obtained using the Looyenga relation and the higher the Wiener relation. These values for $A_s$ are consistent with an arrangement of the particles in multi-row columns rather than single row chains [7,8,9].

Neither $\beta^*$ based entirely on the ratio of conductivities nor entirely on the ratio of the dielectric constants $K'$ can however account for the decrease in $\tau_E$ of the zeolite/oil suspensions which ultimately occurs at high temperatures and possibly also at high water contents, since both ratios increase with these factors. In a previous paper [6] it was suggested that such decreases in $\tau_E$ could result from a leakage of the charge on the particles as the current density becomes large. A model is here proposed for this leakage in terms of the resistance loss which occurs along a chain of particles. It will be seen that the model is in qualitative agreement with experimental observations.

Referring to Fig. 30, let us assume that an electric current $I_c$ flows between the electrodes along the chain consisting of spherical particles, which are separated by a gap containing the host oil. There also exists an interface region in which the charge transfer from the particle to the oil gap occurs. The voltage drop V between electrodes with unit length separation is

$$V = NI_c\left(R_p + R_f + R_i\right) \tag{27}$$

where N is the number of particles in the chain and $R_p$, $R_f$ and $R_i$ are the resistances of the particle, oil gap and interface, respectively. Taking $R_p = \rho_p d_p/A_p$, $R_f = \rho_f x_f/A_f$ and $R_i = 2\rho_i x_i/A_i$ and taking $\rho = 1/\sigma$, Eqn. 27 becomes

$$V = NI_c\left(\frac{d_p}{\sigma_p A_p} + \frac{x_f}{\sigma_f A_f} + \frac{2x_i}{\sigma_i A_i}\right) \tag{28}$$

where $d_p$ is the diameter of the particles, $x_f$ the oil gap spacing, $x_i$ the thickness of the interface region and $A_p$, $A_f$ and $A_i$ the respective areas through which the current flows. Since according to Eqn. 24 the maximum in $\tau_E$ occurs at $R/a = 2.05$, we will take $x_f = 2.5 \times 10^{-2}\, d_p$. Also, we will assume that $2x_i \approx 0.1\, x_f$ and that $A_p = A_f = A_i \approx \pi a^2$. Inserting these values into Eqn. 28, one obtains

$$E = \frac{V}{Nd_p} = j_c\left(\frac{1}{\sigma_p} + \frac{2.5\text{x}\,10^{-2}}{\sigma_f} + \frac{2.5\text{x}\,10^{-3}}{\sigma_i}\right) \tag{29}$$

Further, taking the number of chains per unit area $N_A \approx \phi/\pi a^2$ and assuming that practically all of the current flows along the chains with very little between chains, we obtain

$$j_c = \bar{j}/\phi \tag{30}$$

where $\bar{j} = E\bar{\sigma}$ is the current density in the suspension with conductivity $\bar{\sigma}$. Inserting $E\bar{\sigma}/\phi$ for $j_c$ into Eqn. 29 gives

$$E = \frac{E\bar{\sigma}}{\phi\sigma_p} + \frac{2.5\text{x}10^{-2}E\bar{\sigma}}{\phi\sigma_f} + \frac{2.5\text{x}10^{-3}E\bar{\sigma}}{\phi\sigma_i} \tag{31}$$

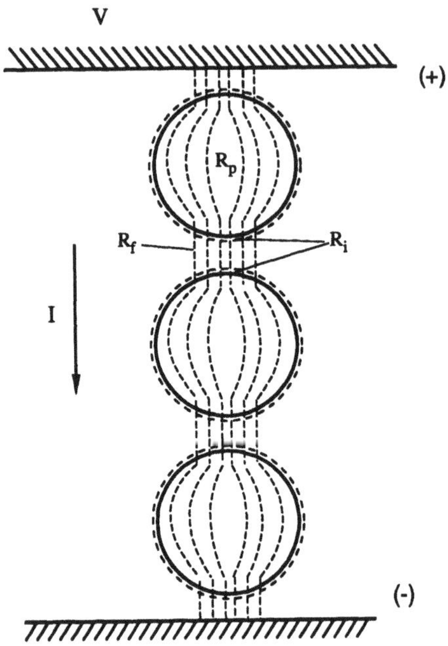


$$V = I\,R = N\,I\,(R_p + R_f + R_i)$$

Fig. 30 Schematic illustrating current flow along a chain of particles.

where the first term on the R.H.S. of Eqn. 31 gives the electric field $E_p$ across the particles, the second $E_f$ that across the oil gap and the third $E_i$ that at the interface between the particle and the oil.

To compare the model with experiment, $\bar{\sigma}$, $\sigma_p$, $\sigma_f$ and $\sigma_i$ are plotted vs 1/T in Fig. 31 for an electric field E = 2 kV/mm applied to a suspension ($\phi$ = 0.28) of Z3125 zeolite particles (6 wt.% $H_2O$) in silicone oil. The value of $\bar{\sigma}$ was that measured for this suspension (Fig. 32); $\sigma_p$ was that derived from the relaxation time $\bar{t}_R$ using Eqn. 15 and taking $\Delta H_{\sigma_p}$ = 16.5 kcal/mole. $\sigma_f$ is shown for various values of the local electric field in the oil gap between the particles, using Eqn. 10 for the effects of electric field and temperature on the conductivity of the silicone oil. $\sigma_i$ was derived using Eqn. 31 and the above-indicated values of $\bar{\sigma}$, $\sigma_p$ and $\sigma_f$ ($E_{local}$ = 38 kV/mm). The reason for this choice of $E_{local}$ will become more evident below.

Focusing on $\bar{\sigma}$ in Fig. 31 it is seen that at temperatures below ~ 110°C the curve of $\bar{\sigma}$ vs 1/T is parallel to $\sigma_p$ (i.e. $\Delta H_{\bar{\sigma}} = \Delta H_{\sigma_p}$), whereas at higher temperatures the curve is parallel $\sigma_f$ (i.e. $\Delta H_{\bar{\sigma}} = \Delta H_{\sigma_f}$). This indicates that at T < 110°C, the *temperature-dependence* of the conductivity of the suspension is determined mainly by that of the particles, whereas at higher temperatures it is governed by the silicone oil. Further, since $\bar{\sigma}$ at T > 110°C lies on the line for $\sigma_f$ with $E_{local}$ = 38 kV/mm, this suggests that the local electric field in the oil gap between the particles is of this magnitude, which is a reasonable value [2].

Of additional interest regarding Fig. 31 is the magnitude and temperature dependence of $\sigma_i$. If we assume that $\bar{\sigma}$ used in Eqn. 31 to derived $\sigma_i$ is that actually measured at all temperatures, the temperature dependence of $\sigma_i$ parallels that of $\bar{\sigma}$ over the entire temperature range with $\sigma_i << \bar{\sigma}$.

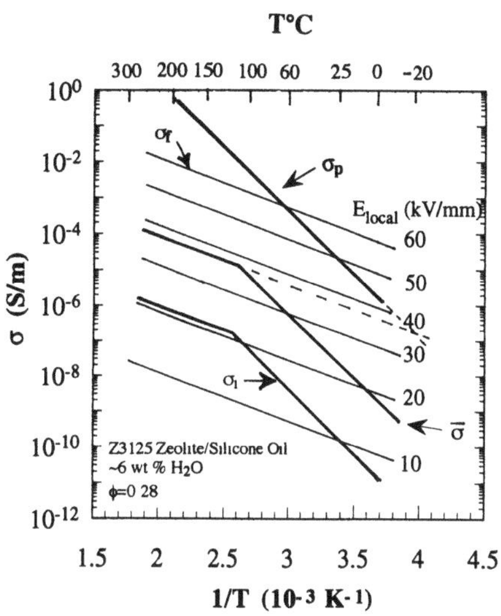


Fig. 31 $\sigma_f$, $\sigma_p$ $\bar{\sigma}$ and $\sigma_i$ vs 1/T for a suspension ($\phi$ = 0.28) of zeolite particles (6 wt.% $H_2O$) in silicone oil with an applied electric field E = 2 kV/mm dc. $E_{local}$ is the increased electric field in the gap between the particles.

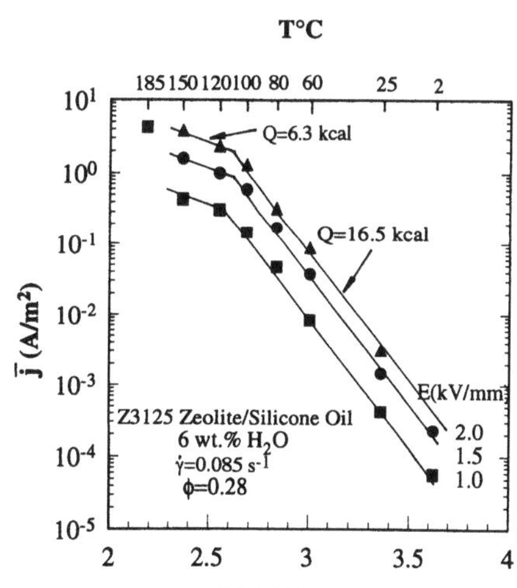


Fig. 32. Log $\bar{j}$ vs 1/T at several dc electric fields for the suspension in Fig. 31. From Conrad et al [4].

Upon comparing the temperature dependence of $\tau_E$ in Fig. 21 with the curves in Fig. 31, we find the following:

a. The temperature (~ –25°C) at which $\tau_E$ approaches zero at low temperatures corresponds to that where $\sigma_p = \sigma_f$ (E = 38 kV/mm).
b. The initial increase in $\tau_E$ with temperature corresponds with the increase in the ratio of $\sigma_p/\sigma_f$ (E = 38 kV/mm) with temperature.
c. The temperature (110°C) at which the maximum in $\tau_E$ occurs corresponds to the temperature of the abrupt change in slope in $\bar{\sigma}$.

These agreements in temperatures between $\tau_E$ and the conductivities gives some credibility to the proposed model.

According to the model it is expected that $\tau_E$ will depend not only on the ratio $\sigma_p/\sigma_f$ defining $\beta^*$ but also on the leakage of the charge from the particles. This leakage is mainly governed by the conductivity of the interface; its degree is reflected in the voltage drop across a particle due to its resistivity. The electric field components associated with the current flow are given in the three terms on the R.H.S of Eqn. 31. A semilog plot of each of these components normalized by the applied electric field is presented Fig.33. It is here seen that with increasing temperature, the normalized ratio for the interface remains relatively constant at approximately unity. In the case of the oil gap, the ratio initially increases with temperature up to ~ 110°C and then remains constant at $3\times10^{-2}$ at higher temperatures. For the particles, the ratio remains constant at $4\times10^{-3}$ up to ~ 110°C and then decreases markedly at higher temperature, attaining relatively low values at 150°C-200°C where $\tau_E$ seems to be approaching zero (Fig. 22). It thus appears that the drop in $\tau_E$ at high temperatures is associated with the decrease in the parameter $\bar{\sigma}/\phi\sigma_p$. Physically, this parameter can be interpreted to reflect a reduction in the electric field responsible for the force between the particles, the reduction being due to the current flow along the chain.

The above thus indicates that the influence of the various factors on $\tau_E$ for $Na_{12}$–A zeolite/oil suspensions considered in the present investigation can be explained qualitatively by the polarization model if we take the following into account:

a. The high conductivity of the particles.
b. The increased local electric field in the oil gap between the particles.
c. The decreased voltage drop across the particles as their conductivity increases.
d. The existence of an interfacial region of high resistivity for charge transfer between a particle and the oil in the gap between the particles.
e. The specific arrangement of the particles, i.e. the so-called structure.

Moreover, it is expected that the model for the role of conductivity in the strength of ER fluids given here applies as well to other hydrous-based ER fluids and also to anhydrous fluids.

An alternative to the polarization model for the ER effect with dc or low frequency ac electric fields is the conduction model proposed by Atten et al [26]. This model gives for the force between spherical particles of radius a with low applied electric field E

$$F \approx 4\pi a^2 K_f K_\Gamma^2 \Gamma^2 E^2 \tag{32}$$

where $\Gamma = \sigma_p/\sigma_f >> 1$, $K_\Gamma = [\pi \ln(a/\delta)]^{-1}$ and $\delta$ is the lateral extend of the conduction zone between the particles. At high electric fields the model gives

$$F = 2\pi a^2 K_f E_c \left\{\ln[(10\Gamma/\pi)(2E/E_c)^{1/2}]\right\}^2 \; E \tag{33}$$

where $E_c$ is a constant (≈ 0.335 kV/mm for polar liquids). Eqn. 32 gives a quadratic electric field dependence and a strong dependence on the ratio $\sigma_p/\sigma_f$, whereas Eqn. 33 is linear in E to a first approximation and exhibits a very weak dependence on the ratio $\sigma_p/\sigma_f$. Recently, Davis [40] has proposed that when considering the yield stress of ER fluids the conductivity model gives a 3/2 power electric field dependence at high fields. This intermediate value of the exponent tended to occur at high electric fields and very low shear rates in the present zeolite suspensions.

$\tau_E$ derived using Eqns. 32 or 33 is in qualitative accord with the increase in conductivity which occurs with water content of the zeolite particles or with increase in temperature. However, these equations do not predict a maximum or peak in $\tau_E$ as the conductivity of the particles increases. Again, this may be accounted for by the increased charge leakage associated with the increase in conductivity of the particles and interface region. However, further work is needed before the conductivity model can be unequivocally accepted as a quantitative description of the strength of ER suspensions with conducting particles.

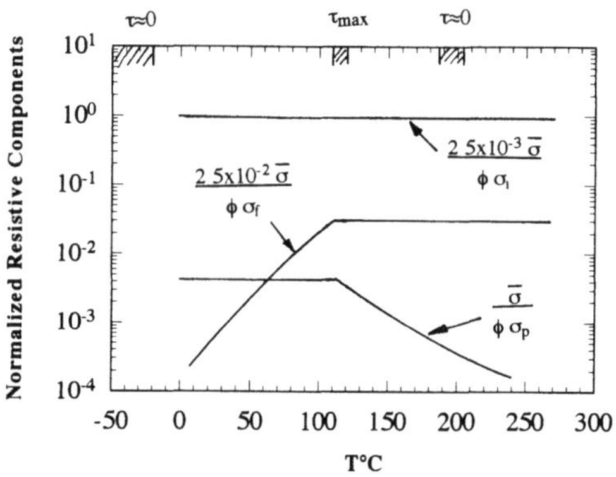


Fig. 33 The variation with temperature of the three normalized resistive components along a chain for the suspension in Fig. 31.

## 4. SUMMARY

Detailed data are presented and analyzed in terms of the Maxwell-Wagner polarization model for the influence of a number of factors on the electrical properties and the ER strength at low shear rates of suspensions of $Na_{12}$–A zeolite particles in various host oils. The conductivity of the particles derived from the relaxation time associated with the dielectric dispersion of the suspensions was in all cases considerably greater than that of the host oil, even when the increased local electric field in the gap between the particles was considered. Qualitative agreement between the measured shear resistance and that predicted occurred when either the conductivity mismatch or the dielectric constant mismatch between the particles and the host oil was considered and allowance was made for the structure or arrangement of the particles. To account for the peak in the strength of the ER suspensions with increase in temperature or water content of the particles (i.e. in the conductivity of the particles) a model is proposed which considers the resistance loss associated with the electric current flowing along a chain of particles. Qualitative agreement with the model is shown. The possibility that the strength of ER fluids may be better described by the conduction model of Atten et al than the polarization model needs further evaluation.

## 5. ACKNOWLEDGEMENTS

This research was jointly sponsored by the NCSU/Industry Consortium on ER Fluids and the NSF Engineering Division under Grant CTS–9313897 with Dr. M. C. Roco as technical monitor.

## REFERENCES

1. D. J. Klingenberg, F. van Swol and C. F. Zukoski, J. Chem. Phys. 94 6160 (1991).
2. Y. Chen, A. F. Sprecher and H. Conrad, J. App. Phys. 70 6796 (1991).
3. L. C. Davis, Appl. Phys. Lett. 60 319 (1992); J. App. Phys. 72 1334 (1992).
4. H. Conrad, Y. Chen and A. F. Sprecher, Proc. 2nd Int. Conf. Electrorheological Fluids, eds. J. B. Carlson, A. F. Sprecher and H. Conrad, Technomic, Lancaster, PA (1990) p. 252.
5. H. Conrad, A. F. Sprecher, Y. Choi and Y. Chen, J. Rheol. 35 1393 (1991).
6. Y. Chen and H. Conrad, ASME AMD–Vol. 175 Developments in Non-Newtonian Flows, eds. D. A. Siginer, W. E. Van Arsdale, M. C. Alton and A. N. Alexandrou, ASME, New York, N. Y. (1993) p. 199.
7. A. F. Sprecher, Y. Chen, Y. Choi and H. Conrad, Proc. 3rd Int. Conf. Electrorheological Fluids, ed. R. Tao, World Scientific, Singapore (1992) p. 142.
8. H. Conrad, Y. Chen and A. F. Sprecher, ibid, p. 195; Int. J. Mod. Phys. B 6 2575
9. Y. Chen and H. Conrad, Proc. 4th Int. Conf. Electrorheological Fluids, ed. R. Tao and G. D. Roy, World Scientific, Singapore (1994) p. 412; Int. J. Mod. Phys. B (1994), in print.
10. H. Uejima, Jap. J. Appl. Phys. 11 319 (1972).
11. Yu. F. Deinega and G. V. Vinogradov, Rheol. Acta 23 636 (1984).
12. H. Block, P. Rattray and T. Watson, Proc. 3rd Int. Conf. Electrorheological Fluids, ed. R. Tao, World Scientific, Singapore (1992) p. 93.
13. H. Conrad, Y. Shih and Y. Chen, "Effect of Prior Heating Zeolite/Silicone Oil Suspensions on Their Subsequent Rheology at 25°C", FED-Vol. 205/AMD –Vol. 190 Developments in Electrorheological Flows and Measurement Uncertainty, eds. D. A. Siginer, J. H. Kim, S. A. Sherif and H. W. Coleman, ASME, New York, N. Y. (1994) p. 83.
14. Molecular Sieves, brochure by UOP Inc. F–1979J–2M (1990).
15. D. W. Breck, Zeolite Molecular Sieves: Structure, Chemistry and Use, John Wiley, New York (1974).
16. A. Dyer, An Introduction to Zeolite Molecular Sieves, Wiley, New York (1988).
17. ANSI/ASTM D2216–80, "Standard Test Method for Laboratory Determination of Water (Moisture) Content of Soil, Rock and Soil Aggregate Mixtures", ASTM, Philadelphia (1980).
18. L. K. H. van Beek, Progress in Dielectrics Vol. 7, ed. J. B. Birks, CRC Press, Chemical Rubber Co., Cleveland, OH (1967) p. 69–114.
19. S. S. Dukhin, Surface and Colloid Science Vol. 3, ed. E. Matijevic, Wiley-Interscience, New York (1971) p. 83–165.
20. K. D. Weiss and J. D. Carlson, Proc. 3rd Int. Conf. Electrorheological Fluids, ed. R. Tao, World Scientific, Singapore (1992) p. 264.
21. K. S. Cole and R. H. Cole, J. Chem. Phys. 9 341 (1941).
22. C. P. Smyth, Dielectric Behavior and Structure, McGraw-Hill, New York (1955).
23. C. W. Wu, Y. Chen and H. Conrad, unpublished research, NCSU (1994).
24. F. E. Filisko and L. H. Radzilowski, J. Rheol. 34 539 (1990).
25. S. Yasufuku and Y. Inuishi, 7th Int. Conf. on Cond. and Breakdown in Dielectric Liquids, Berlin (1981) p. 444.

26. N. Félici, J.-N. Foule and P. Atten, Proc. 4th Int. Conf. Electrorheological Fluids, eds. R. Tao and G. D. Roy, World Scientific, Singapore, p.139.
27. L. Onsager, J. Chem. Phys. 2 599 (1934).
28. T. Garino, D. Adolf and B. Hance, Proc. 3rd Int. Conf. Electrorheological Fluids, ed. R. Tao, World Scientific, Singapore (1992) p. 167.
29. D. Klingenberg and C. Zukoski, Langmuir 6 15 (1990).
30. Yu. F. Deynega, K. K. Popko and N. Ya. Kovganich, Heat Transfer-Soviet Res. 10 50 (1978).
31. G. Schwarz, J. Phys. Chem. 66 2636 (1962).
32. R. W. Sillars, J. Inst. Elect. Engr. (London) 80 378 (1937).
33. D. A. G Bruggeman, Ann. Physik (Leipzig) 24 636 (1935).
34. T. Hanai, Kolloid-Z, 171 23 (1960); 177 57 (1961); Bull. Inst. Chem. Res., Kyoto Univ. 39 3 & 1 (1961).
35. H. Looyenga, Physica 31 401 (1966).
36. O. Wiener, Abh. Sachs Ges. Wiss. Math. phys. 32 509 (1912).
37. R. D. Miller and T. B. Jones, J. Phys. D: Appl. Phys. 21 527 (1988).
38. T. Ohgushi, K. Nonaka, H. Yoshida and T. Takaishi, Bull. Chem. Soc. Jpn. 62 2998 (1989).
39. A. F. Sprecher, Y. Chen and H. Conrad, Proc. 2nd Int. Conf. Electrorheological Fluids, ed. J. D. Carlson, A. F. Sprecher and H. Conrad, Technomic, Lancaster, PA (1990) p. 82.
40. L. C. Davis, ACS Conf. Electrorheology (ER) Materials and Fluids, Amer. Chem. Soc. Meeting, Washington, DC, Aug. 20, 1994.

# CONDUCTION PROCESSES IN ER FLUIDS

Douglas A. Brooks

Advanced Fluid Systems Limited
10/14 Pensbury Industrial Estate
Pensbury Street
London SW8 4TJ
England

## INTRODUCTION

There has been very little in the way of published data or theoretical explanation of the current flow observed in electro-rheological (ER) fluids. Simultaneous measurements of rheological and electrical properties are commonly taken and basic observations customarily made. Namely, that the current is non-ohmic, that it tends to fall with increasing shear rate but increases dramatically with temperature and to a lesser extent with volume fraction of the dispersed phase. Publications that exist are generally vague, alluding to current being either proportional to the field[1], raised to the power n, with n varying from less than 1 to 5,[2] or exponential with the field[3].

As to whether a current is necessary for the existence of the ER effect is a moot point; but regardless of this philosophical aspect in practice a current will flow. This stems from the observation that whilst solid insulators may have negligible conductivity, even pure dielectric liquids, and especially polar ones, always attain some conductivity. In practice dielectric liquids are not pure and are usually contaminated by a variety of charge carriers, mostly ions. In addition impurities in the form of trace water, solid impurities, and dissolved gases make their presence felt. In the case of ER fluids, however, significant quantities of these impurities, for example water, are actually required to achieve the rheological changes associated with the ER effect and consequently the conductivity of an ER fluid will always be greater than that of a "pure" dielectric liquid.

In practice the engineer is interested in the minimum and maximum power requirements of the device. This is characterised by the volume of fluid between the electrodes of the device, the applied voltage, and the current drawn[4]. This reduces effectively to an assessment of the relationship between the current density (J) and the voltage acting across the electrodes, that is, the field strength ($E_0$). Accepting that the power requirement of an ER device, and hence it commercial viability, is governed by the relationship between current density and field strength, it is surprising that so little attention has been focused on the characteristic form, and by implication, the origins

*Progress in Electrorheology*, Edited by K.O. Havelka
and F.E. Filisko, Plenum Press, New York, 1995

of this current. This omission is further compounded by the observation that with increasing temperature, the current density often rises dramatically. This can be a serious limitation at high temperatures. It is not unusual for current densities to double every 5 to 20°C, with the exact rate depending upon the fluid composition. For example, a typical polymer based ER fluid drawing, at room temperature, less than 1 $\mu A/cm^2$ will experience a rise in current density to the order of $10mA/cm^2$ at 100°C, a thousand fold increase.

## APPROACHES

Investigators tackling a complex problem, such as the ER effect, will adopt an approach in keeping with their background. Engineers prefer the top-down approach, measuring the performance of the system and explaining it in terms of know variables and expected models. Hence systems are often described in terms of springs, dashpots, and masses. Scientists, and physicists in particular, tend to adopt the bottom-up approach, specifying everything in terms of basic laws and simple particles. Consequently, in ER fluids for example, the dispersed phase is always spherical and monodisperse.

Although, little work has been published on the conductivity in ER fluids a similar division is already becoming apparent. The engineers prefer to describe the conductivity in terms of resistances and capacitances connected in some form of circuit, which gets progressively more complicated as additional factors are taken into account. A good example of this is the work of Bullough[1,5]. Such an approach, whilst possibly being suitable for the design of power supplies, provides limited insight into the mechanisms behind the transfer of charge and hence the current. The physics community is less well represented in the literature with the only known detailed publication being by Frood[2] who conducted measurements on ER fluids in a large parallel plate condenser. Unfortunately no temperature control was evident and no clear indication of the dimensions of the apparatus were given. The data was presented as a number of log-log plots and therefore it is difficult to extract the original current versus voltage (I-V) data. He concluded that the data could be best described by a power law expression of the form $J=CE_0^n$. Unfortunately no single exponent would adequately cover the entire field strength range, 0 to 4.0 kV/mm, and in practice three were used. This observation led to a fractal based argument for current flow and resulted in a series of equations that to solve would require significant computational time.

In both the examples cited there seems to be one serious omission, the effect of the chemical nature of the components on the ER fluid. In the case of the ER effect, the chemistry is usually reduced to a simple dielectric difference effect with little or no explanation of, and implications for, the chemical nature of the components employed. In the case of conductivity, where so little has been published, the chemistry is seldom even considered. However, given that an ER fluid is well populated by a variety of charge carriers, ions and electrons, stemming directly from the chemical nature of the components, the evolution and transport of these charges is fundamental to the understanding of the current flowing. Therefore attention will be directed to a possible explanation.

## AN ELECTRO-CHEMICAL APPROACH

This approach uses the macroscopic details of the electrode area and applied voltage to define the current density and field strength. The product of the total transported charge and the drift velocity of the charges being the net current. Neither the charge nor the charge carrier need be specified, and at this stage there is no requirement to be specific as to there nature, they can be combined and referred to as a "particle-charge". The particle-charge is subjected to a number of influences, being charged it is field dependent, that is it will preferentially drift with respect to the external field. Similarly the particle-charge is influenced by the chemical system that it is part of, it will have a chemical energy component, in particular some activation energy will be required to move the particle-charge. Finally the particle-charge is subject to thermal energy. Provided that the sum of the thermal and electrical energies exceed the activation energy the particle-charge will move with a net drift velocity in the direction of the applied electric field and a current will flow. It is apparent that if additional fields are applied to the system, specifically a perpendicular hydrodynamic field typified by the shear field in a Couette cell, the net current flow will be modified.

Empirically the current temperature profile is approximately a Boltzmann curve which is in broad agreement with the mechanism discussed above. If it is assumed that the energy is distributed amongst the particle-charges on the basis of a Boltzmann distribution with an associated charge on an electron of q, the particle-charge in an electric field $E_0$, will gain energy proportional to $qE_0$. The particle-charge will have to travel some mean free path, $\delta$, which will depend on the structure of the system. In moving distance, $\delta$, the energy gained will be $\delta qE_0$. As stated before if this energy plus the thermal energy exceeds the activation energy, $\epsilon$, after the particle-charge has moved distance $\delta$ the particle-charge will continue in the conduction process and a net current will flow. It is assumed that no electrical energy is required to charge the particle-charge, if such energy is required it would simple reduce the total electrical energy available and as such represents a reduction in the magnitude of the $\delta qE$ term.

The probability that a particle-charge will move up the potential gradient, $E_0$, will be proportional to $\exp\{(-\epsilon - \delta qE_0)/kT\}$ and the probability that the particle-charge will move down the potential gradient will be $\exp\{(-\epsilon + \delta qE_0)/kT\}$ where k is Boltzmanns constant and T is the absolute temperature. The net current will be the difference of these two probabilities. This can be written as,

$$J = A_o\, e^{-\epsilon/kT} \sinh(\delta qE_0/kT) \qquad (1)$$

where J is the current density and $A_o$ is a constant of proportionality.

The sinh function has two limiting cases. At low values of $E_0$, $\sinh(\delta qE_0/kT)$ is approximately $\delta qE_0/kT$ and therefore:-

$$J = \frac{A_0\, \delta qE_0}{kT} \exp^{-\epsilon/kT} \qquad (2)$$

That is the conduction is ohmic. However, at high fields $\sinh(\delta qE_0/kT)$ is approximately $e^{(\delta qE_0)}/2$ and the conduction is exponential. This explains some of the differences in the literature regarding the nature of the current flow in ER fluids. If the formulation of the ER fluid is such that the operative field results in a charge transfer mechanism operative in the low field region, a linear relationship will be observed. Correspondingly, when operated at the high field limit, an exponential relationship will be observed. Equation (1) is in fact a specific case of the Butler-Volmer equation for interfacial charge transfer where the symmetry factor, $\beta$, has been assigned the commonly accepted value of 0.5, this being the accepted practice for most single electron exchange reactions[6]. Consequently, conductivity in ER fluids is firmly placed within the domain of electrochemistry and its associated charge transfer mechanisms.

## EXPERIMENTAL SET-UP

### Measurement System

Static dc conductivity tests were carried out on a range of ER fluids. The conductivity cell used, shown in cross section in figure 1, was a purpose built cell with the usual precautions of guard rings and high impedance measuring equipment. Coolant was circulated around the cell to control the sample temperature. Actual fluid temperature was monitored via a thermocouple sited within the ER fluid as shown. The power supply used was a Brandenberg photomultiplier unit 872B with an output from 0 to 2.1 kV at 5mA with a 1mV peak-to-peak ripple. Voltage was applied to the cell through a high voltage relay. Current was measured on a Keithly 5½ digit multimeter.

The current sensing electrode was 2.54 cm in diameter, giving an area of 5.060 $cm^2$. The gap used throughout the experiments was maintained at 0.635 mm.

### Measurements

The measurement technique adopted was to fix the lowest temperature and apply the first voltage, 100 volts, through the high voltage relay to the cell. A small surge current was occasionally observed before the current attained a steady value which was noted. This initial current surge has been observed by Bullough[5] who explained it in terms of a capacitance charging effect. After noting the current the voltage was removed. The voltage was increased by 100 volts and the sequence was repeated. Twenty one current measurements were taken at 100 volt increments.

Having collected a set of data the temperature was increased by approximately 5°C and the experiment repeated. Initially a fresh sample was used for each temperature increment although this procedure was later discontinued as no significant differences were noted in the current measurements due to the temperature cycling or the chosen direction of the temperature increment. That is starting with a low temperature and increasing the temperature or starting with a high temperature and reducing the temperature had negligible effect.

The following experiments were conducted:

i) Current - Voltage (I-V) measurements for eight water content fluids at two volume fractions, 30% and 40%. Temperature range of 5°C to 50°C in 5°C increments.

ii) I-V measurements for nine volume fractions fluids from 5% to 45% at three water contents. Temperature range of 5°C to 50°C in 5°C increments.

iii) I-V measurements for five oils. Temperature range of 5°C to 50°C in 5°C increments.

iv) I-V measurements for six fluids manufactured with different milling times. Temperature range of 5°C to 50°C in 5°C increments.

v) A number of miscellaneous tests were conducted using different electrode materials, reverse polarity of the field and repeatability tests.

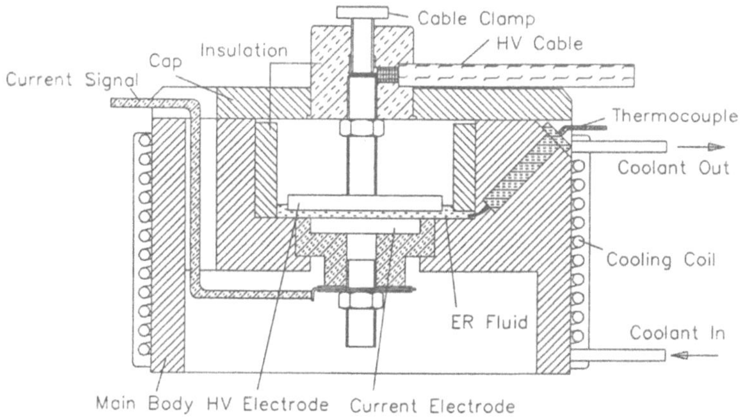


**Figure 1. Cross Section of Conductivity Cell. Details include accurate temperature measurement and temperature control.**

Some overlap in samples took place and these were used to give comparative data. In total some fifty one samples were manufactured and evaluated.

### ER Fluids

The ER fluids manufactured consist of two main components, a continuous phase and a dispersed phase, the latter component containing a quantity of water. The continuous phase used for the majority of the work was a chlorinated paraffin supplied by ICI and can be obtained under the proprietary name of Cereclor 50LV. It was a pale yellow liquid, stable up to 140°C, and compatible with most metals. It had a density of 1.19 gm/cm$^3$ at 25°C and a viscosity of 0.80 poise at the same temperature.

The dispersed phase was the product of the reaction of lithium hydroxide and polymethacrylic acid. The lithium hydroxide was obtained from BDH Chemicals in the form of 99.5% pure lithium hydroxide monohydrate. The polymethacrylic acid was supplied by Allied Colloids under the name of Versicol K13 and was in the form of an 20% aqueous solution. The number average molecular weight was approximately 26,000. After neutralisation, pH 9.00, the solution was spray dried and a fluffy white powder obtained.

The manufacture of the ER fluid is a relatively straightforward process. After obtaining the required water content the required volume of dispersed phase was mixed into the chosen continuous phase. The water content being measured after 40 minutes at 150°C. This being an AFS standard set of conditions for the measurement of water content using an infra red dryer. The mixture was then processed in a bead mill for a defined period to reduce the dispersed phase to an acceptable size. As will be discussed later, the duration of the milling process alters the particle size and particle size distribution thereby effecting the conductivity. Measurements were also made of the particle size using a Coulter Counter.

Each fluid so manufactured was assigned a fluid code for example, a code of 50LV/L35155 indicated that the fluid used Cereclor 50LV as the continuous phase, (50LV), and the alphanumerics L35155 designated a polymer, lithium polymethacrylate, (L), at a volume fraction of 35% and water content 15.5%.

## PRELIMINARY MEASUREMENTS

Obviously with this number of individual sets of data only a small portion can be shown in full. Primary measurements were those of current against temperature, water content, and volume fraction and secondary measurements were those of current against continuous phase, milling time, electrode material, and the polarity of the cell. Primary measurements were defined as those with the most significant effect.

The present interest lies in the electrochemistry of the conduction process and, after detailing some of the data the remainder of the work will concentrate on a reduction of the data based on this approach.

### Variation due to temperature

Temperature was the most convenient probe for delineating the characteristics of the conduction process. It is well known that an increase in the temperature will result in an increase in the current. This is clearly shown in figures 2 and 3 which show the temperature dependence in both linear and logarithmic forms respectively. Figure 3 indicates that the current becomes exponential at high fields and this straight line, observed on the logarithmic plot, is often referred to as a Tafel[7] line. The Tafel line is

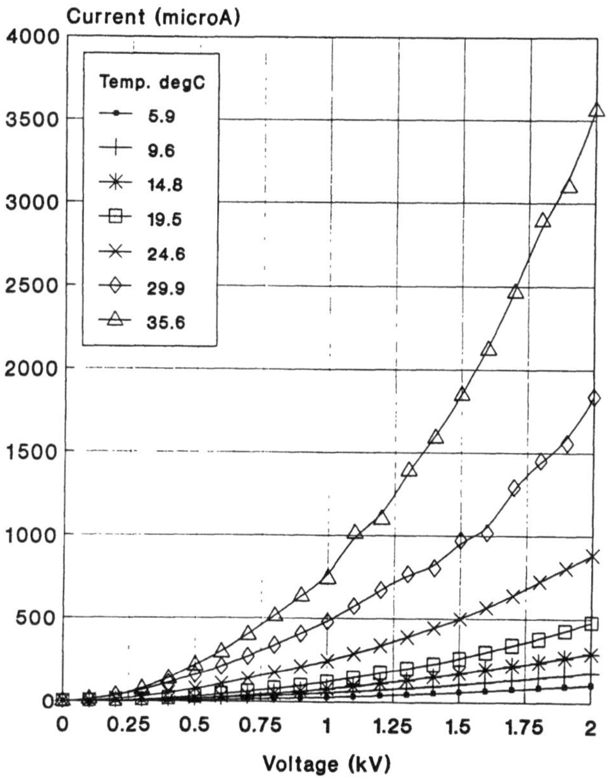


Figure 2. The Effect of Temperature on Current. Basic I-V data collected from conductivity cell. Fluid designated 50LV/L30206 is a 30% volume fraction fluid of 20.6% water content.

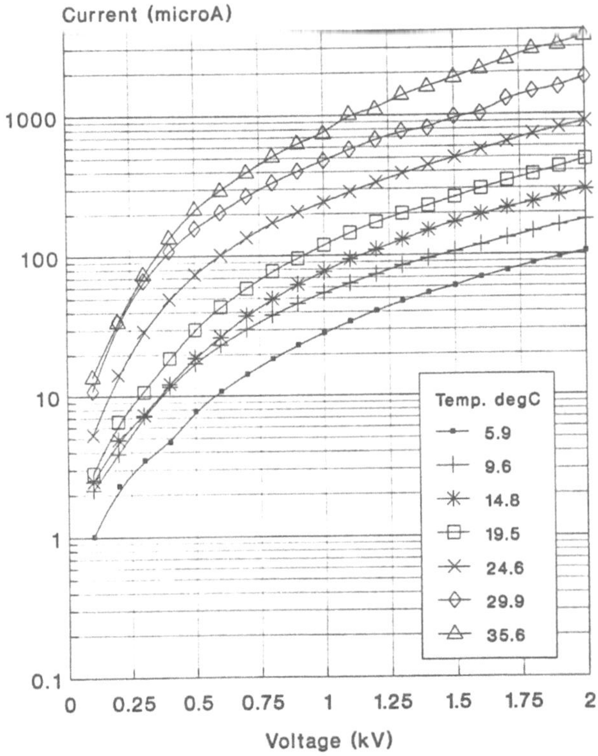


Figure 3. Logarithmic Representation of Figure 2. The high field limit tends towards a straight line and is useful to assess any Joule heating.

the high field strength exponential approximation to the conduction mechanism. This approximation can therefore be used to identify and correct the current measurements at elevated temperatures where Joule heating can be a problem. In addition the logarithmic plot of the results is more revealing of any anomalies in the current data than the linear plot.

### Variation Due to Water Content

It is generally observed that an increase in the water content of the dispersed phase results in an increase in the conductivity of the fluid. The trend is clearly shown in figure 4 which shows the current voltage characteristic, in logarithmic form, for five different water content fluids. The definition of the water content is in terms of the weight loss under and infrared dryer. It should be noted that this is an arbitrary, but repeatable and well quantified, definition for the free water in the polymer.

### Variation due to Volume Fraction

An increase in the volume fraction of the dispersed phase results in an increase in the current. The increase is less than that observed with both temperature and water content but is still significant. To avoid repetition, data is not shown in graphical form although the curves are similar to figure 4. Typically the current doubles, all other conditions being equal, from 5% volume fraction to 35% volume fraction. However, the increase between the commonly used volume fractions of 30% and 40% is in the order of 50%.

### Variation due to the Continuous Phase

The effect of different continuous phases on the current drawn in an ER fluid was assessed by manufacturing fluids with the same water content polymer, 15.5% but different oils. All other manufacturing parameters were held constant. The oils used were a representative cross section of possible oils for an ER fluid. They were, Cereclor 50LV, the standard oil used in current ER fluid formulations. A high temperature gas turbine oil from Burmah-Castrol, V342; two dielectric liquids currently used as replacements for polychlorobipheynls (PCBs) in transformers, Silicone 561 from Dow and Midel 7131 a pentaaerythritol ester from the Micanite and Insulators Company. All the fluids were tested at 30.2±0.2°C. The results indicate that the current is independent of the continuous phase.

### The Effect of Milling Time

The next set of experiments were carried out to assess any changes brought about by the milling process. Previous experimental work on the manufacture of ER fluid has shown that a minimum milling time of 5 minutes was required to reduce the spray dried polymer to a size suitable for a functional ER fluid. The milling time being related solely to the development of a reproducible ER shear stress. These tests were performed to see if there was any advantage in additional milling time with respect to the conductivity of the ER fluid. The current drawn appears to stabilise at a lower value after a milling time of 15 minutes.

### Electrode Material

Four sets of electrodes were made and tested with a 30% volume fraction 10.8% water content fluid. The materials were mild steel, aluminium, stainless steel and Niflor.

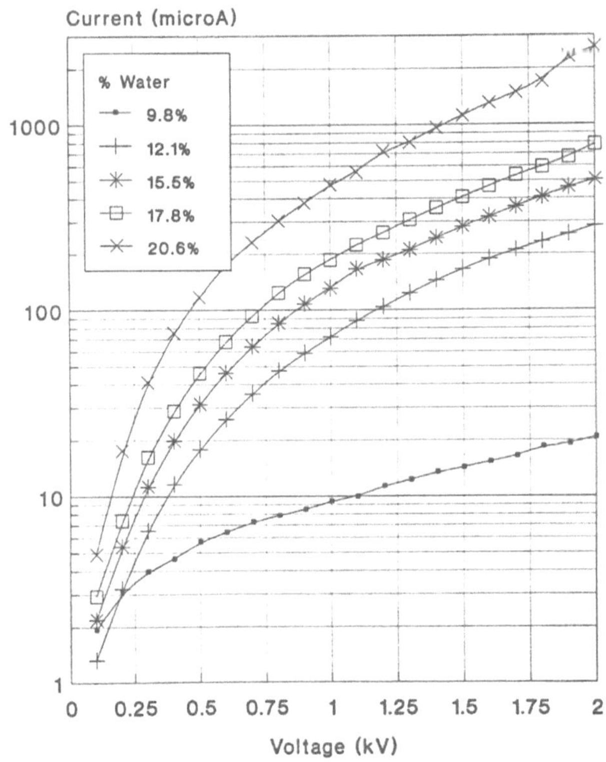


**Figure 4.** The Effect of Water Content. As the water content increases, the current drawn also increases.

The later material was a sprayed on metallic/PTFE coating with an effective 40% concentration of PTFE. With this it was anticipated that the effective surface area of the electrodes could be reduced and consequently the current density. In all case the current drawn was independent of the material used.

**Reverse Polarity**

The output of the high voltage power supply under most tests was -ve high, +ve earth. This can be readily changed by internal jumper to +ve high, -ve earth. Changing of the polarity had no effect on the current drawn.

## DISCUSSION OF RESULTS

The preliminary overview of the data indicated that the only significant variables are the temperature, the water content, the volume fraction, and the milling time. The remaining variables have little or no effect on the conductivity.

The proposed model of the conductivity, based on an ion transport rate limiting process, enables the reduction of the current density versus field strength data to three material parameters. These are an activation energy, $\epsilon$, a mean free path or jump distance, $\delta$, and a charge concentration, $A_o$.

The data can readily be fitted to the expression

$$J = A \sinh B E_0 \tag{3}$$

where $A = A_o e^{-\epsilon/kT}$ and $B = \delta q E_0/kT$ by using a least squares fitting routine. The data shown in figure 5 being typical of the results achievable. The fit is extremely good at intermediate and high field strengths but has a slight tendency to over estimate the current at low field strengths. Further reduction of the data is possible from the measurements taken at different temperatures. The charge concentration and the activation energy can be calculated from the slope and intercept of Ln A plotted against 1/T. T in °K. Furthermore from B the mean free path, $\delta$, can be evaluated. For example if we take a typical value for B of 0.8 mm/kV ($0.8*10^{-6}$ m/V) the mean free path at 30°C is given by $\delta = BkT/q$ and results in a value of $2.08*10^{-8}$ m or 200 Å.

Before assessing the effect of the primary variables on the conductivity it is useful to consider the effect of each of the parameters. Taking the charge concentration, $A_o$, first it is noted that as it is increased the total current will be increased. It is simply a multiplying factor probably linked to the total number of charges available to take part in the conduction process. The activation energy term, $\epsilon$, is in the exponential and consequently the current will be decreased as the activation energy is raised. That is basically saying that the resistance to movement is increased by increasing the depth of the potential well. Finally, the mean free path, $\delta$, will give an increased conductivity as it is increased. This is because the current is essentially the product of the concentration and the charge velocity. The velocity is the product of the mean free path and the frequency, which can be taken as a constant at a given fixed temperature. Hence when the mean free path is increased at fixed frequency the velocity must increase and with it the current.

***Temperature***

The initial step in the reduction of the data is to fit the basic I-V data to the

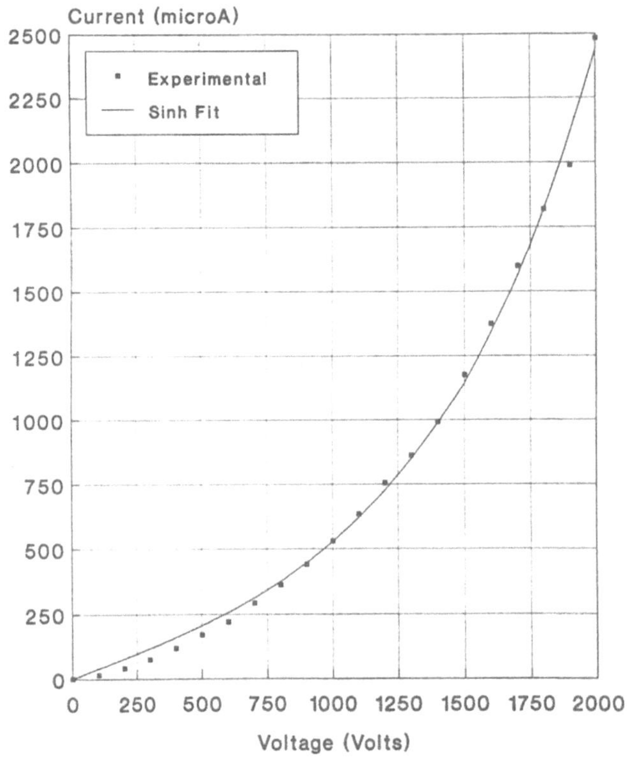


Figure 5. Sinh Fit of Experimental Data. Fluid was designated 50LV/L40155 measured at a temperature of 45°C.

expression $J = A \sinh BE_0$ and extract the A and B term. Once this has been carried out for a number of fluid formulations and temperatures further reduction of these terms can take place. Typical data are shown in figures 6 and 7 which show the two coefficients, A and B respectively. Here data for five of the eight water content fluids are plotted against 1/°K. As can be seen in figure 6, coefficient A is very temperature sensitive and largely accounts for the temperature dependence observed in practice. The scatter observed with coefficient B is significant and clear trends are difficult to detect.

### Water Content

Shown in figure 6 are data for five water content fluids. By using a least squares regression the intercept, $A_0$, and slopes, $\epsilon$, of all eight water content fluids can be obtained. This data is plotted in figures 8 and 9 where it is seen that both the charge concentration and the activation energy increase with water content. It should be noted that in the case of charge concentration we are dealing with a logarithmic quantity so that a doubling of water content from 10% to 20% corresponds to a thousand fold increase in available charges. The activation energy shows a modest increase, but as noted above, such an increase will be reflected in a reduction of current. Therefore, the two effects tend to cancel each other. However, it is clear from the experimental data that the increase in the charge concentration dominates any increased resistance to movement.

On first sight the data in relation to coefficient B, effectively the meam free path is difficult to interpret due to the scatter. Figure 7, the coefficient plotted against 1/°K, shows the best fit straight line through the data. Further detailed analysis of the individual water contents indicates that the slight fall derived from a least squares fit of the data is not universally observed. Some of the individual water contents show a slight positive slope, some virtually zero slope, and some a slight negative slope. Weighing up the body of evidence it is tempting to conclude that coefficient B is independent of temperature. The average value for B is 0.744 mm/kV which gives a mean free path at 30°C of 194 Å.

### Volume Fraction

A similar reduction of the I-V data can be carried out for the different volume fraction fluids. The initial reduction to A and B indicated that values of both increased with volume fraction. The reduced data of $A_0$, $\epsilon$, and $\delta$ are shown in figures 10 and 11. All the variables indicate an increase with the volume fraction. Again the increase in $A_0$ is the most significant and would appear to be described by a straight line given by $A_0 = 18.34 + 41.31 * \phi$, where $\phi$ is the volume fraction. Both the activation energy, $\epsilon$, and B are probably linearly related to the volume fraction.

### Milling Time

It was noted that as the milling time was increased the current fell and reached a steady level after some 15 minutes milling; the reduction in current being approximately 20%. The initial variables, A and B, have been plotted for three volume fractions. However is was noted that B is essentially independent of milling time with an average value of 0.846 (220 Å). However, the coefficient A does vary with milling time as shown in figure 12. Data on the reduction of the particle size and particle size range was available, the particle size analysis being conducted on a Coulter Counter.

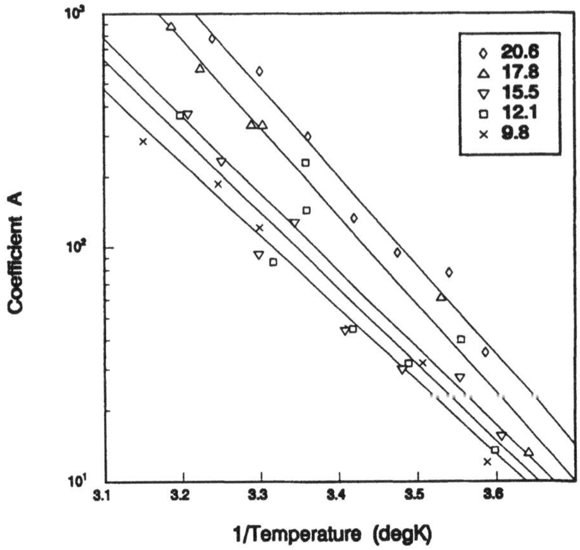


Figure 6. Variation of Coefficient A with Temperature and Water Content. Five 30% volume fraction fluid shown.

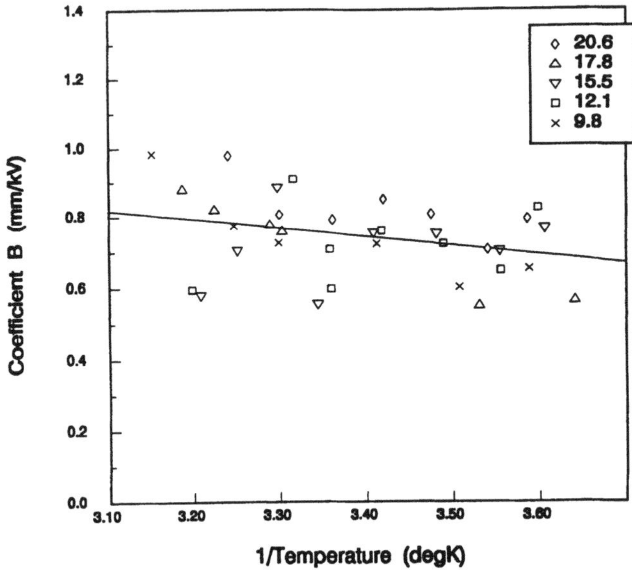


Figure 7. Variation of Coefficient B with Temperature and Water Content. Five 30% volume fraction fluid shown. The average value of B is 0.744 mm/kV corresponding to 194 Å.

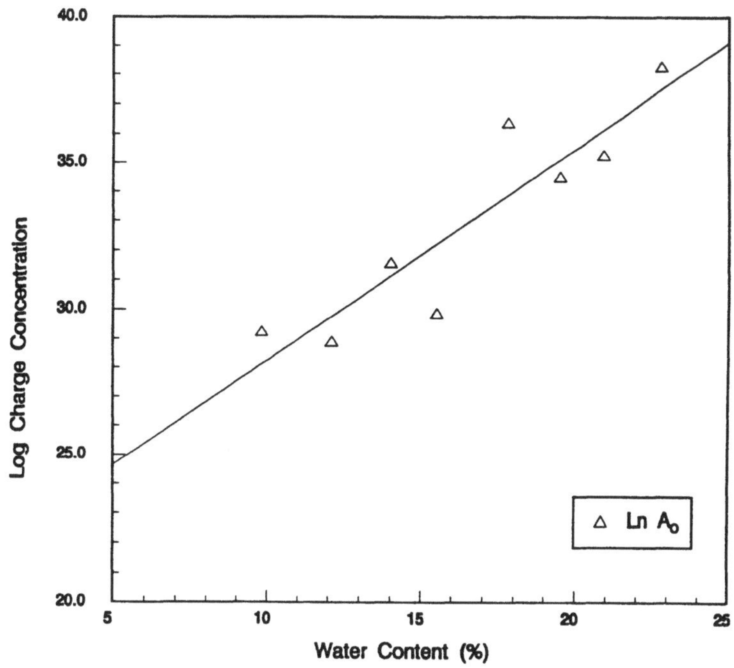


Figure 8. Variation of Charge Concentration with Water Content. Further reduction of A from figure 6 enables the parameter $A_0$ to be evaluated. Linear regression gives a line of the form $LnA_0 = 21.04 + 0.723 * H_2O$.

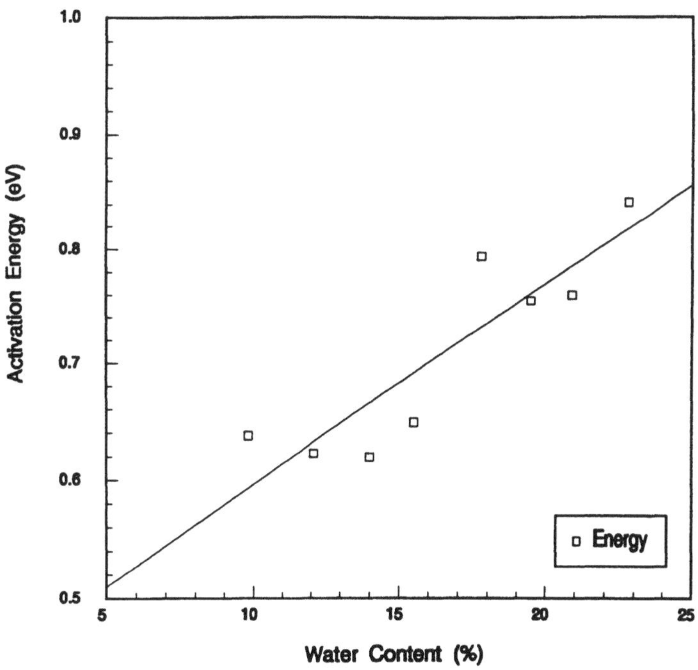


Figure 9. Variation of Activation Energy with Water Content. Further reduction of A from figure 6 enables the parameter $\epsilon$ to be evaluated. Linear regression gives a line of the form $\epsilon = 0.423 + 0.0174 * H_2O$.

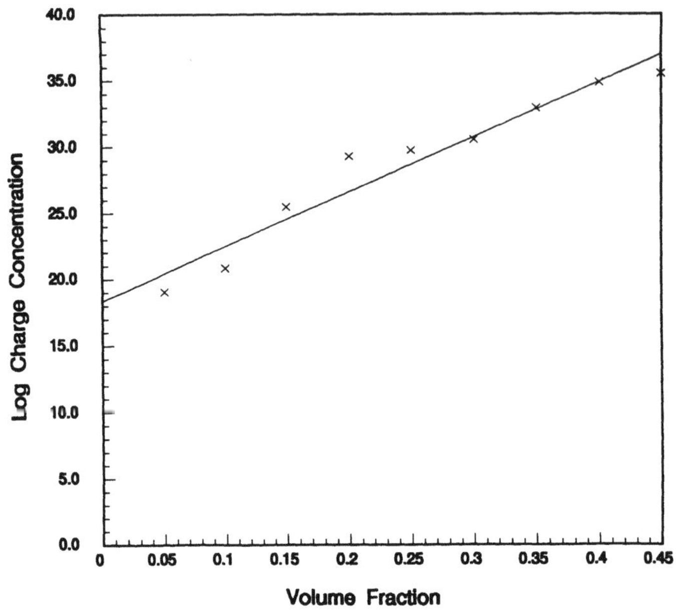


**Figure 10.** Variation of Charge Concentration with Volume Fraction. Data falls along a line described by $A_o = 18.34 + 41.31\ \phi$. Cereclor 50LV based fluid with a 15.5% water content polymer, tested at 30°C.

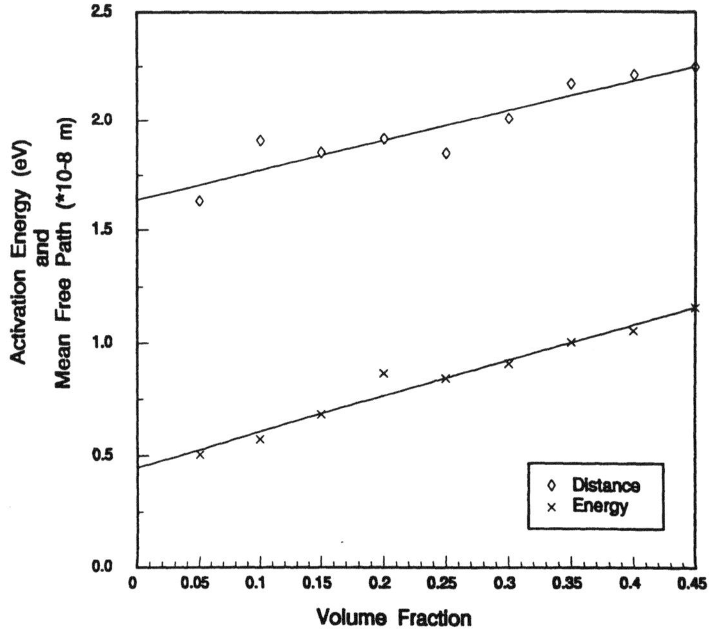


**Figure 11.** Variation of Activation Energy and Mean Free Path with Volume Fraction. Linear regression gives equations $\epsilon = 0.448 + 1.57 * \phi$ and $\delta = 1.64 * 10^{-8} + 1.36 * 10^{-8} * \phi$

What was noted from the analysis of these data was that increasing the milling time brings about a reduction in the particle size range by reducing the number of larger particles in the sample. The lower end of the size range remains around the 1 to 2 micron whereas the higher end pulls in from approximately 25 microns to 11 microns. This change is also reflected in a reduction of the mean particle size from 10 to 7.5 micron. Hence increased milling time reduces the sample polydispersity and the total number of particles must increase without altering the volume fraction.

## Interpretation of Observations

A number of useful comments can be made concerning the model and the interpretation of the data within the framework offered by the model. Firstly, the use of log-linear plots and in particular the high field straight line portion, the Tafel line, is very useful in observing secondary effects, such as heating, and any anomalous data points that would be hidden by a conventional linear-linear presentation. For certain fluids, a high water content and high volume fraction fluid for example, at high fields the power input into the fluid is significant, some 30 watt/cm$^3$, and can present problems with temperature stability. Use of a Tafel line allows such data to be corrected.

The quality of the sinh fit is very good considering the symmetry assumptions of the charge transfer mechanism and the assumption that no energy is required to charge the carriers. A deviation from either assumption would explain the difficulty in approximating the low field values. A further useful observation to make is the role of additional fields, typically hydrodynamic fields. It would be expected that such fields would tend to be dominant at low electrical fields and consequently fitting of the data to a sinh curve would become progressively less accurate as the shear rate in a Couette cell is increased. This is indeed the case. To accommodate such effects additional terms would be required, possibly by altering the symmetry factor, $\beta$. However, this would then introduce additional material parameters and reduce the exercise to a unsophisticated curve fitting approach.

Is there any correspondence between this approach and the RC model approach? Conceivably there is, in that energising the charges can be considered as capacitor charging and charge transfer can be seen in some fashion as analogous to resistance. However, there is at present no clear comparisons.

It is also useful to consider the effect of the main variables of temperature, water content, and volume fraction on the material parameters of $A_0$, $\epsilon$, and $\delta$. If $A_0$ is considered first it is observed that it increases with all three, temperature, water content, and volume fraction. This is not unexpected as $A_0$ is assumed to be the charge concentration and would clearly increase with these variables. The activation energy, $\epsilon$, is seen to be essentially independent of temperature but increases with both water content and volume fraction in a linear fashion. The intercepts, ie 0% water content and 0% volume fraction are 0.423 and 0.447 respectively and are close enough to be considered the same. Finally the mean free path, $\delta$, is seen to be independent of temperature and water content but increases, also in a linear fashion, with volume fraction. Typical values for the three material parameters are shown in Table 1 at a temperature of 30°C.

The possible sources of error on these values can be assessed. For $A_0$ the possible errors are potentially large as measurements were taken at temperatures between 0°C and 90°C but extrapolated back to 0°K. The values for the activation energy, $\epsilon$, are of a realistic nature being typical for the energy required to force the charge across a potential barrier. Also given that it appears independent of

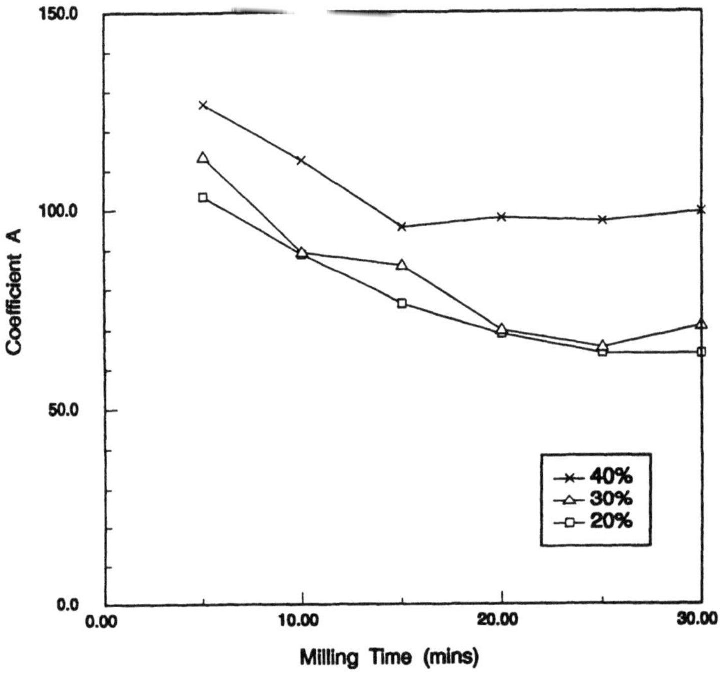


Figure 12. Variation of Charge Carriers with Milling Time. Water content 9.8%, continuous phase 50LV, temperature 30°C.

temperature one can conclude that it is a reasonably accurate value. The mean free path presents some problems. Superficially one might expect this to be the interparticle distance. However, the magnitude is such as to suggest that it is probably more associated with the surface characteristics of the particle.

Data was available concerning the current - voltage characteristic of the oil 50LV alone, that is at 0% volume fraction. The temperature was 27°C. The results show, not surprisingly, that the oil was an ohmic conductor. It is therefore possible to use the low field approximation of the sinh expression, equation 2, to obtain a value for $A_o$ and compare it with the intercept value of figure 10. Values for $\epsilon$ and $\delta$ are required at zero volume fraction and these can be obtained from the intercepts of figure 11. The values are 0.448 and 1.64 * $10^{-8}$ respectively. Inserting the relevant values in equation 2 and accounting for the units indicates that $J = A_o * 2.23 * 10^{-14} E_o$. Data taken form measurements on the oil indicates that $J = 6.56 * 10^{-6} E_o$. Therefore, $A_o$ is approximately 2.9 * $10^8$. This compares favourably with the intercept value of figure 10 of approximately 1 * $10^8$. The logarithmic values are 19.5 and 18.34 respectively.

Another supporting observation relates to the effect of changing the particle size through milling. As noted this caused a slight reduction in the current. When analyzed this shows that the reduction is caused by either a change in $A_o$ or $\epsilon$ and not a change in $\delta$. Unfortunately it was not possible to identify which of the two parameter was responsible, but the most probable candidate is $A_o$. However, as a result of the increased milling time, for a fixed volume fraction fluid, the actual number of particles increases considerably. They are also reduced in size. Therefore, the distance between particles must be reduced. As $\delta$ is independent of milling time and hence particle size and distribution it must be concluded that $\delta$ is not associated with any macroscopic physical property of the fluid. This allows two interpretations, either the mechanism reflects the freeing of charge carriers from the structure of the polymer and details motion within the particle or it reflects the freeing of charge carriers from the surface of the particle and transferring then between particles.

**TABLE 1 Typical Values of Parameters for different Volume Fraction Fluids**

| Fluid | $\phi$(%) | $A_o$ | $\epsilon$(eV) | $\delta$(m) |
|---|---|---|---|---|
| 50LV/L10180 | 10 | $1.08e^9$ | 0.572 | $1.38e^{-8}$ |
| 50LV/L20180 | 20 | $5.09e^{12}$ | 0.864 | $1.77e^{-8}$ |
| 50LV/L30180 | 30 | $1.80e^{13}$ | 0.904 | $1.91^{e-8}$ |
| 50LV/L40180 | 40 | $1.31e^{15}$ | 1.053 | $2.04e^{-8}$ |

## CONCLUSION

Current flow in ER fluids is assumed to be due to a charge transfer mechanism occurring either within the particle or at the particle-oil interface. The use of a simplified Butler-Volmer expression allows three material parameters that are realistic and useful in interpreting the nature of the current in ER fluids. These parameters allow the current to be defined against such variables as temperature, water content, and volume fraction. It is also possible to reconcile some of the differences reported in the literature as to the characteristic form of the conductivity curve.

## REFERENCES


1. Bullough, W.A., Peel, D.J., and Firoozian, R. "Electrical characteristics and other considerations for electro-rheological fluids". Prod 1st Int Conf on ER Fluids, North Carolina State University, Engineering Publications.

2. Frood,D.G."Preliminary Measurements of I-V Characteristics in Starch Based ER Fluids". Prod 2nd Int Conf on ER Fluids, Lancaster PA Technomic Publishing.

3. Conrad, H., Chen, Y., and Sprecter, A.F., "Electro-rheology of Suspensions of Zeolite Particles in Silicone Oil". Prod 2nd Int Conf on ER Fluids, Lancaster PA Technomic Publishing.

4. Duclos, T. G., "Design of Devices using Electrorheological Fluids". SAE Paper No 884434, 1988.

5. Whittle, M., Firoozian, R., Peel, D.J., and Bullough, W.A., "A Model for the Electrical Characteristics of an ER Valve". Prod of Int Conf on ER Fluids, 15-16 Oct 1991, Carbondale, Illinois, p343-386. World Scientific Publishing.

6. Bockeris, J O'M., Reddy, A.K.N. Modern Electrochemistry 2, Plenum/Rosetta NY. pp 862-878.

7. Tafel, J., Z Physik. Chem. (Leipzig), 50, p641 (1905)

# ELECTROSTATIC FORCES IN ELECTRORHEOLOGICAL FLUIDS

L. C. Davis and J. M. Ginder

Research Laboratory
Ford Motor Company
20,000 Rotunda Drive, MD 3028
Dearborn, MI 48121-2053

## INTRODUCTION

Electrorheological (ER) fluids generally consist of highly polarizable or conductive particles in insulating fluids. Application of electric fields of order 1 kV/mm significantly alters the rheology of these fluids, thus making possible a wide range of electro-mechanical devices, such as automotive shock absorbers. One of the critical issues in the performance of ER fluid devices is the strength of the fluid. In this paper, we summarize our current understanding of the electrostatic forces in ER fluids and how they affect fluid properties. This understanding is based on detailed theoretical analysis of the electric fields between particles, including finite element analysis (FEA), in addition to measurements of shear modulus and yield strength of model systems. The effects of dielectric constant mismatch, conductivity mismatch, insulating layers on metallic particles, and fluid breakdown are considered.

## MODULUS AND YIELD STRENGTH

The geometric model of ER fluids we analyze consists of uniformly spaced, single chains which span the distance between the electrodes. If a shear strain $\gamma = \tan\theta_0$, where $\theta_0$ is the shear angle, is applied slowly to the model chain structure, the required shear stress $\tau$ is determined as follows. For small $\theta_0$, the shear stress is linear with slope G, which is the shear modulus. At larger $\theta_0$, the stress levels off at the static yield stress $\tau_y$.

A standard model[1] consists of the geometry described plus the assignment of dielectric constants to the particle ($\epsilon_p=K_p\epsilon_0$, where $\epsilon_0=8.85419\text{x}10^{-12}$ F/m) and to the fluid ($\epsilon_f=K_f\epsilon_0$). This picture, although it is almost universally accepted, really only holds for ac electric fields of sufficiently high frequency (above 10-300 Hz) such that conductivity effects can be neglected. In dc fields, it is sometimes useful to realize that particle and fluid have conductivities $\sigma_p$ and $\sigma_f$. In this case, it is the conductivity

*Progress in Electrorheology*, Edited by K.O. Havelka
and F.E. Filisko, Plenum Press, New York, 1995

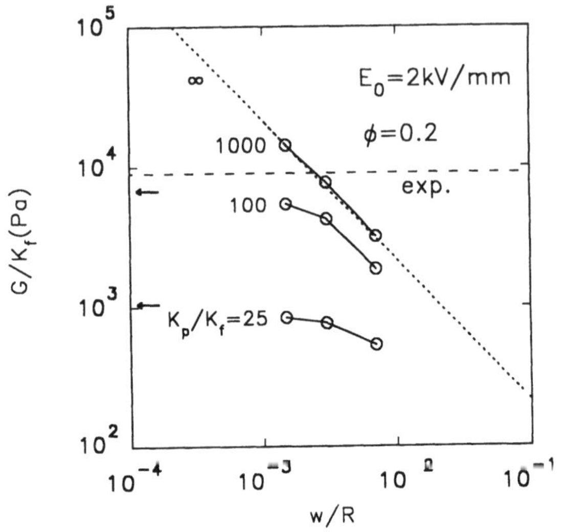


Figure 1. Calculated shear modulus vs interparticle gap. The dotted line is an analytic approximation for particles with infinite dielectric constant. The dash-dotted line is the experimental value for $BaTiO_3$/dodecane (unknown gap). The arrows are an analytic approximation from Ref. 7 for $K_p/K_f=25$ and 100.

contrast that is important in determining the spatial variation of the electric field in the sample and thus in controlling the particle-particle forces. Careful treatments[2,3] of conductivity effects have appeared elsewhere and will not be discussed here. In this section, we consider ac fields only.

The linear shear regime where $\tau = G\ \theta_0$ is interesting because the chains presumably undergo small, reversible deformations. Chain breaking and reforming, which are not as well understood, should not occur to any great extent. Thus one expects to observe the particle-particle interactions most clearly in this regime. Experimentally this can be approached by observing the response to small oscillating shear strains. A model fluid[4] of barium titanate particles having 1.6$\mu$m diameter with $K_p \approx 2000$,[5] suspended at a volume fraction $\phi = 0.2$ in dodecane with $K_f = 2.0$, was utilized in these studies. Measurements were performed with a controlled-strain rheometer using the Couette geometry. Details are given in Ref. 6.

It was found that the shear stress varies with the applied strain and the electric field E(t); in the latter case, the response is at twice the frequency since all inter-particle polarization forces are expected to depend upon $E^2(t)$. For the smallest values of the amplitude of the applied shear strain, $\gamma_{peak}$, the stress is nearly independent of applied field, suggesting that non-polarization particle-particle forces such as van der Waals forces dominate. Since a double modulation technique is used, the stress due to the electrical component can be isolated. The expected linear behavior with strain was observed for small $\gamma_{peak}$. Likewise, the stresses increase roughly as $E_{rms}{}^2$, where $E_{rms}$ is the root-mean-square applied electric field, as expected for a polarization mechanism. The magnitude of the measured moduli (slope at low strain) are consistent with finite element analysis (FEA) calculations for this system (Fig. 1). The moduli are somewhat sensitive to the assumed smallest particle-particle separation. The arrows are the approximate (low field) limiting values determined from Felici, Foulc, and Atten[7], which the FEA results appear to approach:

$$G = 6\epsilon_f \phi \left(\frac{R}{\delta}\right)^2 E_{rms}^2, \qquad (1)$$

$$\pi \frac{R}{\delta} \ln\left(\frac{R}{\delta}\right) = \frac{\epsilon_p}{\epsilon_f}$$

where $\delta$ is an effective radius of the contact zone between particles of radius R.

The yield stress, which is the stress that is approached asymptotically as $\gamma_{peak}$ increases, is also in fair agreement with FEA results. Our measured results and some literature values[5] for particles of $TiO_2$ ($K_p$ = 70) and $SrTiO_3$ ($K_p$ =270) as well as $BaTiO_3$ are compared to theory in Fig. 2. The theory tends to overestimate the yield stress, probably because the assumed ideal chain structure is not completely attained in real fluids. Nonetheless, it appears that for these model fluids we have a good understanding of the ER mechanism for ac fields.

The trend established in Fig. 2 suggests that increasing the particle dielectric constant increases the yield stress. The limit in this regard is metal particles, since they essentially possess an infinite dielectric constant. Metal particles might be undesirable because of abrasion or settling problems, but it is useful to examine how large a yield stress they might produce. To avoid electrical short circuits from metal particles that are in contact, a thin insulating oxide layer must cover the particles. Let the oxide thickness be $\delta_{ox}$ and its dielectric constant be $\epsilon_{ox}$. From an analysis previously published,[8] the yield stress can be determined:

$$\tau_y = \frac{\epsilon_f}{2} 3\phi E_{rms}^2 \sqrt{\frac{\epsilon_{ox} R}{2\epsilon_f \delta_{ox}}}. \qquad (2)$$

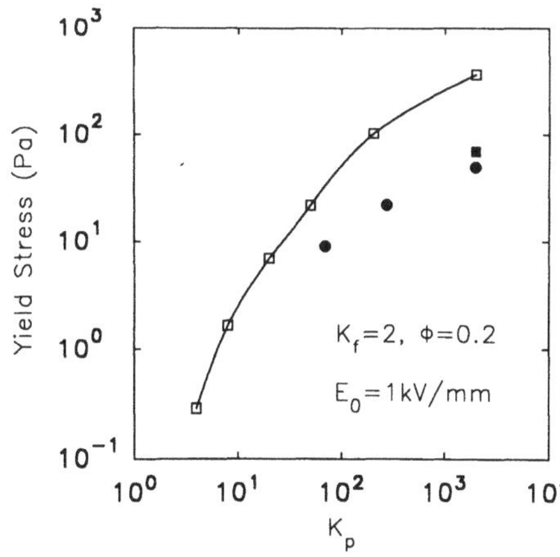


Figure 2. Yield stress vs particle dielectric constant. The open squares are FEA values; the filled circles are experimental results from Ref. 5; and the filled square is the measured value for $BaTiO_3$/dodecane.

The same analysis shows that the maximum field in the fluid between the particles is

$$E_{max} = \frac{\epsilon_{ox}}{\epsilon_f} \frac{R}{\delta_{ox}} \sqrt{2} E_{rms}. \tag{3}$$

Although Eq. (2) shows that reducing $\delta_{ox}$ improves the yield stress at a given electric field, Eq. (3) shows that the field between the particles can become so large that the dielectric breakdown field of the fluid (or the oxide) is exceeded. Breakdown effects will be treated in the next section.

## THEORETICAL LIMITATIONS

In this section, we present an analysis of the theoretical limitations on the ultimate shear strength of ER fluids. The idea is based upon the work of Felici *et al.*[7], who considered the effects of field-induced conductivity enhancement in the high field regions between nearly touching particles. These nonlinear conductivity effects limit the electric field between particles and thus limit the ER activity. We consider two particles in an infinite chain of particles. Only single chains, not multiple-chain columns, are again assumed. We take the particles to be highly conducting so that $\sigma_p >> \sigma_f$, that is, the particle conductivity greatly exceeds the fluid conductivity. Here we take the electric field to be static. The local electric field in the z direction (applied field $\mathbf{E_0} = E_0 \hat{z}$) in the gap region is given by

$$E_z = \frac{E_0 2R}{w + \rho^2/R}, \quad \rho > \delta \tag{4}$$

where $\rho$ is the radial distance from the axis of symmetry, w is the gap between particles and R is the particle radius. At some value of $\rho$, which is denoted by $\delta$, $E_z$ reaches a breakdown field of value $E_m$. We assume that the local conductivity increases to ensure that

$$E_z = E_m, \quad \rho < \delta. \tag{5}$$

Neglecting w since the particles are nearly touching, we can equate the two expressions at $\rho = \delta$ to obtain

$$\frac{\delta}{R} = \sqrt{2 \frac{E_0}{E_m}}. \tag{6}$$

From Felici *et al.*[7], [or from Eq. (1) with $\delta$ given by Eq. (6)] we can determine the shear modulus:

$$G=3\epsilon_f \phi E_0 E_m. \tag{7}$$

Note that the modulus no longer depends on $E_0^2$, but is linear in applied electric field. The parameter $E_m$ depends upon the composition and purity of the suspending fluid. Felici *et al.* suggest that $E_m$ is of order 30 kV/mm. Similar field intensifications and conductivity enhancements are expected in the gaps between high dielectric constant particles in low frequency ac excitation. Applied to the $BaTiO_3$/dodecane system, Eq. (6) gives G = 637 Pa at $E_0$ = 2 kV/mm, which is considerably below the measured value of 18 kPa.[6] Maximizing the shear stress with respect to $\theta_0$ (as in Ref. 8.), we find the yield stress has an $E_0^{3/2}$ dependence:

$$\tau_y=\frac{4}{\sqrt{3}}\epsilon_f \phi E_0^{3/2}\sqrt{E_m}. \tag{8}$$

For $BaTiO_3$/dodecane, Eq. (8) predicts that $\tau_y$(2kV/mm) = 127 Pa compared to 140 Pa obtained experimentally.

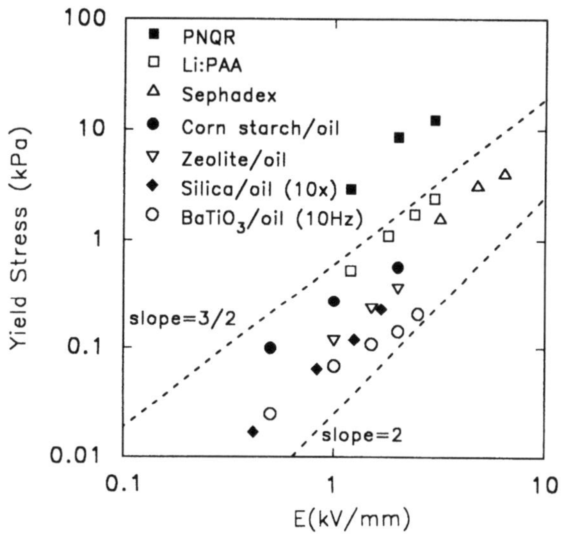


Figure 3. Yield stress vs applied electric field (dc, except as noted otherwise, room temperature). ■PNQR poly(acenequinones)/Cereclor ($\phi$ = 0.35) [Ref. 9]; △ Sephadex QAE/paraffin oil, 100°C [Ref. 10]; □ hydrated particles of lithium salt of poly(methacrylate)/chlorinated paraffin oil (Cereclor, ICI, $\phi$=0.35) [Ref. 11]; ● corn starch (7.4 wt.% $H_2O$)/corn oil ($\phi$=0.28) [Ref. 12]; ▽ Sigma Z3125 zeolite particles/silicone oil ($\phi$=0.25) [Ref. 13]; ◆ $10\tau_y$ for hollow silica spheres/corn oil ($\phi$=0.35) [Ref. 14]; ○ $BaTiO_3$/dodecane ($\phi$=0.2) at 10 Hz [present work].

In Fig. 3, the yield stress as function of applied field $E_0$ is shown for several model fluids reported in the literature. These fluids include particles that are comprised of conducting polymer (PNQR), ferroelectric ($BaTiO_3$), hydrous, and anhydrous materials. Results for commercial fluids described in product literature and other industrially prepared fluids (with references to the appropriate literature) are displayed in Fig. 4. The detailed composition of these fluids is not revealed in most instances. It can be seen that the trend is for $\tau_y$ to be proportional to $E_0^2$ at low fields and to approach $E_0^{3/2}$ at high fields in most, but not all, of the fluids. A notable exception (not shown) is the data for PAnQR (35% volume fraction) in Cereclor (See

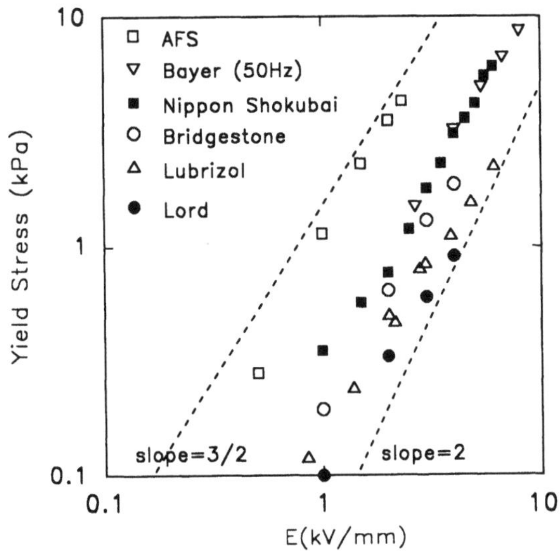


Figure 4. Yield stress vs applied electric field (dc, except as noted) for industrially prepared fluids. □ Advanced Fluid Systems (hydrated particles of lithium salt of poly(methacrylate)/chlorinated paraffin oil), 50°C; ▽ RHEOBAY® TP AI 3565, 40°C, 50 Hz (polyurethane/silicone oil); ■ Nippon Shokubai NSP(5.5) (styrene-co-divinylbenzene particles reacted with chlorosulfonic acid, 5.5 wt. % $H_2O$/silicone oil) [Ref. 15]; ○ Bridgestone's ERF (carbonaceous particles); △ Lubrizol (cellulose, 15 wt.% $H_2O$, hydroxypropyl functionalized polydimethlsiloxane, 2 wt.% $H_2O$/silicone oil), 60 °C [Ref. 16]; ● Lord VersaFlo™ ER-100.

Fig. 9 of Ref. 6.), which show an approximate $E_0^2$ dependence over the range 0.8 to 3.6 kV/mm. There is no evidence that the yield stress will eventually approach an $E_0^{3/2}$ behavior in this fluid. It is not clear why some fluids appear to follow the predicted scaling and others do not. The limited range of electric fields and difficulties with measuring yield stresses at both low and high values of $E_0$ may be partially responsible.

Some predicted values of yield stress are given in Table I for various assumed values of $E_m$. Larger, non-linear current flow in the fluid will occur for increasing values of $E_0$, although a detailed model of $\sigma_f(E)$ in the gap region is required to predict the current.

Table I. Predicted yield stresses for various values of $E_m$ for $\phi = 0.4$, $E_0 = 4$ kV/mm, and $\epsilon_f = 12\ \epsilon_0$.

| $E_m$(kV/mm) | $\tau_{max}$(kPa) |
|---|---|
| 30 | 4.3 |
| 300 | 13.6 |
| 1000 | 24.8 |

## CONCLUSIONS

Calculations of the shear modulus and yield strength on model ER systems (e. g., $BaTiO_3$/dodecane) excited by ac electric fields show reasonable agreement with measurements. In this case conductivity effects can be neglected, which simplifies the analysis. For dc excitation, replacing the dielectric constant ratio $\epsilon_p/\epsilon_f$ by the conductivity ratio $\sigma_p/\sigma_f$ in force expressions is an approximate method to predict the rheological response of ER fluids in low fields. At higher fields, field-induced conductivity enhancement is an important effect and may be incorporated into the model as suggested by Felici *et al.*[7] However, their suggested value for the critical local field at which conductivity enhancement occurs is too small to account for some measurements in alternating fields[6]. The model predicts a $E_0^{3/2}$ dependence of the yield stress, which tends to agree with results on a broad range of fluids reported in the literature and in product information brochures.

## REFERENCES


1. L. C. Davis, Finite-element analysis of particle-particle forces in electrorheological fluids, *Appl. Phys. Lett.* 60:319 (1992).
2. L. C. Davis, Polarization forces and conductivity effects in electrorheological fluids, *J. Appl. Phys.* 72:1334 (1992).
3. R. A. Anderson, Effects of finite conductivity in electrorheological fluids, *in*: "Proceedings of the Third International Conference on Electrorheological Fluids", R. Tao, ed., World Scientific, Singapore (1992), p. 81.
4. Prof. P. P. Phule (private communication) has suggested that small $BaTiO_3$ particles may not behave as linear dielectrics and may possess a conducting, hydroxyl surface layer. However, the present experiments were done with large particles that we expect to be multidomain and thus to show a linear response. Some conductivity effects could occur at low frequencies, but larger frequency excitation eliminates these effects.
5. T. J. Garino, D. Adolf, and B. Hance, The effect of solvent and particle dielectric constants on the electrorheological properties of water-free ER fluids, *in*: "Proceedings of the Third International Conference on Electrorheological Fluids," R. Tao, ed., World Scientific, Singapore (1992), p. 167.
6. J. M. Ginder and L. C. Davis, Viscoelasticity of electrorheological fluids: Role of electrostatic interactions, *in*: "Proceedings of the Fourth International Conference on Electrorheological Fluids," R. Tao, ed., World Scientific, Singapore, (1994), p. 267.
7. N. Felici, J.-N. Foulc, and P. Atten, A conduction model of electrorheological effect, *in*: "Proceedings of the Fourth International Conference on Electrorheological Fluids," R. Tao, ed., World Scientific, Singapore, (1994), p. 139.
8. L. C. Davis, The metal-particle/insulating oil system: An ideal electrorheological fluid, *J. Appl. Phys.* 73:80 (1993).
9. H. Block, J. P. Kelly, A. Qin, and T. Watson, Materials and mechanisms in electrorheology, *Langmuir* 6:6 (1990).

10. F. E. Filisko, Rheological properties and models of dry ER materials, *in*: "Proceedings of the Third International Conference on Electrorheological Fluids," R. Tao, ed., World Scientific, Singapore (1992), p. 116.
11. W. S. Yen and P. J. Achorn, A study of the dynamical behavior of an electrorheological fluid, *J. Rheol.* 35:1375 (1991).
12. Y. Chen and H. Conrad, Effects of water content on the electrorheology of corn starch/corn oil dispersions, *in*: ASME AMD-Vol. 175, "Developments in Non-Newtonian Flows 1993," D. A. Siginer, W. E. Van Arsdale, M. C. Altan, and A. N. Alexandrou, eds., ASME, New York (1993), p. 199.
13. H. Conrad, Y. Chen, and A. F. Sprecher, Electrorheology of suspensions of zeolite particles in silicone oil, *in*: "Proceedings of the Second International Conference on Electrorheological Fluids," J. D. Carlson, A. F. Sprecher, and H. Conrad, eds., Technomic, Lancaster, PA (1990), p. 252.
14. D. J. Klingenberg, F. van Swol, and C. F. Zukoski, The small shear rate response of electrorheological suspensions. II. Extensions beyond the point-dipole limit, *J. Chem. Phys.* 94: 6170 (1991).
15. Y. Asako, S. Ono, R. Aizawa, and T. Kawakami, Properties of electrorheological fluids containing numerously sulfonated polymer particles, *Polymer Preprints* 35:352 (1994).
16. J. W. Pialet and D. R. Clark, The dependence of shear stress and current density on temperature and field for a model electrorheological fluid, *Polymer Preprints* 35:367 (1994).

# SURFACTANT-ACTIVATED ELECTRORHEOLOGICAL SUSPENSIONS

Y. D. Kim and D. J. Klingenberg

Department of Chemical Engineering
and Rheology Research Center
University of Wisconsin
Madison, WI 53706

**Abstract**

Applications of the electrorheological (ER) response are limited by a lack of effective fluids. Surfactants may be used to tailor ER suspension properties. We report investigations of the influence of nonionic surfactants on the field-induced dynamic yield stress of various alumina suspensions. We explore the dependence of the yield stress on surfactant type and concentration, particle type, water content, and electric field strength and frequency. Suspension dielectric properties are also investigated. Experimental results at small surfactant concentrations are qualitatively reproduced by a simple model based on Maxwell-Wagner polarization of the disperse phase. This suggests that surfactants act primarily to increase particle conductivity, and hence polarizability, leading ultimately to an increased yield stress. Possible explanations are proposed for deviations from this model at large surfactant concentrations.

## INTRODUCTION

The electrorheological (ER) response is defined as the dramatic change in rheological properties of a suspension of small particles due to the application of a large electric field transverse to the direction of flow. ER suspensions are typically composed of nonconducting or semiconducting particles dispersed in a nonconducting continuous phase. The observation of a large ER effect was first reported by Winslow in 1949[1]. The

*Progress in Electrorheology*, Edited by K.O. Havelka
and F.E. Filisko, Plenum Press, New York, 1995

simplicity of engineering designs based on ER materials has facilitated the development of specifications for a broad range of devices, such as dampers, clutches, and adaptive structures[2].

Although many ER devices have been brought successfully to the prototype stage, and despite much industrial activity in the U.S. and abroad, there are currently no commercially available devices. The main limitation of ER technology development is a lack of effective fluids[3,4]. Most applications require fluids that possess a large field-induced yield stress, are stable to settling and irreversible aggregation, are environmentally benign, and draw limited current. Our inability to design such acceptable fluids stems largely from a lack of a fundamental understanding of the mechanisms that control ER behavior.

Surfactants are added to ER suspensions for a variety of reasons[1,5,6,7,8,9,10,11], and can be used to tailor suspension properties. They are often used to promote colloidal stability, which is necessary to keep particles from irreversibly flocculating, and to control rheological properties in the absence of the electric field. Surfactants are also used to "activate" suspensions. Some suspensions display little or no ER activity unless a small amount of water or surfactant is added, while other suspensions exhibit a significantly enhanced response with activators present[9,10,11,12,13]. Enhancing ER activity with surfactants offers advantages over other approaches, such as adding water which severely limits the allowable temperature range of operation, promotes corrosion, and also increases suspension conductivity and power consumption. Furthermore, additional independent variables (*i.e.*, type and amount of surfactants) give flexibility to designing desired properties that is not possible by simply varying the materials of the disperse and continuous phases.

In this paper, we report experimental studies of the influence of nonionic surfactants on the ER properties of various alumina suspensions. In the following section, sample preparation, and rheological and dielectric measurement methods are briefly described, followed by a discussion of experimental results. The dependencies of the dynamic yield stress on surfactant type and concentration, particle type, water content, and electric field strength and frequency are presented, along with suspension dielectric properties. A model based on Maxwell-Wagner polarization of the disperse phase is then developed, and is found to qualitatively reproduce both the rheological and dielectric data at small surfactant concentrations. Possible explanations for deviations from this model at large surfactant concentrations are then discussed.

## EXPERIMENT

### Suspension Preparation

Three different types of activated alumina particles were employed: neutral, acidic, and basic (Aldrich, $\rho_p = 3970$ kg/m$^3$, average pore diameter = 58 Å). The alumina particles were approximately spherical and sieved to obtain diameters in the range of 63 - 90 $\mu$m. Nonionic surfactants investigated were glycerol monooleate (GMO, Chemical Service), glycerol trioleate (GTO, Chemical Service), and Brij® 30 ($C_{12}H_{25}(OCH_2CH_2)_4OH$, Aldrich).

In order to investigate the influence of water content on ER activity, particles were processed three different ways. The particles were either used as received ("nondried" suspensions), were dried for 3 hours under vacuum (< 0.5 torr) at 58 °C ("dried" suspensions), or were dried for 4 hours under vacuum (< 0.5 torr) at 160 °C ("highly dried" suspensions) to remove free water. The water contents of the particles were

**Table 1.** Water contents of particles, silicone oil, and surfactants determined by Karl Fisher titration.

| Material | Water content (wt %) |
|---|---|
| Nondried neutral alumina | 3.47 ±0.12 |
| Nondried acidic alumina | 4.26 ±0.12 |
| Nondried basic alumina | 1.51 ±0.12 |
| Dried neutral alumina | 1.20 ±0.09 |
| Dried acidic alumina | 1.28 ±0.09 |
| Dried basic alumina | 1.18 ±0.09 |
| Highly dried neutral alumina | 0.53 ±0.09 |
| Silicone oil | 0.113 ±0.003 |
| GMO | 0.460 ±0.003 |
| Brij® 30 | 0.239 ±0.003 |
| GTO | 0.022 ±0.003 |

determined by Karl Fisher titration. The results of this analysis are presented in Table 1, along with the water contents of the silicone oil and surfactants.

Alumina suspensions were prepared by first adding the desired amount of surfactant to silicone oil (SF96, General Electric, $\eta_c$ = 0.0968 Pa·s, $\rho_c$ = 968 kg/m$^3$). The surfactants were soluble in the silicone oil for all concentrations reported. The particles were then added to the surfactant solution, sealed in a vial, and stored in a desiccator to minimize contact with air. Suspensions were allowed to equilibrate for at least 24 hours, and rheological experiments were performed within 48 hours. Samples prepared in this manner gave yield stresses and dielectric data ($\epsilon'$ and tan$\delta$) reproducible with 5 % and 1 %, respectively.

## Rheological Measurements

Rheological experiments were performed at 23 °C on a Bohlin VOR rheometer fitted with parallel plates, and modified for the application of large electric fields (a schematic diagram of the modified rheometer is presented in Fig. 1). Potential differences were supplied by a function generator (Stanford Research Systems, model DS345) and amplified with a Trek amplifier (model 10/10). Experiments were conducted with an electric field frequency of 500 Hz (except for the frequency sweep experiments).

The influence of surfactants on ER activity was determined by their effect on the dynamic yield stress. Samples were placed between the parallel plates and sheared for one minute at a large shear rate (> 40 s$^{-1}$) and zero field strength to insure a uniform distribution of particles. The desired electric field was then applied for one minute with no shear prior to measurements. Rheological measurements were performed by shearing the suspensions at a constant shear rate under the applied electric field, and recording the shear stress transmitted by the suspension. Experiments were performed with decreasing and then increasing shear rates, to obtain plots of shear stress as a function of shear rate. Values for the dynamic yield stress, $\tau_0$, were determined by extrapolating the shear stress-shear rate data to zero shear rate, using data over the range of shear rates, 0.01 s$^{-1}$ < $\dot{\gamma}$ < 0.1 s$^{-1}$.

### Dielectric Measurements

Suspension capacitance and loss were measured using a General Radio GR 1689M RLC Digibridge, which probes frequencies in the range of 12 Hz to 100 kHz, and operates with potential differences in the range of 0.01 - 1.0 V (rms).

A three terminal, guarded dielectric cell was employed. Experiments were performed with decreasing and then increasing applied field frequencies, to obtain plots of dielectric constant and loss tangent as a function of field frequency.

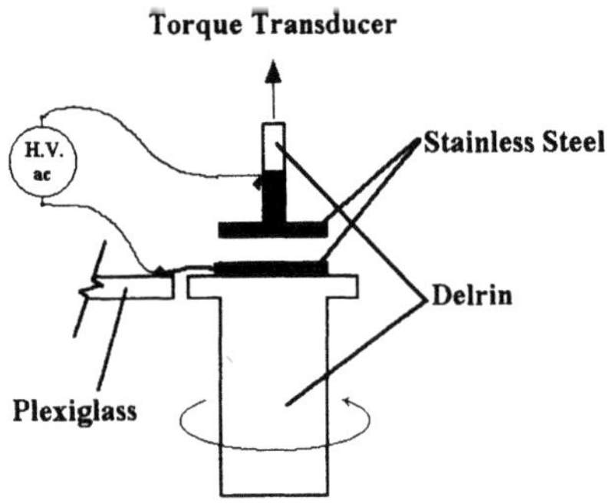


Figure 1. Schematic diagram of the rheometer modified for the application of large electric fields.

## RESULTS

### Rheological Data

The dependence of the yield stress on Brij® 30, GMO, and GTO concentration is presented in Fig. 2 for 20 wt% dried neutral alumina suspensions. For all cases, the yield stress initially increases with surfactant concentration and then passes through a maximum; the surfactant concentration at the maximum is insensitive to the applied electric field strength $E$, especially at large field strengths, but does depend on the surfactant type.

The dependence of the yield stress on the particle type is presented in Fig. 3 for 20 wt% dried alumina suspensions (acidic, neutral, and basic alumina; $E = 1.0$ kV/mm and $f_E = 500$ Hz), using Brij® 30 as the surfactant. All particles show similar trends, with the maximum yield stress varying with particle type. Acidic alumina shows the largest yield stress and basic alumina shows the smallest yield stress; neutral alumina suspensions show the largest relative increase in yield stress, approximately 1000 % increase at 2 kV/mm. We believe that the differences observed among the different particle types are associated with different adsorption characteristics on the various surfaces.

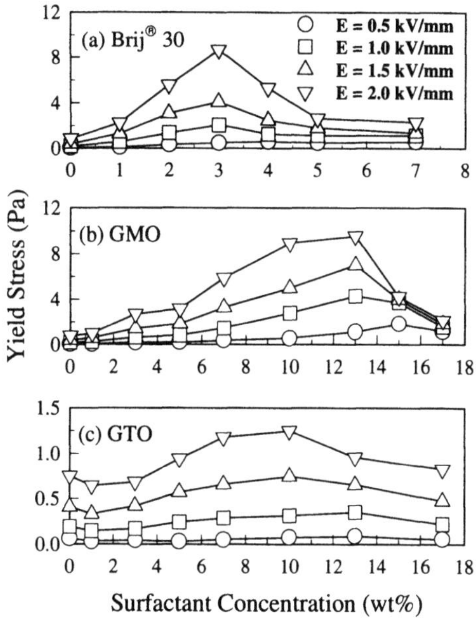


Figure 2. Yield stress as a function of surfactant concentration for 20 wt% dried neutral alumina suspensions in silicone oil: (a) Brij® 30, (b) GMO, and (c) GTO.

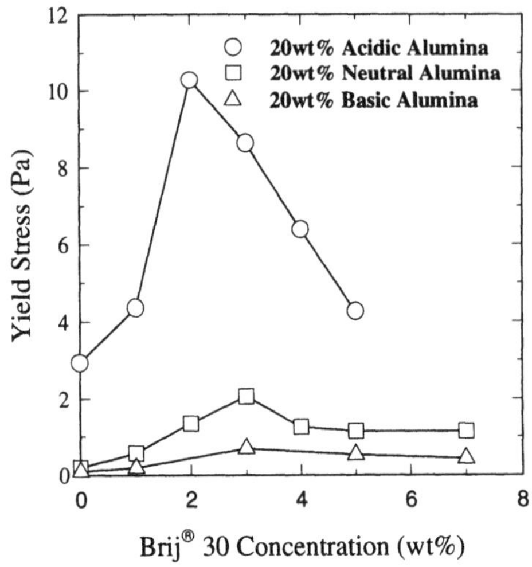


Figure 3. Yield stress as a function of Brij® 30 concentration for 20 wt% dried alumina suspensions in silicone oil.

The dependence of the yield stress on water content is presented in Fig. 4 for neutral alumina suspensions containing various amounts of Brij® 30 ($E$ = 1.5 kV/mm and $f_E$ = 500 Hz). The yield stress again passes through a maximum at ~ 3 wt% Brij® 30 at all water contents. The ER enhancement increases with water content and is almost negligible for the highly dried particles. For the nondried neutral suspensions, there appears to be a threshold surfactant concentration for yield stress enhancement.

The dependence of the yield stress on the electric field strength is depicted in Fig. 5. Here, the yield stress divided by the field strength squared is plotted against Brij® 30 concentration ($f_E$ = 500 Hz). At small Brij® 30 concentrations, the curves at different electric field strengths superpose and the yield stress scales with the electric field strength squared. However, at larger Brij® 30 concentrations (> 3 wt%), the yield stress deviates from the field squared dependence, increasing approximately linearly with field strength. We also note that at the large surfactant concentrations, the sample conductance increases significantly, inferred from the significant increase in sample temperature (at which point the experiments were stopped).

## Dielectric Data

The yield stress and suspension dielectric constant for 20 wt% dried neutral alumina suspensions, with and without surfactant, are plotted as a function of electric field frequency in Fig. 6. The yield stresses in this figure were measured at a field strength of 1.5 kV/mm, while the dielectric constants were measured under a small field strength (< 2 V/mm).

In the absence of surfactant, both the yield stress and suspension dielectric constant decrease with frequency over the range of 10 - 10,000 Hz. When surfactant is added, the yield stress and dielectric constant increase, and the decrease with increasing frequency becomes more pronounced.

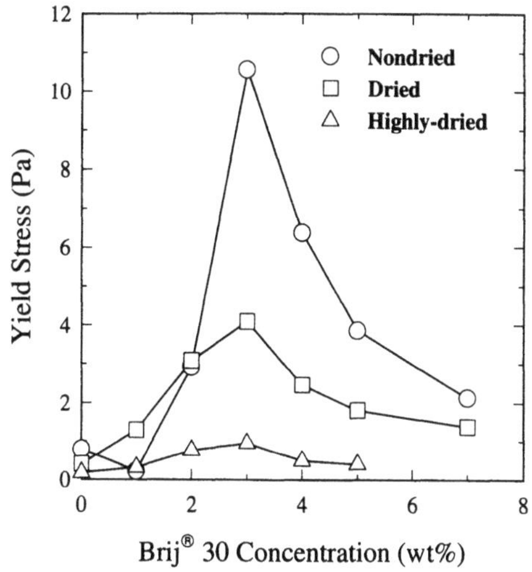


Figure 4. Yield stress as a function of Brij® 30 concentration for 20 wt% neutral alumina suspensions in silicone oil, with varying water contents. Nondried, 3.47 wt% (g $H_2O$/100 g dried alumnia); dried, 1.20 wt%; highly dried, 0.53 wt%.

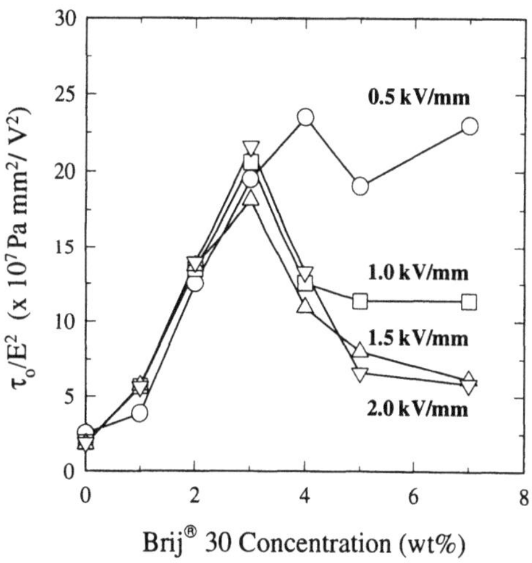


Figure 5. Yield stress divided by the electric field strength squared as a function of Brij® 30 concentration for 20 wt% dried neutral alumina suspensions in silicone oil.

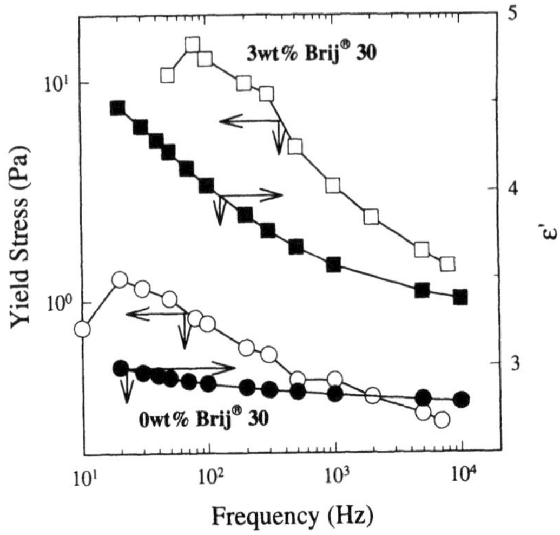


Figure 6. Yield stress and dielectric constant as a function of electric field frequency for 20 wt% dried neutral alumina suspension with 3 wt% Brij® 30 ($E$ = 1.5 kV/mm).

The surfactant also has a pronounced effect on the dielectric loss. Dielectric loss data for dried neutral alumina suspensions is presented in Table 2 for various Brij® 30 concentrations, along with data for surfactant/silicone oil solutions. Also included, for reference, are values for the real part of the complex conductivity, $\sigma' = \epsilon_0 \epsilon'' \omega$, where $\epsilon_0$ is the permittivity of free space, $\epsilon''$ is the imaginary part of complex dielectric constant, and $\omega$ is the applied electric field frequency.

## DISCUSSION

The similar frequency dependencies of the suspension dielectric properties and yield stress (Fig. 6) suggest that they are intimately related. In fact, both of these macroscopic properties have been related to the polarization of the disperse phase in existing models[14,15,16,17,18,19]. The dielectric constant of dilute, nonaqueous suspensions can be accurately described by volume averaging the electric polarization, leading to an effective suspension dielectric constant that increases linearly with the polarizability of the disperse phase[19,20]. The electrostatic polarization model for electrorheology[14,15,16,17,18,21] describes the yield stress in terms of the electrostatic forces acting between particles, induced by the applied electric field—this force increases quadratically with the particle polarizability for weakly polarizable particles. A frequency dependent particle polarizability would thus lead to similar frequency dependencies in the dielectric constant and yield stress, as observed in Fig. 6. Furthermore, the low frequency dispersion seen in this figure suggests that the particle polarization is controlled by migration polarization[5,6,8,19,20].

Hence, the activation of these ER suspensions appears to be related to the enhancement of the particle polarizability. Preliminary data indicates that the surfactants adsorb strongly to the alumina particles. An obvious hypothesis is that the surfactants

**Table 2.** Dielectric losses and conductivities for dried neutral alumina suspensions and silicone oil solutions ($c_s$ is the Brij® 30 concentration and $E = 2.0$ V/mm).

| Surfactant/silicone oil solutions | | | | |
|---|---|---|---|---|
| $c_s$ | $\tan\delta$ (x$10^3$) | | $\sigma'$ (x$10^{11}$ mho/m) | |
| (wt%) | 500 Hz | 1.0 kHz | 500 Hz | 1.0 kHz |
| 0 | 0.2 ±0.1 | 0.2 ±0.1 | 0.4 ±0.2 | 0.8 ±0.4 |
| 2 | 0.7 ±0.2 | 1.3 ±0.4 | 1.4 ±0.4 | 5.0 ±1.5 |
| 3 | 1.6 ±0.2 | 3.0 ±0.4 | 3.2 ±0.4 | 12.1 ±1.6 |
| 4 | 2.7 ±0.2 | 5.2 ±0.4 | 5.6 ±0.4 | 21.5 ±1.6 |
| 5 | 4.5 ±0.2 | 8.7 ±0.4 | 9.7 ±0.4 | 37.4 ±1.7 |
| 20 wt% dried neutral alumina suspensions | | | | |
| $c_s$ | $\tan\delta$ (x$10^3$) | | $\sigma'$ (x$10^{11}$ mho/m) | |
| (wt%) | 500 Hz | 1.0 kHz | 500 Hz | 1.0 kHz |
| 0 | 13 ±1 | 12 ±1 | 12.9 ±1.0 | 22.6 ±1.9 |
| 1 | 55 ±1 | 47 ±1 | 73.1 ±1.4 | 122.5 ±2.4 |
| 3 | 73 ±1 | 60 ±1 | 93.4 ±1.4 | 149.6 ±2.6 |
| 4 | 80 ±1 | 82 ±1 | 150.5 ±1.8 | 297.8 ±3.6 |
| 5 | 102 ±1 | 104 ±1 | 235.9 ±2.4 | 454.0 ±4.3 |

increase the surface conductivity; as the particle are porous, they will appear to display an increased bulk conductivity and polarizability. This is indeed consistent with the data presented in Table 2, where the suspension conductivity increases with surfactant concentration. Below, we develop a model to simultaneously describe the role of enhanced polarizability on the suspension dielectric and rheological properties.

The simplest treatment of frequency-dependent particle polarization controlled by migration polarization is the Maxwell-Wagner model[19,20]. In this model, the complex dielectric constant of each phase, $k$, is written in terms of its real dielectric constant, $\epsilon_k'$, and bulk conductivity, $\sigma_k$,

$$\epsilon_k^*(\omega) = \epsilon_k' - j\frac{\sigma_k}{\omega\epsilon_0}, \tag{1}$$

where $\omega$ is the angular frequency, $\epsilon_0$ is the permittivity of free space, and the $\epsilon_k'$'s and the $\sigma_k$'s are assumed to be independent of both the electric field strength and frequency. Charge neutrality is assumed to be maintained everywhere within the suspension except within an infinitesimally thin layer near the fluid-particle interface (hence the alternate descriptor "interfacial polarization"). Further assuming that the average electric field, averaged over a volume containing many particles, is uniform, the effective medium dielectric constant can be obtained by dividing the volume-averaged dielectric displacement by the volume-averaged electric field strength[19,20]. Maxwell's approximate expression, which can be obtained from a heuristic mean field argument, is given by[20]

$$\epsilon^* = \epsilon_c^*\frac{1 + 2\beta^*(\omega)\phi}{1 - \beta^*(\omega)\phi}, \tag{2}$$

where $\epsilon_c^*$ is the complex dielectric constant of the continuous phase, $\phi$ is the particle volume fraction, $\omega$ is the electric field frequency, and $\beta^*$ is the complex relative polarizability defined as

$$\beta^*(\omega) = \frac{\epsilon_p^* - \epsilon_c^*}{\epsilon_p^* + 2\epsilon_c^*}, \tag{3}$$

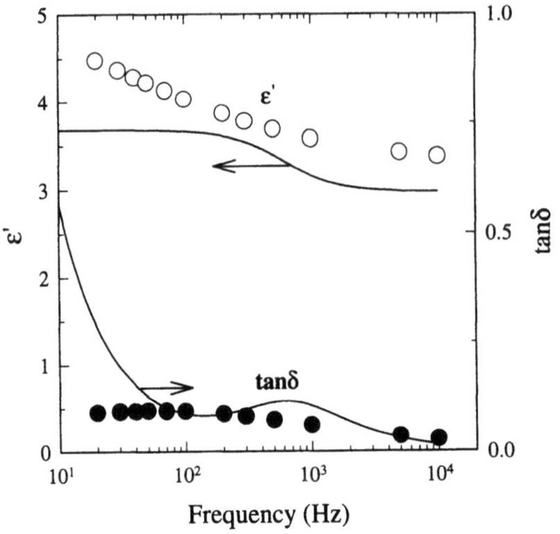


Figure 7. Dielectric constants as a function of electric field frequency for 20 wt% dried neutral alumina suspension with 3 wt% Brij® 30 (the symbols represent experimental data and the solid curve represent the model predictions).

where $\epsilon_p^*$ is the complex dielectric constant of the particle. Thus, even though the permittivities of the phases are frequency independent, the suspension permittivity is not, described in terms of the frequency dependent complex relative polarizability, $\beta^*(\omega)$.

We now extend this model to qualitatively describe the influence of surfactant on the dielectric properties. We assume that the surfactant adsorption may be described simply by the Langmuir adsorption isotherm[22],

$$\theta = \frac{bc_{s,eq}}{1 + bc_{s,eq}}, \tag{4}$$

where $\theta$ is the fraction of the surface area covered by surfactant, $b$ is the adsorption coefficient, and $c_{s,eq}$ is the equilibrium surfactant concentration in the fluid phase.

As low frequency dispersion is commonly attributed to the relatively large-distance migration of charged species, and since we find that the addition of surfactant increases this dispersion, we assume initially that the surfactant only significantly influences the dc conductivity of each phase. Linear relationships are assumed initially to describe this dependence:

$$\sigma_p = \sigma_{p0} + k_p\theta, \tag{5}$$

$$\sigma_c = \sigma_{c0} + k_c c_{s,eq}, \tag{6}$$

where $\sigma_{p0}$ and $\sigma_{c0}$ are the conductivities of the particulate and continuous phases in the absence of surfactant, and $k_p$ and $k_c$ are the proportionality constants describing the composition dependence of the conductivity in each phase.

Dielectric data for 20 wt% dried neutral alumina suspensions with 3 wt% Brij® 30 are presented in Fig. 7. Included in this figure is the best fit of the simple model described above. The agreement is not perfect, but certain features are captured. The magnitude of the dielectric constant and loss tangent are within 20 % (above ~ 50

Hz) and the low frequency dispersion is reproduced by the model, as expected. The assumption of frequency-independent phase properties in the Maxwell-Wagner model produces a sharper dispersion than observed experimentally. Finally we note that this fit requires a much larger increase in the particle conductivity than the continuous phase conductivity ($k_p\theta/k_c c_{s,eq} = 659$), consistent with the hypothesis that the surfactants act to increase the conductivity of the particles, as well as our preliminary results indicating that the surfactants adsorb strongly. As adsorbed species are known to increase the surface conductivity of hydrophilic particles dispersed in hydrophobic liquids[5,6,20,23], and since these alumina particles are very porous, an increase in the bulk particle conductivity is not surprising.

Rheological properties can be related to dielectric properties within the framework of the electrostatic polarization model[14,15,16,17]. Once the many particle electrostatic problem is solved, the force on individual particles can be determined as[24]

$$\mathbf{F}_i^{el} = \int_{S_i} \boldsymbol{\sigma}^M \cdot \mathbf{n} dS. \tag{7}$$

where $\boldsymbol{\sigma}^M = \epsilon_0\epsilon_c(\mathbf{EE} - \frac{1}{2}\boldsymbol{\delta}|\mathbf{E}|^2)$ is Maxwell's electrostatic stress tensor.

The electrostatic contribution to the (mechanical) stress tensor is given by

$$\boldsymbol{\sigma} = -\frac{1}{V}\sum_i^N \mathbf{r}_i \mathbf{F}_i^{el}. \tag{8}$$

Regardless of the level of approximation used in calculating the electrostatic forces, the stress, and hence the yield stress, must scale as

$$\boldsymbol{\sigma} \propto |\beta^*(\omega)|^2 E^2 \mathbf{f}(\beta^*(\omega)) \tag{9}$$

where $E$ is the magnitude of the applied electric field. For weakly polarizable particles, the function $\mathbf{f}(\beta^*)$ is only weakly dependent on $\beta^*$, and the yield stress is proportional to $|\beta^*|^2$ for the point-dipole approximation. Note that within the framework of this model, the frequency and surfactant concentration dependence of the yield stress lies entirely within the "normalized" yield stress, $|\beta^*(\omega)|^2$. The $\mathbf{f}(\beta^*)$ term becomes more important as the polarizability increases, producing a more rapid increase in yield stress with particle polarizability[14].

Measured yield stresses are compared to predicted normalized values ($|\beta^*|^2$) in Figures 8 and 9. The experimental data is for 20 wt% dried neutral alumina suspensions under large electric fields. The normalized stresses are calculated using model parameters obtained from a best fit of the model to the dielectric data. Thus, this comparison represents a critical evaluation of the original premise—that suspension dielectric and rheological properties are both determined by the surfactant-activated particle polarizability.

The yield stresses are plotted as a function of surfactant concentration in Fig. 8 ($f_E$ = 500 Hz). The predicted yield stress reproduces the increase with surfactant concentration at low values, although the rate of increase is somewhat smaller. This is not surprising, since the point-dipole approximation underpredicts the yield stress, more so as the particle polarizability increases. We also find that the prediction does not capture the maximum in the yield stress. This also is not surprising since the yield stress beyond the maximum does not scale as $E^2$, a feature that is not captured in this model.

In Fig. 9, the yield stresses are plotted as a function of electric field frequency. Here, the measured values are for 20 wt% neutral alumina suspensions with 3 wt% Brij® 30 (near the maximum in the yield stress), and again, the predicted values are calculated

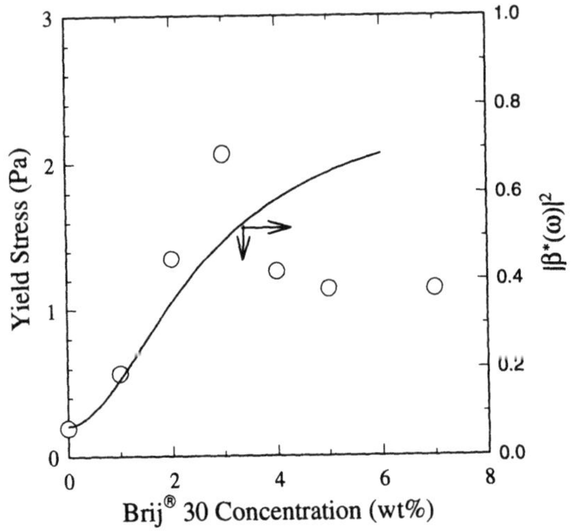


Figure 8. Experimental yield stress (symbols, $E = 1.0$ kV/mm) and predicted normalized yield stress (solid curve) as a function of Brij® 30 concentration for 20 wt% dried neutral alumina suspension, $f_E$ = 500 Hz.

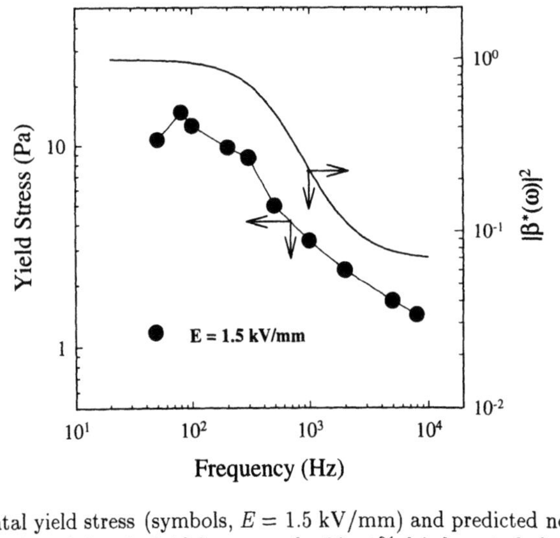


Figure 9. Experimental yield stress (symbols, $E = 1.5$ kV/mm) and predicted normalized yield stress (solid curve) as a function of electric field frequency for 20 wt% dried neutral alumina suspension with 3 wt% Brij® 30.

using parameters obtained from the dielectric data. The predicted values capture the decrease with frequency observed experimentally, but the dispersion occurs over a more narrow range of frequencies, as expected for the Maxwell-Wagner model.

Thus for surfactant concentrations below the maximum in the yield stress, this model qualitatively reproduces the experimental data, supporting the hypothesis that surfactants act to increase the particle conductivity (and polarizability), and hence the yield stress and suspension dielectric constant. Quantitative agreement will require a more sophisticated model that captures the broader dielectric dispersions and includes a more accurate calculation of the electrostatic interactions between particles.

Unfortunately, this level of analysis does not provide any insight into the microscopic mechanism or mechanisms by which the surfactant molecules might enhance the particle polarizability. This is due in part to the complicated microscopic structure envisioned for this system. Single surfactant molecules exist alone in solution and adsorbed on the particle surfaces, perhaps associated with adsorbed water molecules. Surfactants may also exist in aggregates in solution (inverted micelles) and on the particle surfaces ("hemimicelles"[25,26,27]). Some of these aggregates may also be swollen with water, and perhaps ions. Thus, enhanced surface conductivity might be realized within a variety of structures that are indiscernible by modeling such macroscopic properties as yield stress and suspension dielectric properties.

At large surfactant concentrations, above the maximum in the yield stress, the yield stress scales as $E^n$ where $n \approx 1$, a feature that is not captured by the simple model described above. However, possible explanations for the non-$E^2$ behavior can be obtained by examining the assumptions within the Maxwell-Wagner model that lead to the $E^2$ scaling of the yield stress. Two assumptions that may be violated at large surfactant concentrations are (1) a uniform average electric field and (2) phase dielectric properties that are independent of electric field strength.

For suspensions with zero conductivity, the electrostatic potential averaged over a volume containing many particles (but small relative to the electrode gap dimensions) will decrease linearly from the high to low potential electrodes. As the suspension conductivity increases, as is observed at large surfactant concentrations, it is possible for space charge layers to build up near the electrodes, screening the electric field near the center of the electrode gap. The screening is likely to be more pronounced as the conductivity or the field strength is increased. Within the framework of the electrostatic polarization model, the decreased field strength leads to a decreased yield stress, and an apparent field strength scaling less than quadratic.

The other possible mechanism for non-$E^2$ scaling of the yield stress is nonlinear, nonhomogeneous conduction. Conductivities of many fluids increase at field strengths comparable to those expected between particles in ER fluids[28,29,30]. Foulc *et al.*[31] have shown that increased conductivity in the fluid between conducting particles leads to electrostatic interactions and yield stresses that increase slower than $E^2$, due to the decreased field strength between the particles. Such a situation might occur in the activated suspensions; at large surfactant loadings, the continuous phase conductivity may increase and become field dependent. It is also conceivable that surfactant aggregates may be able to extend or travel across the interparticle gap, providing another nonlinear charge transport mechanism.

## CONCLUSION

The nonionic surfactants, Brij® 30, GMO, and GTO, increase the field-dependent yield stress of various alumina suspensions. For most cases, the yield stress initially increases with surfactant concentration and then passes through a maximum. The surfactant concentration at the maximum is insensitive to electric field strength. The yield stress is proportional to $E^2$ at low surfactant concentrations, but at large concentrations, increases slower than $E^2$. The enhancement appears to require the presence of water, and increases with water content. The enhancement is also associated with changes in suspension dielectric properties, with the surfactants enhancing low frequency dispersion.

A simple model, based on Maxwell-Wagner polarization of the disperse phase, is found to qualitatively reproduce the rheological and dielectric data at small surfactant concentrations. Our observations are consistent with the hypothesis that surfactants act primarily to increase the apparent particle conductivity.

Deviation from the linear ($E^2$ dependent) Maxwell-Wagner model at large surfactant concentrations, where the yield stress increases approximately linearly with field strength, may arise from a variety of mechanisms. Possible mechanisms include (1) nonuniform electric fields arising from excessive conduction, and (2) nonlinear conduction across interparticle gaps.

## ACKNOWLEDGEMENTS

Acknowledgement is made to the Donors of the Petroleum Research Fund, administered by the American Chemical Society, for partial support of this research. This work was also partly supported by the National Science Foundation (CTS-9401293).

## REFERENCES

[1] W. M. Winslow, Induced fibration of suspensions, *J. Appl. Phys.*, 20:1137 (1949).

[2] Z. P. Shulman, R. G. Gorodkin, E. V. Korobko, and V. K. Gleb, The electrorheological effect and its possible uses, *J. Non-Newt. Fluid Mech.*, 8:29 (1981).

[3] K. D. Weiss and J. D. Carlson, Material aspects of electrorheological systems, *J. Intell. Sys. and Struct.*, 4:13 (1993).

[4] D. L. Hartsock, R. F. Novak, and G. J. Chaundry, ER fluid requirements for automotive devices, *J. Rheol.*, 35:1305 (1991).

[5] Y. F. Deinega and G. V. Vinogradov, Electric fields in the rheological disperse system, *Rheol. Acta*, 23:636 (1984).

[6] H. Block and J. P. Kelly, Electro-rheology, *J. Phys. D: Appl. Phys.*, 21:1661 (1988).

[7] A. P. Gast and C. F. Zukoski, Electrorheological fluids as colloidal suspensions, *Adv. Coll. Int. Sci.*, 30:153 (1989).

[8] T. C. Jordan and M. T. Shaw, Electrorheology, *IEEE Trans. Elect. Insul.*, 24:849 (1989).

[9] G. G. Petrzhik, O. A. Chertkova, and A. A. Trapeznikov, Electrorheological effect in nonaqueous dispersions of various compositions in relation to the electric field parameters, *Dokl. Akad. Nauk SSSR*, 253:73 (1980)

[10] A. A. Trapeznikov, G. G. Petrzhik, and O. A. Chertkova, Electrorheological properties of nonaqueous dispersions of titanium dioxide and silicone dioxide in relation to concentration and moisture content of filler, *Koll. Zhurn.*, 43:83 (1981).

[11] O. A. Chertkova, G. G. Petrzhik, and A. A. Trapeznikov, Influence of nature of surfactant on the electrorheological effect in nonaqueous dispersions, *Koll. Zhurn.*, 44:83 (1982).

[12] H. Uejima, Dielectric mechanism and rheological properties of electro-fluids, *Jap. J. Appl. Phys.*, 11:319 (1972).

[13] N. Sugimoto, Winslow effect in ionic exchange resin dispersion, *Bull. JSME*, 20:1476 (1977).

[14] D.J. Klingenberg and C.F. Zukoski, Studies on the steady-shear behavior of electrorheological suspensions, *Langmuir*, 6:15 (1990).

[15] D. J. Klingenberg, Frank van Swol, and C. F. Zukoski, Dynamic simulation of electrorheological suspensions, *J. Chem. Phys.*, 91:7888 (1989).

[16] D. J. Klingenberg, Frank van Swol, and C. F. Zukoski, The small shear rate response of electrorheological suspensions. I. Simulation in the point-dipole limit, *J. Chem. Phys.*, 94:6160 (1991).

[17] D. J. Klingenberg, Frank van Swol, and C. F. Zukoski, The small shear rate response of electrorheological suspensions. II. Extension beyond the point-dipole limit, *J. Chem. Phys.*, 94:6170 (1991).

[18] P. M. Adriani and A. P. Gast, A microscopic model of electrorheology, *Phys. Fluids*, 31:2757 (1988).

[19] H. A. Pohl, "Dielectrophoresis," Cambridge University Press, Cambridge (1978).

[20] S. S. Dukhin, Dielectric properties of disperse systems, *J. Surf. Coll. Sci.*, 3:83 (1970).

[21] R. T. Bonnecaze and J. F. Brady, Yield stresses in electrorheological fluids, *J. Rheol.*, 36:73 (1992).

[22] R. J. Hunter, "Foundations of Colloid Science," Vol. 1, Oxford University Press, Oxford (1989).

[23] Y. D. Kim and D. J. Klingenberg, Surfactant-activated ER suspensions, *in:*"Electrorheological Fluids," R. Tao and G. D. Ray, World Scientific Pub., New Jersey (1994).

[24] J. D. Jackson, "Classical Electrodynamics," 2nd ed., John Wiley and Sons, Inc., New York (1975).

[25] M. J. Rosen, "Surfactants and Interfacial Phenomena," 2nd. ed., Wiley, New York (1989).

[26] B.-Y. Zhu and T. Gu, Surfactant adsorption at solid-liquid interface, *Adv. Coll. Int. Sci.*, 37:1 (1991).

[27] C. C. Nunn, R. S. Schecter, and W. H. Wade, Visual evidence regarding the nature of hemimicelles through surface solubilization of pinacyanal chloride, *Langmuir*, 1:251 (1985).

[28] C. Brosseau, Electrical conduction in impregnants for all-film power capacitors, *J. Appl. Phys.*, 70(10):5544 (1991).

[29] Z. Randriamala, A. Denat, J. P. Gosse, and B. Gosse, Field-enhanced dissociation, the validity of Onsagar's theory in surfactant solutions, *IEEE Trans. Elect. Insul.*, 1:167 (1985).

[30] N. Felici, High-field conduction in dielectric liquids revisited, *IEEE Trans. Elect. Insul.*, 1:233 (1985).

[31] D. Atten, J.-N. Foulc and N. Felici, A conduction model of electrorheological effect, *in*: "Electrorheological Fluids." R. Tao and G. D. Ray, World Scientific Pub., New Jersey (1994).

# POLY(ORTHO-TOLUIDINE HYDROCHLORIDE) AND POLY(ANILINE HYDROCHLORIDE) AS SEMI-CONDUCTING POLYMERS FOR ELECTRORHEOLOGY SUBSTRATES

Keith M. Blackwood[1], George Tsangaris[1], Dmitri N. Vorobiev[1], Harry Block[1] and Jacqueline Akhavan[2]

[1]Centre for Molecular Electronics
School of Industrial and Manufacturing Science
Cranfield University
Cranfield
Bedford, MK43 0AL, UK

[2]School of Mechanical, Materials and Civil Engineering
Cranfield University
Royal Military College of Science
Shrivenham, Swindon
Wiltshire, SN6 8LA, UK

## INTRODUCTION

The conductance range for semi-conductors as ER fluid substrates matches well with the typical values for rather poor semi-conducting polymers[1-9]. Such polymers tend to be insoluble powdery materials with densities similar to common dispersing liquids used in ER formulations. They are also soft and therefore do not abrade the device. Polyaniline when treated to reduce its conductance from that of the emeraldine hydrochloride form is of particular interest because it is simple to prepare[10-13]. Therefore polyaniline and its analogues are of technological interest. This paper reports on the conductive, and ER behaviour of dispersions of polyaniline and poly(o-toluidine).

## EXPERIMENTAL

### Production of Anhydrous ER Fluids

Polyaniline and poly(o-toluidine) were synthesised by oxidising the appropriate monomer with acid persulphate at temperatures of $<5^{o}C$[14]. The amine in 2M hydrochloric acid was

*Progress in Electrorheology*, Edited by K.O. Havelka
and F.E. Filisko, Plenum Press, New York, 1995

added dropwise to the vigorously agitated ammonium persulphate in 2M hydrochloric acid. The resultant fine particles of emeraldine hydrochloride were initially washed with copious amounts of hydrochloric acid and then converted to the partially neutralised states by equilibrating for 24 hours with phosphate or borate buffers to a predetermined pH. After filtration and washing with de-ionised water, the samples were dried at 50°C under vacuum to a constant weight.

Fluids for dielectric and ER measurements were made up using 20cS silicone fluid, previously dried over Drierite.

### Conductivity Measurements

Polymer conductance was measured in a pressurised conductivity cell[2]. The sample was held between two platinised anvils in a dye type cell. This was placed on top of a compressive load cell and the unit was put in a hydraulic press. A voltage supply and an electrometer were connected across the anvils and the conductivity of the samples were recorded at varying voltages and pressures. Utilising this apparatus allowed the values for the bulk conductivity of the polymeric powders to be determined before dispersion in the oil.

Thus the effects of different conductivities on the bulk properties of ER fluids, as well as observing the effects on the electrical properties when altering the pH of the buffer baths could to determined.

The electrorheology of dispersed powders were measured for static yield using the yield test meter, and under flow with a modified Carri Med controlled stress electro-rheometer. For the latter a Couette cell was used with a 0.2 mm gap.

## RESULTS AND DISCUSSION

From the range of polymer derivatives prepared during this work, it was found that the toluidine derivative (R = $CH_3$) gave the most promising results. Poly(o-toluidine) was found

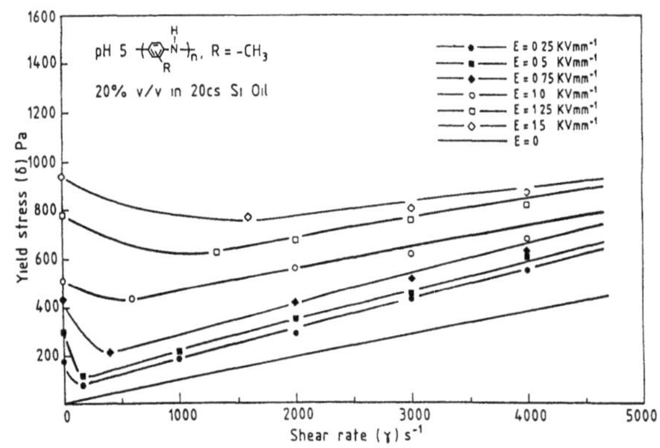


**Figure 1.** The change in yield stress with increasing shear rate and field strength for poly(o-toluidine) dispersed in 20cS silicone oil.

to have similar conductivity values to base treated polyaniline with a two fold increase in the achievable yield stress as shown in Figures 1 and 2

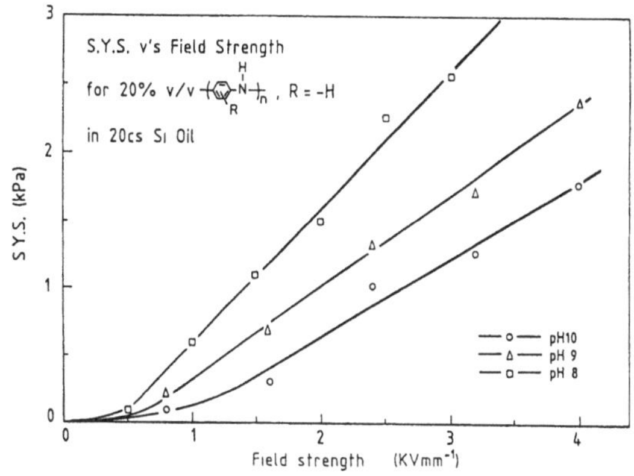


**Figure 2.** The change in static yield stress with increasing field strength for polyaniline dispersed in 20cS silicone oil.

Results of the dynamic measurements, up to 5000s$^{-1}$ of poly(o-toluidine) dispersed in 20cS silicone oil are presented in Figure 1. From Figure 1, it can be seen that the yield stress holds up well under flow. The curves shown in Figure 1 exhibit no real low shear rate hysterisis and there is seen to be only a limited drop off from the static value.

The effect of altering the pH buffering treatment of the sample on the conductivity is shown in Figures 3 and 4. A decrease in the pH of the sample was found to increase the conductivity.

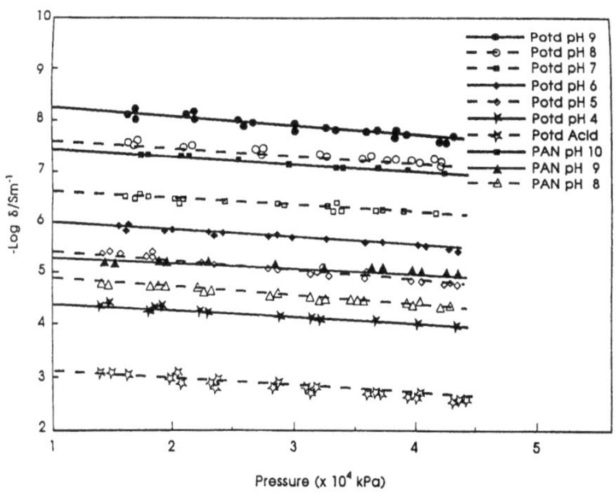


**Figure 3.** The effect on conductivity of poly(o-toluidine) as a function of pressure.

However reducing the pH in order to increase the yield stress is clearly done at the expense of the materials conductivity (see Figure 4) and thus the fluids bulk current density. From Figure 4 it can be seen that the two curves are parallel with poly(o-toluidine) being displaced towards the lower pH values. This suggests that methyl substitution has reduced the conductivity of poly(o-toluidine) and that it is necessary to produce a more acid form for the poly(o-toluidine) to achieve comparable conductivities to polyaniline.

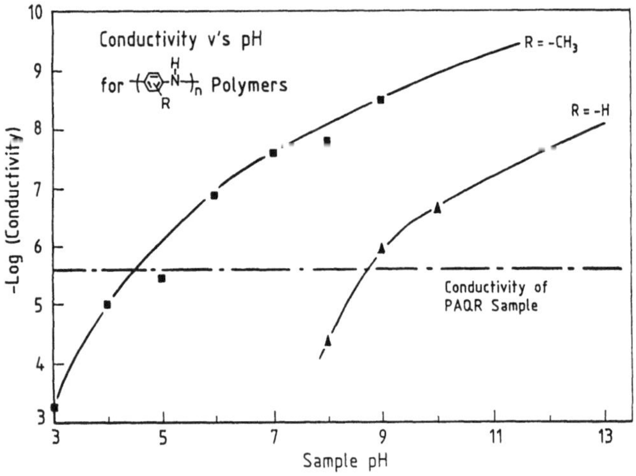


**Figure 4.** The effect on conductivity of poly(o-toluidine) and polyaniline dispersed in 20cS silicone oil as a function of the buffer pH.

Poly(o-toluidine) at pH values below 4 was of little use for ER fluids as the current densities were prohibitively high. There is also evidence to suggest that increased conductivity leads to a reduction of the achievable stress by inhibiting the formation of structures.

The maximum achievable yield stress of this material has not yet been fully investigated, but altering the base oil can have dramatic effects; a 20% v/v dispersion in a fluorinated silicone oil was found to have a yield stress in excess of 30kPa at a field of ~$2kVmm^{-1}$. Although it must be pointed out that this fluid exhibited a very high current density (approximately $100\mu Acm^{-2}$) it clearly demonstrates the forces achievable with 'dry' ER fluids.

## CONCLUSION

This work has shown that 'dry' ER fluids are a commercially viable alternative to those materials currently finding applications in industry. We have shown that while dry ER fluids can provide comparable mechanical characteristics they have overcome the temperature dependence problem associated with promoted fluids.

Combining this with the low abrasiveness of the organic materials which form the basis of our fluids, the low sedimentation rate, (due to a close density match of the solid and the oil), and the relative ease of redispersion, could make these materials the new generation of ER fluids.

The results of this and previous work[2] also demonstrates the existence of a 'window' of

The results of this and previous work[2] also demonstrates the existence of a 'window' of conduction for effective ER materials. Substrates with low conductivities and consequent small polarisations, form weaker structures and thus give a low static yield stress. When flow is also present there is the added effect of flow on the rate of polarisation, also making such substrates rather poor ER materials[2-5]. With the more acid and therefore more conductive materials, the fibrillation in the absence of the flow leads to larger currents, making the establishment of a field difficult and reducing its level from that estimated in terms of the applied voltage. There has also been the suggestion that, even in flow when chains are broken, too high a conductance for the substrate is bad for large shear stress enhancement under field. The cause then resides in the ability for the polarisation of particles to realign under the rotational action of shear flows, destroying incipient structuring[3-5].

Clearly there is much work still to be done in the field of fluid formation, but anhydrous ER materials demonstrates the way forward for this technology.

## REFERENCES


1. H. Block and J.P. Kelly, *US Patent* 4687589 (1987); *UK Patent* 217501B (1986).
2. H. Block, J.P. Kelly, A. Qin and T. Watson, *Langmuir* 6:6 (1990).
3. H. Block, J.P. Kelly and T. Watson, *in*: "High Value Polymers," A.H. Fawcett, ed., RSC Special Pub No 87, (1991).
4. H. Block, The nature, *in*: "Actuator 92 Conference Proceedings," K. Lenz and H. Borgmann, ed., VDI/VDU-Technologiezentrum Informationstechnik Gmbh, Berlin (1992).
5. H. Block, P. Rattray and T. Watson, Semi-conducting polymers as ER fluid substrates, *in*: "Electrorheological Fluids - Mechanism, Properties, Structure, Technology, and Application," R. Tao, ed., World Sci. Pub. Co., Singapore - New Jersey - London - HongKong (1992).
6. K.M. Blackwood and H. Block, *Trends in Polymer Science* 1:98 (1993).
7. K.M. Blackwood, H. Block, P. Rattray, G. Tsangaris and D.N. Vorobiev, *in*: "Proc. Int. Conf. Electrorheological Fluids, Feldkirch, Austria," R. Tao, ed., World Sci. Pub. Co., Singapore - New Jersey - London - HongKong in press.
8. H. Block, J. Chapples and T. Watson, *UK Patent App.* 8900825.6 (1989).
9. C.J. Gow and C.F. Zukoski, *J. Colloid. Interf. Sci.* 136:175 (1990).
10. A.G. MacDiarmid, J.C. Chiang, A.F. Richter and A.J. Epstein, *Synth. Metals* 18:285 (1988).
11. A.G. MacDiarmid, J.C. Chiang, A.F. Richter, N.L.D. Somasiri and A.J. Epstein, *in*: "Conducting Polymers," L. Alcacer, ed., D. Reidel Pub Co., (1987).
12. A.G. MacDiarmid and A.J. Epstein, *Faraday Disc. Chem. Soc.* 88:317 (1988).
13. E.M. Genies, A. Boyle, M. Lapkowski and C. Tsintavis, *Synth. Metals* 36:139 (1990).
14. K.M.Blackwood and H.Block, "Redox Polymerisation Coating Process", *UK Patent Application* 9313408.8 (1993).

# ANHYDROUS ELECTRORHEOLOGICAL FLUID USING CARBONACEOUS PARTICULATE AS DISPERSED PHASE

Yuichi Ishino, Takayuki Maruyama, Toshiyuki Ohsaki,
Shigeki Endo, Tasuku Saito and Norio Goshima

Research & Development Division
Bridgestone Corporation
3-1-1 Ogawahigashi-cho,
Kodaira-shi, Tokyo, 187, Japan

## Introduction

An Electrorheological Fluid(ERF) is a suspension essentially composed of fine solid particulates and an electrically insulating oil. Without an electric field, ERF exhibits the rheological properties of a Newtonian fluid whose shear stress is proportional to the shear rate. When an electric field is applied to ERF, the rheological properties of the fluid are instantaneously changed to a Bingham plastic which has the yield stress given by the intercept of the shear stress and the same constant of the proportionality between shear stress and shear rate as the Newtonian viscosity without an electric field. The yield stress rises with increasing voltage and no shear occurs until the shear stress exceeds the yield stress. This phenomenon is called the ER effect which gives rise to the instantaneous change of the apparent viscosity of the fluid.

Since the discovery of ERF in the 1940's[1], many types of fluid composition have been proposed[2]. As the continuous phase, typical electrically insulating oils, such as mineral oil, silicone oil, perfluoropolyether, dibutyl sebacate, and chlorinated paraffin has been used. As long as the electrical resistivity of those oils is sufficiently high, it works as the continuous phase of ERF, although the effect of its density and dielectric constant have been considered by some researchers[3,4] in order to avoid the sedimentation of the dispersed phase and to increase the ER effect respectively. On the other hand, the solid dispersed phase has been recognized to play very important role in the ER phenomenon. So far many kinds of solid particulate have been proposed. As the organic solid, starch, cellulose, ion exchange resin, phenol resin, and hydrophilic polymer such as lithium polymethacrylate were proposed. As the inorganic solid, silica gel, alminosilicate, zeolite, titanium dioxide, calcium titanate, alumina were also used as the dispersed phase of ERF. Most of these solid dispersed phases were hydrophilic and their adsorbed water content very much affects the electrorheological

*Progress in Electrorheology*, Edited by K.O. Havelka
and F.E. Filisko, Plenum Press, New York, 1995

properties of ERF[5,6]. The ERF containing these hydrophilic dispersed phase is normally called a hydrous ERF and it essentially requires water content in the dispersed phase to generate the ER effect. In other words if water is evaporated from the dispersed phase, hydrous ERF does not show any ER effect. Existence of water also limits the operating temperature range and the strength of the ER effect is very much affected by the quantity of water in the dispersed phase. These hydrous ERFs have been widely tested in device applications[7,8] such as engine mounts, dampers, clutches, and actuator. Since hydrous ERF has several problems about durability, operating temperature and stability of ER effect, so far hydrous ERF has not been applied to an actual use.

Because of the unsatisfactory characteristics of the hydrous ERF, recently some anhydrous ERFs, which do not need any water content to generate the ER effect, have been proposed. It was demonstrated that the completely dried zeolite still works as the dispersed phase of ERF[9]. Some semiconducting polymers, such as polyacenequinone and polyaniline, have been proposed as the dispersed phase of anhydrous ERF[10,11]. It was reported that metallic particles covered by an insulating layer works as the dispersed phase of ERF under AC field[12]. However, those anhydrous ERF have not yet seen an actual use.

The authors have been engaged in working on the application of ERF for automobile parts for several years. Since we realized that hydrous ERF cannot be applicable to automobile parts because of their poor durability and very strong temperature dependence, we have developed new anhydrous ERF using carbonaceous particulate as the dispersed phase (Carbonaceous ERF for short)[13]. In this paper, we would like to show the superior characteristics of Carbonaceous ERF, such as high temperature stability, stable temperature dependence, durability and other properties compared to common hydrous ERF.

## Experiment

Anhydrous ERF used in this experiment basically consists of carbonaceous particulate as dispersed phase and silicone oil as continuous phase. The material characteristics of carbonaceous particulate are described elsewhere[13]. In this paper, two types of carbonaceous particulate were used. One is the standard type and the other is the low current type which has higher volume resistivity compared to the standard type. The viscosity of silicone oil used is approximately 10 centistokes. The ratio of carbonaceous powder and silicone oil was changed to fit each measurement. In order to compare the characteristics of anhydrous ERF to that of hydrous ERF, the fluid consisting of sodium polyacrylate and silicone oil was used. Sodium polyacrylate used in this experiment is commercial powder and its average particle diameter is approximately 10μm.

Rheological properties of ERF were measured by a modified Rheometrics RDS-II dynamic spectrometer for a high voltage use. In this instrument, the outer cup is rotating and a high voltage is applied to the inner bob which is electrically isolated from the viscometer. The diameter and height of the inner bob is 50mm and 20mm respectively. The gap between the inner bob and outer cup is 1mm. Using steady shear mode, the apparent viscosity(shear stress divided by shear rate) at certain shear rate with consecutively increasing voltage and the shear stress with changing shear rate at several levels of voltage were measured. The high voltage supply used in this experiment is Trek model 610C where an input voltage is generated by a personal computer through a D/A converter. Rheological parameters and current densities were calculated by personal computer connected to RDS-II and 610C. For the temperature dependence measurement, liquid nitrogen was used for the temperature control below room temperature. For the response measurement the apparent viscosity and the input voltage were monitored by a digital oscilloscope. The water content of a solid dispersed phase was measured by Karl-Fischer method. We also conducted some durability tests of ERF using the following two methods:

(1)After stirring ERF at 120°C for several days or several ten days, rheological properties of a small portion of the heated fluid were measured by RDS II at room temperature.

(2)Using a specially designed rotational viscometer for durability test, the fluid was continuously exposed to $750s^{-1}$ of shear rate, 2kV/mm of electric field and 60°C. The apparent viscosity and current density of the fluid were intermittently measured at the same condition.

## Results and Discussion

Since Winslow reported in his paper[14] that semiconductive solid works as the dispersed phase of ERF, we have tested many kinds of semiconductive powders. Especially we paid attention to the hydrophobic material to avoid the hydrophilic characteristics of the material which absorbs moisture and shows the ER effect as the hydrous ERF. Some of the semiconductive materials such as carbon black, which is normally used in rubber compound, is too conductive for the dispersed phase of ERF. After many trials of material selection, finally we found that a certain type of carbonaceous powder works as the dispersed phase of ERF. Actually the color of the Carbonaceous ERF is totally black, the fluid can be held between the electrodes like a solid when DC voltage is applied, but it becomes normal liquid state when the voltage is turned off. When we see the structure change of the dilute suspension of carbonaceous powder and silicone oil under an optical microscope, carbonaceous particles form chains between the electrodes when a DC voltage is applied. Thus, at room temperature the macroscopic and microscopic behavior of Carbonaceous ERF is very similar to the common hydrous ERF.

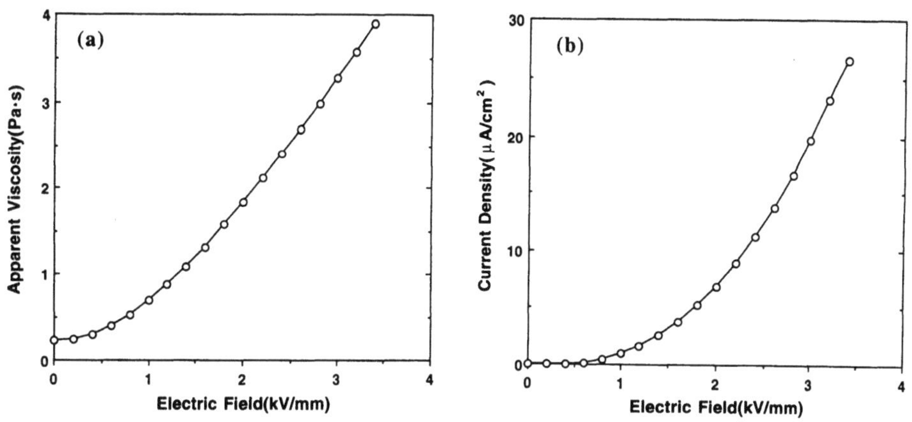


**Figure 1.** ER effect(a) and current density(b) of Carbonaceous ERF(standard type) at $366s^{-1}$ of shear rate.

In order to quantify the characteristics of Carbonaceous ERF, rheological and electrical characteristics, such as the ER effect and current density were measured by a rotational viscometer under an electric field. Figure 1 is the typical example of the ER properties of Carbonaceous ERF. In this figure, the apparent viscosity at a certain shear rate represents the ER effect in order to compare directly with the initial viscosity. As shown in figure 1, when 3kV/mm is applied, the viscosity change is almost ten times of the initial viscosity at

$366s^{-1}$ and 20 μA/cm² of current density is required. Figure 1 is one example of a Carbonaceous ERF, and it is possible to obtain a larger ER effect or lower current density when the type of carbonaceous particulate and continuous phase and their mixing ratio are selected in compliance with each purpose. For example low current type fluid shows a slightly smaller ER effect but it requires only 2 μA/cm² of current density at 3kV/mm.

The rheological characteristics of Carbonaceous ERF under changing shear rate and electric field were measured at 25°C(Figure 2). As shown in this figure when an electric field increases, the yield stress at very low shear rate increases consecutively and the yield stress is maintained at an almost constant value up to $1000s^{-1}$ of shear rate. Accordingly Carbonaceous ERF shows a typical Bingham plastic behavior when an electric field is applied and its ER effect is maintained at high shear rate. In Figure 2 the initial viscosity and yield stress of Carbonaceous ERF was about 0.2Pa·s and 1800Pa at 4kV/mm respectively. In this experiment current density tends to decrease slightly as shear rate increases. Thus, Carbonaceous ERF shows a large ER effect which can be maintained at a wide shear rate range.

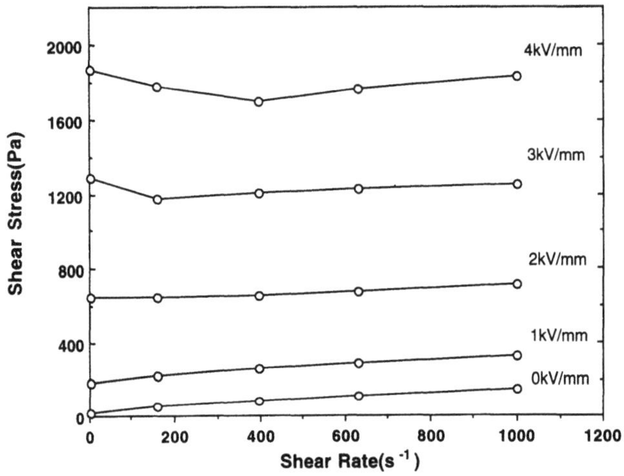


**Figure 2.** Shear stress versus shear rate for Carbonaceous ERF(standard type) with and without electric field at room temperature.

In order to compare the characteristics of Carbonaceous ERF to a typical hydrous ERF, hydrous ERF containing a sodium polyacrylate particulate as dispersed phase was used. Figure 3 shows the result of a heat resistance test at 120°C. The ER effect is represented by the difference of apparent viscosity at 1.2kV/mm and initial viscosity. As shown in figure 3, when the hydrous ERF was heated at 120°C, its ER effect was completely lost within 7 days. On the other hand, Carbonaceous ERF maintains its ER effect even after heated at 120°C for 40 days. Thus, Carbonaceous ERF can be continuously used above 100°C. The reason for the excellent heat resistance of Carbonaceous ERF will be explained in figure 4. This figure shows how much water content influences the ER effect. In this figure the ER effect is represented by the difference of apparent viscosity at 2kV/mm and initial viscosity. As shown in figure 4(a), in the case of hydrous ERF, when the water content is nearly 10%, hydrous ERF shows substantial ER effect. However, when the water content is decreased to 3%, hydrous ERF almost loses its ER effect. On the other hand, because of its hydrophobic nature the carbonaceous particulate usually contains less than 0.5% of water. As shown in figure 4(b), even when the water content is decreased to almost 0%, the ER effect of

Carbonaceous ERF is still maintained. Therefore, we can say that Carbonaceous ERF is truly anhydrous. The difference of the mechanisms between hydrous and Carbonaceous ERF is explained as follows: In the case of hydrous ERF, when an electric field is applied, positive and negative ions in the solid dispersed phase are attracted to the negative and positive electrodes respectively and the particle of solid dispersed phase is electrically polarized. Since the water content is required for ionizing ions in solid dispersed phase, if water is evaporated, hydrous ERF does not show any ER effect. On the contrary, in the case of Carbonaceous ERF, migration of electrons or positive holes in the solid dispersed phase occurs when electric field is applied. Since electrons or positive holes in carbonaceous particulate are utilized to generate the ER effect, it does not require any water content in the dispersed phase. Therefore, since Carbonaceous ERF shows truly anhydrous characteristics, it can be operated above 100°C.

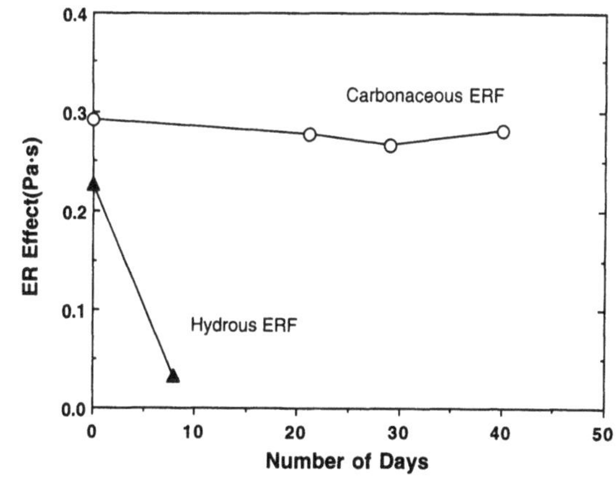


**Figure 3.** Heat resistance of hydrous and Carbonaceous ERF at 120°C. Shear rate and electric field is $366s^{-1}$ and 1.2kV/mm respectively.

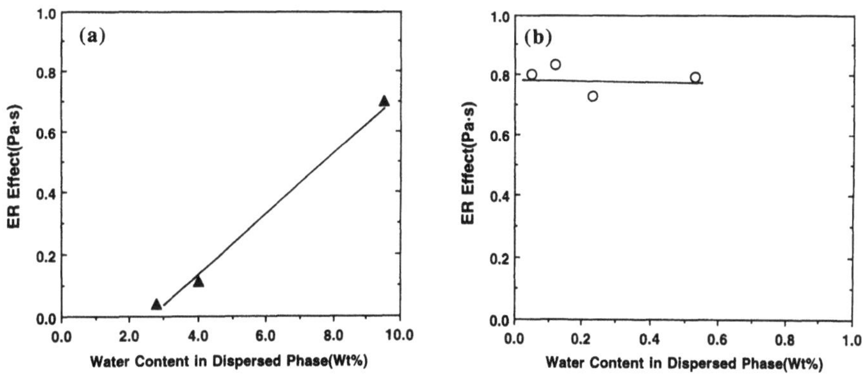


**Figure 4.** Effect of water content in hydrous ERF(a) and Carbonaceous ERF(b). Shear rate and electric field is $366s^{-1}$ and 2kV/mm respectively.

In theory, the mechanism of anhydrous ERF is explained by the Maxwell-Wagner type interfacial polarization[15]. Since ERF is a suspension in which there are boundaries between the dispersed solid particles and non-conducting oil, electrons or positive holes, which migrate under the electric field, are restricted by the boundaries. This interfacial polarization causes the particle polarization under electric field. When the dielectric spectra of Carbonaceous ERF is measured, Maxwell-Wagner type relaxation can be observed.

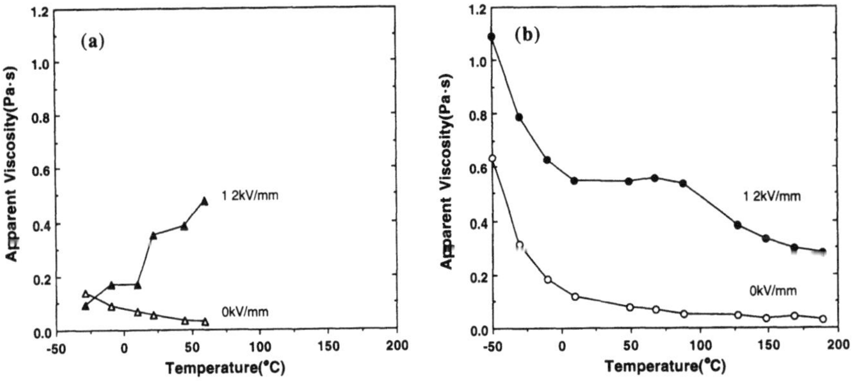


**Figure 5.** Temperature characteristics of hydrous ERF(a) and low current type Carbonaceous ERF(b). Apparent viscosity at 1.2kV/mm and without electric field was measured at $366s^{-1}$ of shear rate.

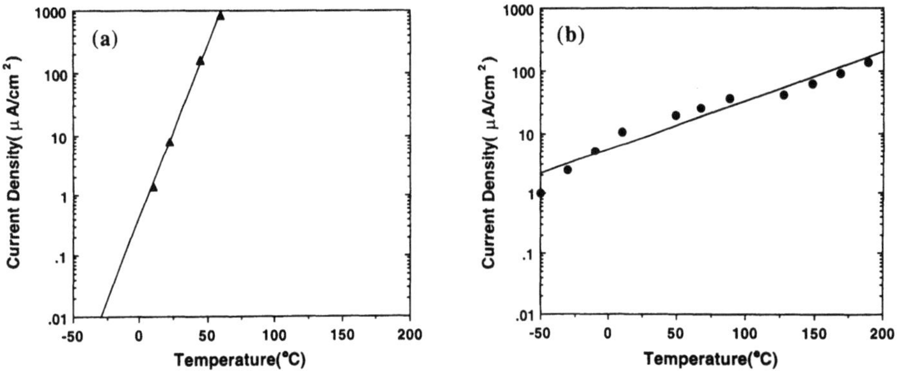


**Figure 6.** Temperature characteristics of hydrous ERF(a) and low current type Carbonaceous ERF(b). Current density at 1.2kV/mm corresponding to the apparent viscosity of figure 5 is shown.

Because of the truly anhydrous nature of Carbonaceous ERF, it also shows stable characteristics when temperature is varied. Figure 5 and 6 shows the comparison of the temperature dependence between hydrous and Carbonaceous ERF. In figure 5 the apparent viscosity at 1.2kV/mm and 0kV/mm are shown. The difference of the apparent viscosity at 1.2kV/mm and 0kV/mm is the ER effect. In figure 6 the current density at 1.2kV/mm are shown as the semilogarithmic plot. As shown in figure 5(a), hydrous ERF shows very little ER effect below 0°C. In the case of hydrous ERF, as temperature is increased, ER effect is also increased. However, as shown in figure 6(a), since the current density of hydrous ERF is drastically increased when temperature is increased, above 60°C the current density becomes excessive and a voltage can not be applied to the fluid. On the other hand, as shown in figure 5(b), when temperature is cooled down to -50°C, the initial viscosity of Carbonaceous ERF goes up, but it still shows ER effect. When temperature is raised above 100°C, even at 180°C, ER effect is observed. As shown in figure 6(b), the current density of Carbonaceous ERF is gradually increased when temperature is increased, however even at 180°C the required current density is only 100μA/cm$^2$.

Although in the case of figure 5 and 6 the low current type fluid is used to demonstrate the characteristics at high temperature, figure 7 shows the temperature dependence of the standard type fluid. As shown in figure 7, the apparent viscosities(open circle) of Carbonaceous ERF without and with electric field(2kV/mm) become increased at lower temperature and they become decreased at higher temperature, because the viscosity of the continuous phase has temperature dependence. However, ER effect, which is the difference between the apparent viscosities with and without electric field and is represented as shear stress(black circle) in Figure 7, keeps almost constant from -40°C to 120°C. These stable temperature characteristics under wide temperature range, which are caused by the truly anhydrous nature, have never appeared in the case of hydrous ERF.

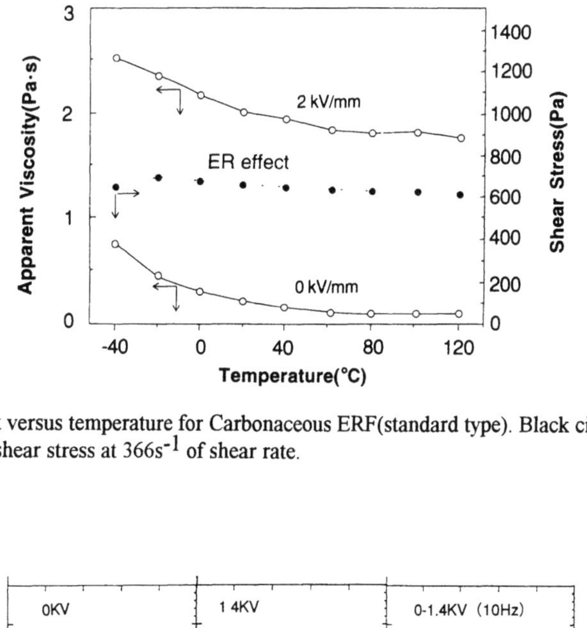


**Figure 7.** ER effect versus temperature for Carbonaceous ERF(standard type). Black circle represents ER effect as shear stress at 366s$^{-1}$ of shear rate.

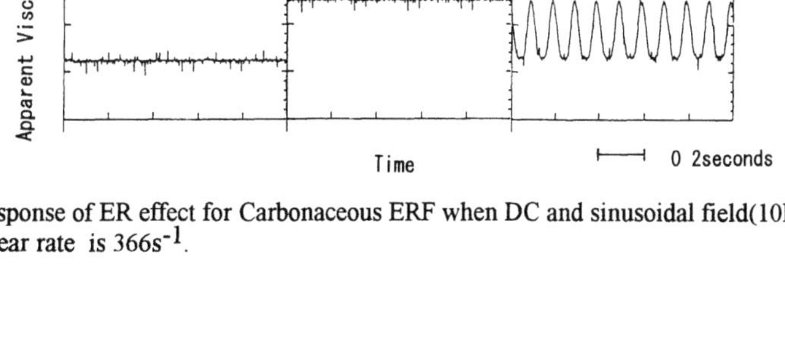


**Figure 8.** Response of ER effect for Carbonaceous ERF when DC and sinusoidal field(10Hz) is applied. Shear rate is 366s$^{-1}$.

In addition, Carbonaceous ERF shows excellent response characteristics. The response of Carbonaceous ERF was tested by a rotational viscometer at $366s^{-1}$ of shear rate(Figure 8). In Figure 8, the upper part of this figure shows an electric field and the lower part is apparent viscosity(arbitrary unit). When 1.4kV of DC field is applied to the fluid, the apparent viscosity goes up as shown in figure 8. When sinusoidal voltage at 10Hz is applied, apparent viscosity follows the sinusoidal voltage and its viscosity change does not show any phase delay. The viscosity at the peak top is almost the same viscosity as the viscosity when 1.4kV/mm of DC electric field is applied. Thus, Carbonaceous ERF has a fast response to the input voltage.

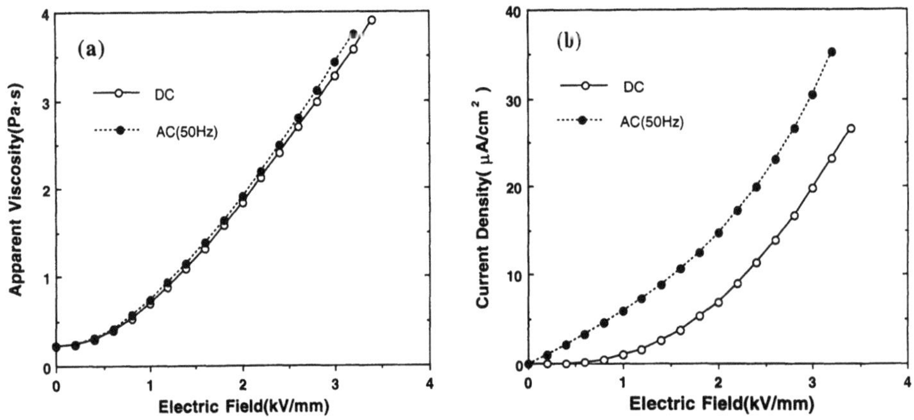


**Figure 9.** Comparison between ER effect of Carbonaceous ERF under AC(black circle) and DC field (open circle). Frequency of AC field is 50Hz. ER effect(a) and current density(b) at $366s^{-1}$ of shear rate were measured.

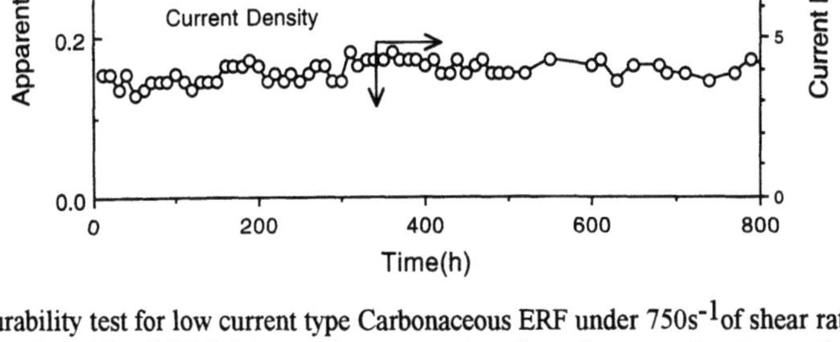


**Figure 10.** Durability test for low current type Carbonaceous ERF under $750s^{-1}$of shear rate, 2kV/mm of electric field and 60°C. Measured apparent viscosity and current density are shown as data points.

Figure 9 is the comparison between the ER effect under DC and 50Hz of AC field. When the DC field is applied to the fluid, no electrophoresis of dispersed phase is observed and the ER effect is completely reversible. As shown in figure 9, when effective voltage of AC field at 50Hz is the same voltage as the DC field, Carbonaceous ERF shows almost the same ER effect except AC requires more current density, because AC allows almost the not only conductive current but also capacitive current. Thus, carbonaceous ERF can be used under both DC and AC fields.

Figure 10 shows the results of a durability test under heat, shear and electric field. In this test low current type Carbonaceous ERF was continuously exposed to $750s^{-1}$of shear rate, 2kV/mm of electric field and 60°C for 800 hours. The apparent viscosity and current density of the fluid were intermittently measured. As shown in Figure 10, Carbonaceous ERF maintains stable ER effect and current density for 800 hours even when the electric field, heat and shear are continuously applied. Since both the low current and standard type carbonaceous particulates are uniform from surface to the center of the particle, if particle is worn, it still shows the same ER effect. Further, the carbonaceous particle is not abrasive and it does not wear conventional metal electrodes. Since hydrous ERF contains water content in its dispersed phase, it is corrosive for the conventional electrodes. On the other hand Carbonaceous ERF is non-corrosive because of its truly anhydrous nature.

## Conclusions

As the summary of the experimental results, since Carbonaceous ERF has a truly anhydrous nature, it has the following superior characteristics:

(1)Since Carbonaceous ERF does not need any water content to show the ER effect, it shows excellent durability at high temperature. It can be operated continuously above 100°C.

(2)Carbonaceous ERF shows constant ER effect under a wide temperature range.

(3)Carbonaceous ERF shows 1.8kPa as ER effect at 4kV/mm.

(4)When electric field is applied, Carbonaceous ERF shows typical Bingham plastic behavior, and its yield stress is maintained under a wide shear rate range.

(5)Its ER effect can be generated by not only DC field but also AC field.

(6)Under shear, heat and electric field, Carbonaceous ERF maintains constant ER effect for long period of time.

## References


1.W.M.Winslow, Method and means for translating electrical impulses into mechanical force, US Patent No.2417850(1947)

2.K.D.Weiss, J.D.Carlson, J.P.Coulter, Material aspects of electrorheological systems, *J.Intelligent Material Systems and Structures*, **4**,13-34(1993)

3.J.E.Stangroom, Electroviscous Fluids, US Patent No.4502973 (1985)

4.A.F.Sprecher, Y.Chen, Y.Choi, H.Conrad, The effect of various dielectric oils and volume fraction of solids on the mechanical properties of hydrous based electrorheological fluids, in "Electrorheological fluids" R.Tao ed., World Scientific Publishing Co. Singapore (1992)

5.H.Uejima, Dielectric mechanism and rheological properties of electrofluids, *Jap.J.Appl.Phys.*, **11(3)**319-26(1972)

6.N.Sugimoto, Winslow's effect in ionic exchange resin dispersion, *Bull. JSME*, **20(149)**1476-1483(1977)

7.T.G.Duclos, D.N.Acker, J.D. Carlson, Fluids thicken electrically, *Machine Design*, **21**, 42-46, January(1988)

8.D.A.Brooks, Fluids get tough, *Physics World*, **2**, 35-38, August (1989)

9.F.E.Filisko, W.F.Armstrong, Electric field dependent fluids, US Patent No.4744914 (1988)
10.H.Block, J.P. Kelly, A.Qin, T.Watson, Materials and mechanisms in electrorheology, *Langmuir*, **6(1)**6-14(1990)
11.C.J.Gow,C.F.Zukoski, The electrorheological properties of polyaniline suspensions, *J.Colloid Interface Sci.*, **136(1)**175-88 (1990)
12.A.Inoue, Study of new electrorheological fluid, in" Proceedings of the Second International Conference on ER Fluid" Technomic Publishing Company Inc. Lancaster PA(1990)
13.Y.Ishino, T.Ohsaki, S.Endo, S.Tomita, T.Maruyama, Y.Fukuyama and T.Saito, Electroviscous fluid, US Patent No.5087382(1992)
14.W.M.Winslow, Induced fibration of suspensions, *J of Applied Physics*,**20(12)**,1137-1140(1949)
15.H.Block, J.P.Kelly, Electro-rheology, *J.Phys.D.Appl.Phys*,**21(12)**,1661-77(1988)

# PROPERTIES OF ELECTRORHEOLOGICAL FLUIDS CONTAINING NUMEROUSLY SULFONATED POLYMER PARTICLES

Yoshinobu Asako, Satoru Ono, Ryuji Aizawa, and Toshihiro Kawakami

Tsukuba Research Laboratory
Nippon Shokubai Co., Ltd.
Tsukuba, Ibaraki, 305
Japan

## INTRODUCTION

Of great interest is that electrorheological (ER) fluids are put in practical use.[1,2,3] In order that an ER fluid can be utilized in a variety of applications, the ER fluid needs to meet several requirements all together.[4,5] At the present time, no ER fluid is likely to fill all of those requirements. It is well-known that ER fluids containing hydrophilic particles with small amounts of moisture induce a relatively large shear stress under an electric field.[6,7,8] However, when the electric field is continuously supplied to the ER fluid, a concurrent current density gradually increases and finally dielectric breakdown occurs. In order to put the ER fluids in practical use, they have to show good durability.

The hydrophilic particles have an ability to absorb moisture in the atmosphere. When the particles are employed as dispersed phase of an ER fluid, the moisture absorbed influences ER properties, e.g. shear stress property, current property, dielectric breakdown strength, durability, etc. Therefore, the ER properties of the fluids containing the hydrophilic particles have been investigated in term of the moisture absorbed in the particles. Sugimoto reported that an ER fluid containing sulfonic acid type ion-exchange resin induced a large shear stress even if the moisture absorbed was very small.[6] In the study, however, the absolute value of the moisture absorbed was not determined. We reported that the value of the moisture absorbed by the sulfonated(styrene-*co*-divinylbenzene) particles (SSD) can be evaluated by use of a Karl-Fisher moisture meter.[9] SSD had the same composition as the ion-exchange resin. Shear stress of a suspension of SSD increased, as the water content in SSD increased in the range between 2 and 5 wt%. The suspensions, however, showed poor durability.

When the water content of SSD decreased up to 1.0wt% or less, a durability of its suspension was found to get better, but the induced shear stress got smaller. In order to give an ER fluid both with outstanding shear stress property and durability, numerously sulfonated polymer particles (NSP) were synthesized.[10] The ER properties of a suspension of NSP were found to be influenced by the contained water as the cases of the ion-exchange resin and SSD. In this paper, effects of the water contained in NSP or SSD on shear stress property, current property and durability of suspensions of NSPs and SSDs are discussed and the ER performance of suspensions of NSPs is compared with those of SSDs.

*Progress in Electrorheology*, Edited by K.O. Havelka
and F.E. Filisko, Plenum Press, New York, 1995

## EXPERIMENTAL SECTION

### Materials

**Synthesis of NSP.** Spherical particles of styrene-*co*-divinylbenzene (380 g) were obtained from a mixture of 364 g of styrene and 36 g of an industrial grade divinylbenzene (a mixture of 55 wt% of divinylbenzene, 35 wt% of ethylstyrene, etc., Wako Pure Chemical) according to the manner similar to that described in the literature.[11] An average particle diameter of the particles was found to be 5 microns by use of a particle size analyzer (Shimadzu SALD-1000). The particles (100 g) were reacted with 1000 g of chlorosulfonic acid at 80 °C for 5 hours. The reaction mixture was poured into large amounts of iced-water, then the solid was separated by filtration and washed with water several times. The obtained particles were poured into 700 g of aqueous solution of 10 wt% sodium hydroxide and the mixture was stirred at 20 °C for 10 hours. After separation by filtration, the particles were washed with water several times. Drying the particles at 80 °C for 5 hours gave 210 g of NSP(10) whose water content was 10 wt%. The water content was determined by use of a Karl-Fisher moisture meter (Kyoto Electric Manufacturing MKS 3P). Hereafter the numbers in the parentheses indicate the water content in the sulfonated polymer particles. In NSP(10), each aromatic ring was found to have 1.24 sulfonic acid substituents on average from an elemental analysis. An average particle diameter of NSP(10) was found to be 5 microns by use of the particle size analyzer.

**Synthesis of SSD.** SSD(10) was obtained by reacting the same styrene-*co*-divinylbenzene particles used in NSP synthesis with 98 wt% sulfuric acid instead of chlorosulfonic acid. An elemental analysis of SSD(10) showed that it had a sulfonic acid substitution level of 0.88 per aromatic ring. An average particle diameter of SSD(10) was found to be 5 microns.

**Preparation of Suspensions of NSPs and SSDs.** 30 g of NSP(5.5) was dispersed in 70 g of dimethyl silicone oil (Shin-Etsu Silicone Oil KF96-50CS) to give a suspension of NSP(5.5). Suspensions of NSP(3.5), NSP(1.5), SSD(3.5), SSD(1.8) and SSD(0.8) were obtained likewise.

### Measurements

**Evaluation of Hygroscopic Property of NSP and SSD.** After NSP(10) was divided into three portions, each portion was heated at 80, 100 or 150 °C for 24 hours in a forced, non-vacuum air oven. The residual water contents of the resultant NSPs were determined by use of the Karl-Fisher moisture meter. The same experiments were carried out for SSD(10).

**Measurements of Values of Induced Shear Stress and Concurrent Current Density of the Suspensions.** In order to evaluate shear stress property and current property of the suspensions, the cup and bob of a rotary viscometer (Contraves Reomat115) were modified to facilitate the supply of an external electric field across 1 mm clearance between the cup and bob. An AC electric source (Tokyo Seiden ITS-5005-A-SP) was used to supply the electric field. Each of suspensions of NSPs and SSDs was placed between the cup and bob. The values of shear stress and current density were measured under 60 Hz electric fields at different field strength. The shear rate investigated was 200 /s.

**Evaluation of Durability of the Suspensions.** Durability of the suspensions was investigated by measuring changes of the shear stress and the current density under the continuous supply of 3000 V/mm of AC electric field by use of the same modified rotary viscometer used in the shear stress and current density measurements. The shear rate investigated was 200 /s.

**Measurements of Zero-Field Viscosity of the Suspensions.** Zero-field viscosity of the suspensions was measured by use of an E type viscometer (Tokyo Keiki Visconic ELD).

## RESULTS AND DISCUSSION

### Characteristic of NSP and SSD

It is well-known that the introduction of sulfonic acid groups into aromatic rings within an aromatic polymer can be followed by sulfonation using a sulfonation reagent.[12] The sulfonation of spherical styrene-*co*-divinylbenzene particles was performed at 80 °C for 5 hours using sulfuric acid and then followed by sulfonated (styrene-*co*-divinylbenzene) (SSD) particles. The vertical sectional view of an SSD particle was followed with scanning electron microscopy (SEM) and showed that the sulfonic acid groups were substantially distributed all over the particle.[9] The elemental analysis showed that each aromatic ring in SSD had 0.88 sulfonic acid substituent on average. These observations support the SSD particle has some aromatic rings derived from component of the copolymer with one sulfonic acid group and the other rings without the substitution, as shown Fig.1, SSD. In order to prepare numerously sulfonated polymer (NSP) particles, the above-said styrene-*co*-divinylbenzene particles were sulfonated at 80 °C for 5 hours using chlorosulfonic acid since the acid is a stronger sulfonation reagent than sulfuric acid. The elemental analysis showed that each aromatic ring in NSP had 1.24 sulfonic acid substituents on average. This indicates that the NSP particle has some aromatic rings with two sulfonic acid groups and the other rings with the one substituent, as shown Fig.1, NSP.

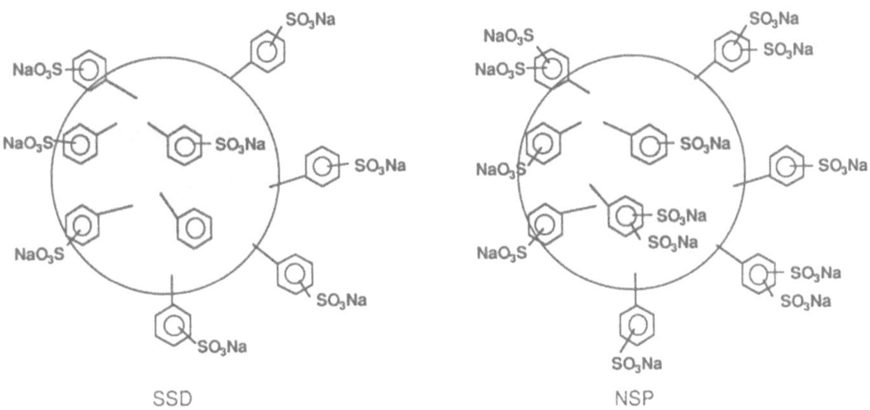


**Figure 1.** Schematic illustrations of SSD and NSP particles.

### Hygroscopic Property of NSP and SSD

In the ER suspensions containing hydrophilic particles, the moisture absorbed in the particles largely influences ER properties of the suspensions. NSP and SSD had exhibited hygroscopic property as in the cases of sulfonic acid type ion-exchange resins and silica gels. Hygroscopic property of NSP and SSD was investigated by evaluating residual water content after heating NSP(10) and SSD(10) at different heating temperature (Fig.2). The water content was determined by use of a Karl-Fisher moisture meter.

After heating SSD(10) at 100 °C for 24 hours, the water content decreased to 1.8 wt%. After heating SSD(10) at 150 °C for 24 hours, the water content was found to be 0.8 wt% and the water could not be removed further at higher temperatures. The 0.8 wt% of water is considered to be strongly bound within SSD as "structural" water.

The content of residual water in NSP was larger than that in SSD at the same heating temperature. After heating NSP(10) at 150 °C for 24 hours, the 1.5 wt% of water remained in NSP and the residual water could not be removed further at higher temperatures. The content of structural water in NSP (1.5 wt%) was larger than that in SSD (0.8 wt%). These results can be explained in term of the higher substitution level of sulfonic acid groups in NSP than that in SSD.

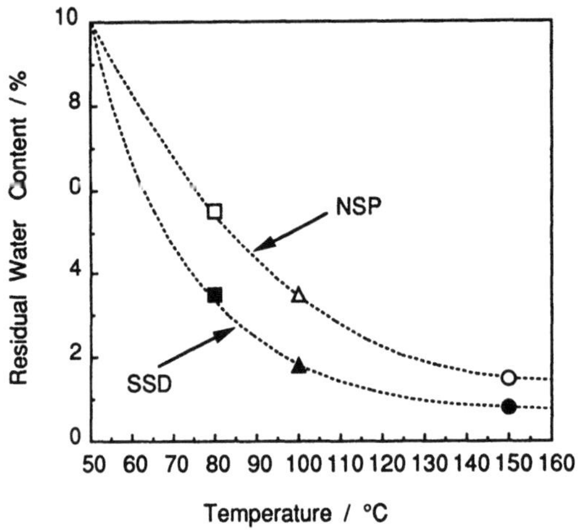


**Figure 2.** Plots of residual water content vs heating temperature for SSD (solid symbols) and NSP (open symbols).

## Zero-Field Viscosity of Suspensions of NSPs and SSDs

Each of NSPs and SSDs with different water content was dispersed in the silicone oil of the kinetic viscosity of 50 CS to form the suspension. All the suspensions had the same viscosity of 90 cP and thus, the residual water level was found to have no influence on the zero-field viscosity.

## Shear Stress Property and Current Property of the Suspensions of NSPs and SSDs

Shear stress property and current property of the suspensions of SSDs at different water content are shown in Fig.3 and Fig.4, respectively. As the applied electric field strength increased, both of the shear stress and the current density increased for all the suspensions. As the water content in SSD decreased, both of the shear stress and the current density decreased at any field strength applied.

Shear stress property and current property of the suspensions of NSPs at different water content are shown in Fig.5 and Fig.6, respectively. Similarly to the suspensions of SSDs, both of the shear stress and the current density increased with the increase of applied electric field strength for all the suspensions. And both of the shear stress and the current density decreased with the decrease of water content at any field strength applied.

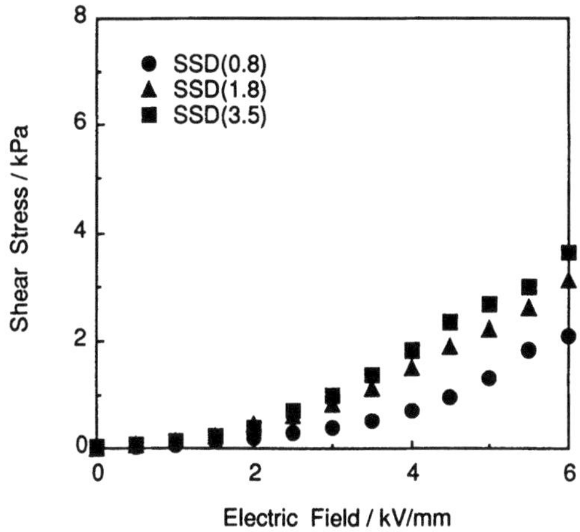


**Figure 3.** Plots of induced shear stress vs applied electric field strength for suspensions of SSDs with different water content in silicone oil. Water content; (●) 0.8 wt%, (▲) 1.8 wt%, (■) 3.5 wt%.

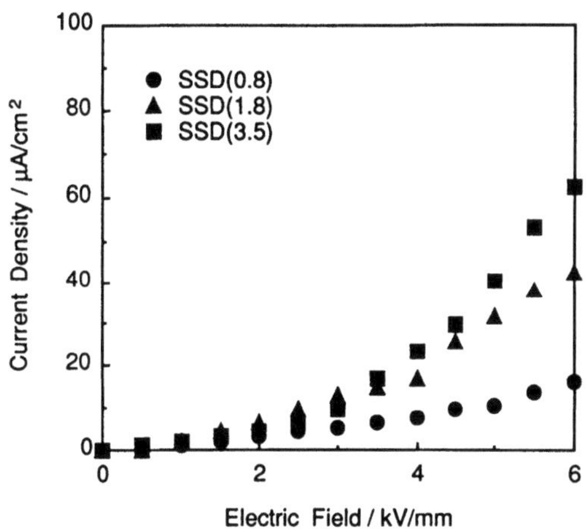


**Figure 4.** Plots of concurrent current density vs applied electric field strength for suspensions of SSDs with different water content. Symbols are the same as Figure 3.

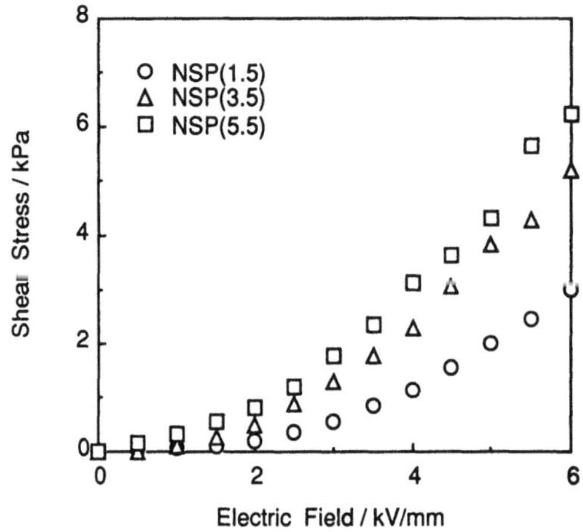


**Figure 5.** Plots of induced shear stress vs applied electric field strength for suspensions of NSPs with different water content in silicone oil. Water content; (○) 1.5 wt%, (△) 3.5 wt%, (□) 5.5 wt%.

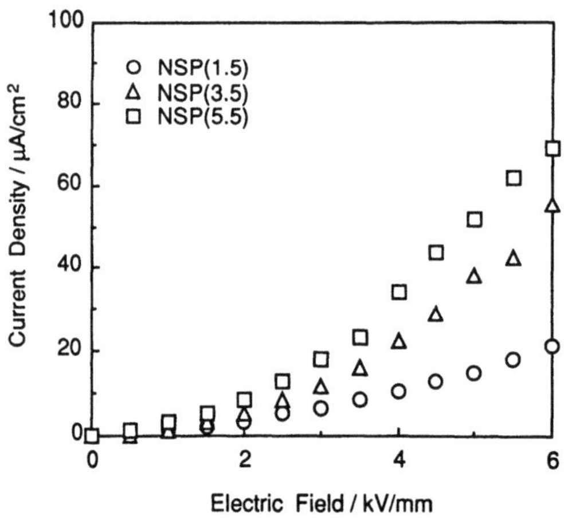


**Figure 6.** Plots of concurrent current density vs applied electric field strength for suspensions of NSPs with different water content. Symbols are the same as Figure 5.

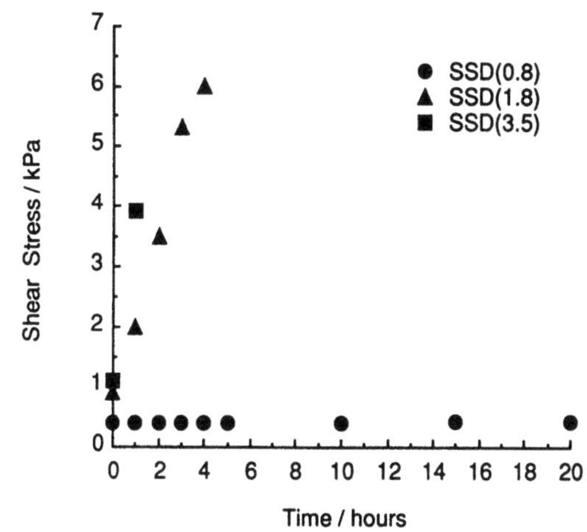


**Figure 7.** Plots of induced shear stress vs passage of time for suspensions of SSDs with different water content under the 3000 V/mm electric field strength. Symbols are the same as Figure 3.

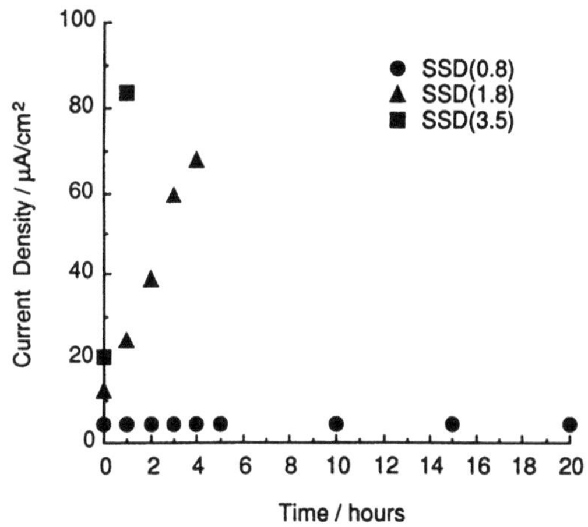


**Figure 8.** Plots of concurrent current density vs passage of time for suspensions of SSDs with different water content under the 3000 V/mm electric field strength. Symbols are the same as Figure 3.

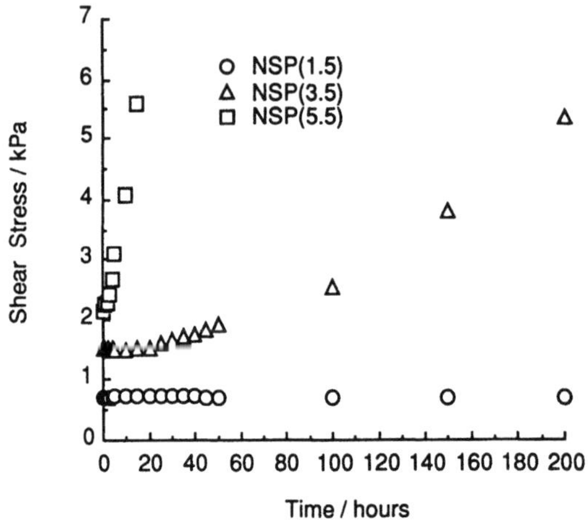


**Figure 9.** Plots of induced shear stress vs passage of time for suspensions of NSPs with different water content under the 3000 V/mm electric field strength. Symbols are the same as Figure 5.

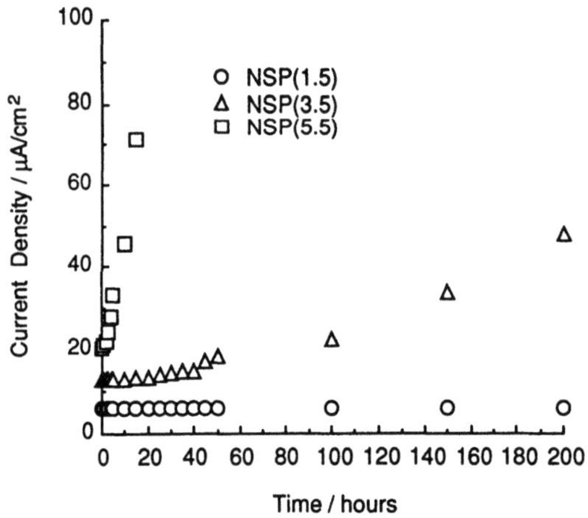


**Figure 10.** Plots of concurrent current density vs passage of time for suspensions of NSPs with different water content under the 3000 V/mm electric field strength. Symbols are the same as Figure 5.

The suspension of NSP(3.5) induced larger shear stress than that of SSD(3.5) at the same field strength. The difference of the induced shear stress became larger with the increase of applied field strength. The current density of NSP(3.5) was smaller than that of SSD(3.5) at the same field strength. The difference of the current density became larger with the increase of applied field strength. The induced shear stress of the suspension of NSP(1.5) was almost same as that of SSD(1.8) at the same field strength. The current density of the suspension of NSP(1.5) was smaller than that of SSD(1.8) at the same field strength and the difference increased with the increase of applied field strength.

### Durability of the Suspensions of NSPs and SSDs

In order to evaluate durability of the suspensions of NSPs and SSDs, changes with the passage of time of the shear stress and the current density were measured under the continuous supply of 3000 V/mm electric field. Fig.7 and Fig.8 show time-dependence of the shear stress and the current density of the suspensions of SSDs at different water content, respectively. In the cases of SSD(3.5) and SSD(1.8), the shear stress and the current density sharply increased during the durability test, and within 5 hours, the dielectric breakdown occurred. Thus, the suspensions of SSD(3.5) and SSD(1.8) exhibited poor durability. The shear stress and the current density of the suspension of SSD(0.8) did not change up to 20 hours. The good durability can be explained based on that SSD(0.8) contained only the structural water.

Changes of the shear stress and the current density of the suspensions of NSPs at different water content are shown in Fig.9 and Fig.10, respectively. In the case of NSP(5.5), the shear stress and the current density sharply increased during the durability test, and within 20 hours, the dielectric breakdown occurred. In the case of NSP(3.5), the shear stress and the current density were constant up to 20 hours and gradually increased after 20 hours. The durability of the suspension of NSP(3.5) is considered to be still insufficient for practical use, but was far better than that of SSD(3.5). The improved durability can be attributed to the higher substitution level of sulfonic acid group in NSP than that in SSD. In the case of NSP(1.5), the shear stress and the current density did not change over 200 hours and the durability was superior to that of SSD(1.8).

Comparing the ER performance of the suspension of NSP(1.5) with that of SSD(0.8), the shear stress property of NSP(1.5) was far better than that of SSD(0.8) and the durability of NSP(1.5) was much better than that of SSD(0.8). The excellent ER properties of the suspension of NSP(1.5) can be attributed to both of the higher sulfonic acid substitution level which exceeds equation ratio per aromatic ring and the higher content of structural water.

## CONCLUSIONS

Two types of sulfonated polymer particles (SSD and NSP) with different sulfonic acid substitution level were synthesized. NSP which had the high sulfonic acid substitution level of 1.24 bound water more strongly than SSD whose substitution level was 0.88. The contents of structural water in NSP and SSD were 1.5 wt% and 0.8 wt%, respectively. The ER properties of the suspensions of SSD and NSP dispersed in silicone oil were investigated at various residual water contents and under varied electric field strengths. The induced shear stress and the concurrent current density of the suspensions increased with the increase of applied field strength. As the residual water content decreased, the shear stress and the current density decreased. The durability was improved with the decrease of residual water content. The suspension of SSD(0.8) with only the structural water showed good current property and durability but the induced shear stress was not large. The suspension NSP(1.5) with only the structural water exhibited outstanding shear stress property, current property and durability. Since the ER fluid containing NSP(1.5) had excellent properties, it could be applicable to practical use.

## REFERENCES


1. J.D.Carlson, A.F.Sprecher and H.Conrad, ed., "Electrorheological Fluids," Technomic, Lancaster (1990).
2. C.A.Rogers and R.C.Rogers, ed., "Recent Advances in Adaptive and Sensory Materials and Their Applications," Technomic, Lancaster (1992).
3. K.Koyama, ed., "Applications & Development of Electro Rheological Fluid," CMC, Tokyo (1994).
4. K.D.Weiss, J.P.Coulter and J.D.Carlson, Electrorheological materials and their usage in intelligent material systems and structures, Part I: Mechanisms, formulatons and properties, ***in*** : "Recent Advances in Adaptive and Sensory Materials and Their Applications," C.A.Rogers and R.C.Rogers, ed., Technomic, Lancaster (1992).
5. D.L.Hartsock, R.F.Novak and G.J.Chaundy, ER fluid requirements for automotive devices, *J. Rheol.* 35:1305 (1991).
6. N.Sugimoto, Winslow's effect of ion-exchange resin suspensions, *Trans. Jpn. Mech. Eng. (Part2).* 43:1075 (1977).
7. J.E.Stangroom, **UK Patent 1570284** (1980).
8. J.Goossens, G.Oppermann, W.Podszun and V. Härtel, **US Patent 4668417** (1987).
9. Y.Asako, I.Okada, M.Aoki, S.Ono and M.Kobayashi, Properties of electrorheological fluid containing sulfonated poly(styrene-co-divinylbenzene) particles, *Nippon Reoroji Gakkaishi.* 20:61 (1991).
10. Y.Asako, S.Ono, M.Aoki, I.Okada, M.Kobayashi and R.Aizawa, **US Patent 5326489** (1994).
11. J.Hiltner and W.F.Bartoe, **US Patent 2264375** (1941).
12. Robert L. Albright and Peter A. Yarnell, Ion-exchange polymers, ***in*** : "Encyclopedia of Polymer Science and Engineering, Vol.8," Jacqeline I. Kroschwitz, ed., John Wiley & Sons, New York (1987).

# PRECERAMIC POLYMER ROUTES TO AMORPHOUS AND CRYSTALLINE ALUMINOSILICATE POWDERS FOR ELECTRORHEOLOGICAL APPLICATIONS. I.

Rita Baranwal,[1] Alexandra Zika,[1] Brian L. Mueller,[1]
and Richard M. Laine[1,2]

Materials Science and Engineering Department[1]
Chemistry Department[2]
University of Michigan
Ann Arbor, MI 48109-2136

## INTRODUCTION

Electrorheological (ER) fluids prepared using suspensions of amorphous and crystalline aluminosilicates in inert, nonpolar liquids are reported to exhibit good-to-exceptional ER activity.[1-3] Unfortunately, the measured ER properties can vary greatly even for compositionally similar fluids. This variation results, in part, because previous efforts to correlate ER behavior with the properties of well-defined aluminosilicates relied on commercially produced aluminosilicates. The commercial sources rarely provided a range of products (e.g. wide variety of compositions, particle sizes, etc.), or detailed quantification of powder properties (phase and chemical purity, surface area and pore size distribution, etc.). In addition, because the synthetic procedures are typically proprietary, it is not possible to identify important differences in methods of preparation. Hence, a detailed understanding of the exact mechanism(s) of ER activity as a function of individual aluminosilicate properties has been difficult-to-impossible to establish. The objectives of the work reported here are to: (1) develop a detailed and general synthetic approach to aluminosilicate powders; (2) develop protocols for characterizing their spectroscopic and physical properties, and (3) conduct preliminary ER studies to demonstrate the proof-of-principle of our chosen approach.

We describe here a general synthetic route to group I aluminosilicate powders, and

*Progress in Electrorheology*, Edited by K.O. Havelka
and F.E. Filisko, Plenum Press, New York, 1995

discuss in detail the preparation and characterization of one member of this group, the 1:1:1 Li:Al:Si composition lithium aluminosilicate.

## EXPERIMENTAL SECTION

### General Procedures

All chemicals were purchased from standard vendors and used as received; ethylene glycol (EG) (Mallinckrodt, Paris, KY), $LiOH \cdot H_2O$ (Johnson Matthey, Ward Hill, MA), $SiO_2$ (Johnson Matthey Electronics, Ward Hill, MA), $Al(OH)_3 \cdot xH_2O$ (Aldrich Chemical Co., Milwaukee, WI), triethanolamine (Aldrich Chemical Co., Milwaukee, WI), and acetonitrile (Baxter Healthcare Corp., Muskegon, MI) were used.

**Syntheses.** A typical synthesis of the lithium aluminosilicate polymer precursor is described below. Further details concerning the general syntheses of the alkali aluminosilicate materials are reported elsewhere.[4,5] In the standard reaction, 1.0 equivalent of MOH (M = Li, Na, K, etc.) is reacted, with 1.0 equivalent of $SiO_2$, 1.0 equivalent of $Al(OH)_3 \cdot xH_2O$ (53 wt. percent $Al_2O_3$) and 3.0 equivalent of triethanolamine (TEA), in excess EG with continuous distillative removal of water, followed by cooling. The resulting polymer is recovered by vacuum evaporation of solvent and/or washing with acetonitrile.

A mixture of 0.5 mole of $LiOH \cdot H_2O$ (17 g, 57 wt % LiOH by TGA), 0.5 mole of $SiO_2$ (30 g), 0.5 mole $Al(OH)_3 \cdot xH_2O$ (38 g, 53 wt % $Al_2O_3$ by TGA), 3.0 moles of TEA (180 g), and 500 mL of EG are placed in a 1 L flask equipped with a still head. The mixture is heated to ≈ 200°C under $N_2$ with constant mechanical stirring. The silica, alumina and alkali hydroxide readily dissolve with continuous distillation of EG and simultaneous removal of $H_2O$ formed as by-product. The liquid reaction goes clear and is complete in 4-5 h, when 350-400 mL of EG have distilled off. The clear solution becomes increasingly viscous as EG is removed. The resulting product is dried under a dynamic vacuum at temperatures of 200-250°C.

**Pyrolysis.** Precursor samples were pyrolytically converted to the target alkali aluminosilicate using a two step pyrolysis at 400°C. The first pyrolysis step required staged heating to 400°C, in air, to decompose and evolve most of the organics from the sample.[6] In this step, a 1500 mL crystallization dish containing a concentrated precursor solution (200 g, 25 wt. percent ceramic yield) was inserted into a Thermolyne Type 6000 furnace (Model No. F6038C, Dubuque, IA). The sample was ramped at 1°C/min to 75°C and held for 2 h. The pyrolysis temperature was then increased to 100° (1°C/min, 2 h hold) and thereafter, the material was heated in stages with 50°C/2h hold increments to 350° and finally held at 400°C for 4 h. After cooling, the powder was ground with an alumina mortar and pestle. The second pyrolysis step was a fixed bed oxidation at 400°C for 8 h in flowing $O_2$ (500 ml/min) to oxidize residual carbon. This step was run with 10 g samples heated in a single zone, Lindberg (Model No. 58114, Watertown, WI), tube furnace, equipped with a Eurotherm programmable temperature controller (Model No. 818P, Northing, England). A total of 50 g of ceramic LAS was made for the studies conducted here. The resulting material consists of large (-200 mesh; 74 μ), porous, aluminosilicate particles. Additional details concerning the processing can be found in reference 7.

The samples (5-10 g) of the resulting 400°C product were then heated to selected temperatures for periods of 2 or 8 h in flowing $O_2$ and then characterized as follows. The samples produced at 400° and 500°C were run using 8 h pyrolysis times to further ensure that carbon was completely removed.

## Characterization

The phase transformations and chemical changes that occur during pyrolytic transformation of the precursors to amorphous and/or crystalline materials were characterized using powder x-ray diffractometry (XRD), diffuse reflectance infrared Fourier transform spectroscopy (DRIFTS), and thermal gravimetric analysis (TGA). Porosimetry was used to determine specific surface area (SSA, $m^2/g$) and pore size distribution.

**XRD.** The XRD patterns were determined using a Rigaku Rotating Anode Goniometer (Rigaku Denki Co. Ltd.,Tokyo, Japan). Powder samples (100-200 mg) were: (1) ground with an alumina mortar and pestle, (2) packed in a glass specimen holder, and (3) placed in the goniometer. Scans were continuous from 5-80° 2θ at a scan rate of 2° 2θ/min using 0.05° 2θ increments and Cu $K_\alpha$ ($\lambda$ = 1.542 Å) radiation. The peak positions and the relative intensities of the powder patterns were identified by comparing with standard JCPDS files for characterization purposes.

**DRIFTS.** The DRIFT spectra were recorded on a Mattson Galaxy Series FTIR 3000 spectrometer (Mattson Instruments, Inc., Madison, WI). Optical grade, random cuttings of KBr (International Crystal Laboratories, Garfield, NJ) were ground using an alumina mortar and pestle and transferred to a flask. DRIFT samples were prepared by mixing 0.3-1.0 weight percent of the sample to be analyzed, with ground KBr.

Sample concentrations were kept below 1 weight percent to ensure adherence to Beer's law. The dilute samples prepared in KBr were subsequently packed firmly in the sample holder, leveled off at the upper edge to provide a smooth surface and transferred to the FTIR sample chamber, which was flushed constantly with $N_2$. A minimum of 64 scans were collected for each sample at a resolution of ± 4 $cm^{-1}$. Diffuse reflectance peak positions were identified using a standard peak searching program.

**Thermal Analytical Techniques.** TGAs of the precursor polymers were performed on a 2950 Thermal Analysis Instruments (TA Instruments, Inc., New Castle, DE). TGA samples (≈ 10-15 mg) were loaded on a Pt pan and heated in flowing (60 cc/min) dry air with a $N_2$ balance (40 cc/min), at 50°C/min to 1000°C, in the TGA "Hi-Res" mode.

**Porosimetry.** SSAs and pore size distributions were obtained using a volumetric technique on a Micromeritics ASAP 2000M sorption analyzer (Norcross, GA). Samples were degassed at 110°C for 2 h, and then at 400°C until the degas rate was less than 10 μm Hg/min. The SSAs were determined by the BET multipoint method with at least 5 data points with relative pressures between 0.001 and 0.20. Pore size distributions were determined using the density functional theory (DFT)[8-12] using the set of isotherms in which $N_2$ is adsorbed on a C substrate (at 77 K), with a model assuming slit-like pore geometries. Analysis was done with the software package supplied with the instrument.

**Electrorheological Measurements.** Samples were ground (-200 mesh; 74 μ) with an alumina mortar and pestle. The samples were then sealed in a Thermolyne Type 21100 Tube Furnace (Model No. F21125) under dynamic vacuum and heated to remove physisorbed water. Samples were typically dried at 600°C, except for the samples heat treated to 400°C and 500°C, which were dried at 400°C and 500°C, respectively. Samples were then cooled to room temperature under vacuum. Mineral oil (light paraffin oil, supplied by Mallinckrodt, Paris, KY) was added to the sample under vacuum in a ratio of 8 ml to 3 g sample. The sample was allowed to absorb mineral oil for
4 h and then removed from vacuum in air and transferred immediately to a rheogoniometer.

A Weissenberg rheogoniometer (Model No. R16) connected to a high voltage DC supply was used to measure the electrorheological behavior of the samples at room temperature. The dispersed samples were loaded into the cup of the couette cell of the rheogoniometer. A constant shear rate of 0.0236 $s^{-1}$ was used for all measurements. An electric field was applied to the couette cell (a 1 mm gap exists between the cup and couette) in increments of 0.5 kV/mm until the ER response reached equilibrium, and a torque was measured on the upper bob. Torque deflection readings were taken every 0.5 kV/mm until the sample experienced dielectric breakdown.[13] ER activity was measured in shear stresses of Pa.

## RESULTS AND DISCUSSION

The key to the work we are undertaking is the development of a very general synthetic route to alkali aluminosilicates. In this regard, we recently identified a simple and direct route to group I silicate precursors via the reaction of the corresponding hydroxide with silica:[6]

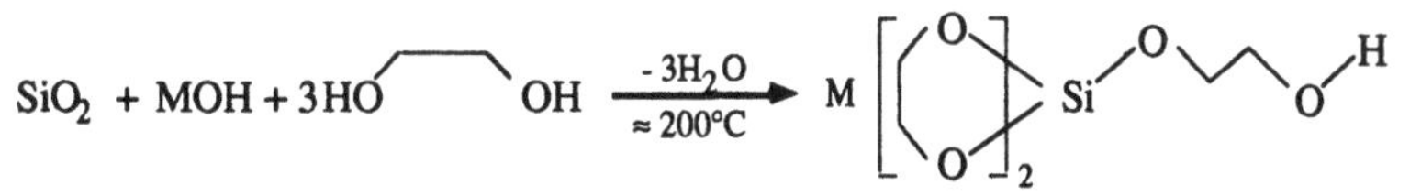 (1)

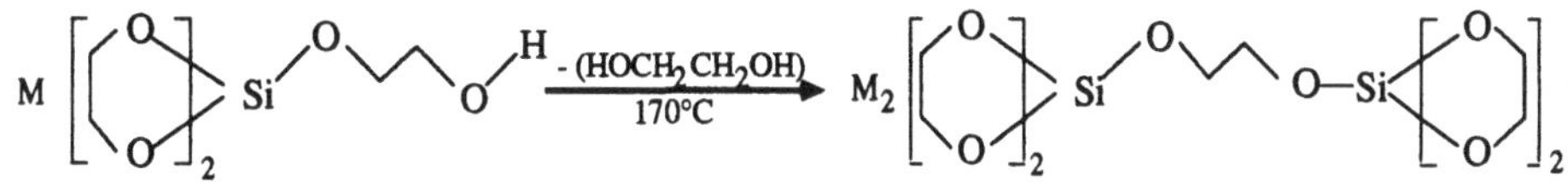 (2)

M = Li, Na, K, Cs

Pyrolysis of the dimeric species shown in reaction (2) provides nearly phase pure silicates, $M_2Si_2O_5$.[4] More recently, we discovered that it is possible to make the neutral alkoxy alanes:[5]

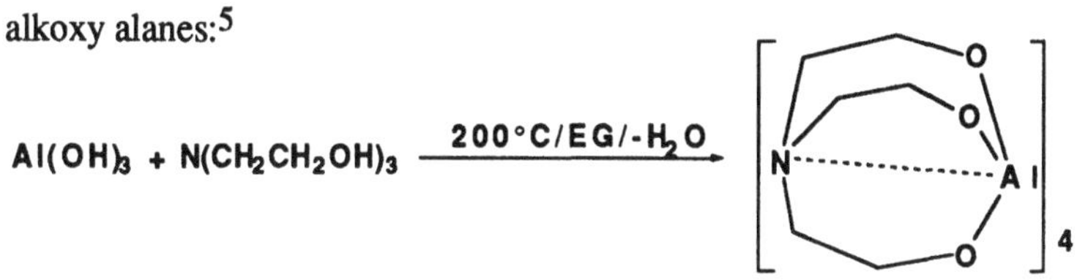

(3)

We now find that combining these reactions provides a novel, "One Pot Oxide Process," wherein it is possible to synthesize alkali aluminosilicates of any stoichiometry, simply by mixing the reactants in the desired stoichiometry:

$$xMOH + ySiO_2 + zAl(OH)_3 + (x+y+z)TEA \xrightarrow{200°C/x\text{'s } EG/10\ h/-H_2O} M_xSi_yAl_z(EG)_{2.5y}(TEA)_z \quad (4)$$

We have not as yet delineated the chemical structures for any of the alkali aluminosilicate precursors produced via reaction (4); although, we have some concept of the bonding environment at Si and Al. Thus, the formula given should be treated only as a generic label for these materials.

All of the precursors to aluminosilicates produced using reaction (4) are viscous liquids or brittle plastics that can be pyrolyzed directly in air or oxygen at temperatures up to 700°C, to form the corresponding inorganic amorphous alkali aluminosilicates, with no apparent change in stoichiometry. Further heating to higher temperatures results in crystallization. Details of the process as conducted for the 1:1:1 Li:Al:Si (LAS) composition aluminosilicate are as follows.

## Pyrolysis of LAS Bulk Precursor

The LAS aluminosilicate precursor was prepared according to the procedure in the experimental section. TGA of the resulting viscous polymer in air demonstrated that most of the organic component was lost on heating to 400°C. This result identifies the lowest bulk pyrolysis temperature as 400°C. Pyrolysis studies conducted at both 400°C and at 500°C showed no obvious difference in the properties of the powders produced at both temperatures. For the work reported below, the initial processing temperature was 400°C. The first step was a staged 4 h, pyrolytic decomposition of the precursor in air. This step removed the majority of the organics but resulted in the formation of some residual carbon containing species. The remaining carbonaceous material was then removed via an 8 h isothermal pyrolysis in $O_2$ at 400°C. Approximately 50 g of amorphous, 400°C powder was produced to serve as dispersed phase material for the ER fluids prepared.

A set of samples was then processed by heating 7-8 g samples of the 400°C powder to 500°, 600°, 700° or 900°C ($O_2$) for 8 h for the two lower temperatures and 2 h for the two higher temperatures. The resulting powders were then characterized by DRIFTS, XRD, and porosimetry prior to being used for ER studies. The results of the characterization studies are presented below.

## DRIFT Studies

Figure 1 records FTIR spectra for samples heated to selected temperatures in $O_2$ followed by a 2 h (8 h) hold at temperature. The broad, intense bands are typical of amorphous materials. The peaks at 980, 700, and 500 $cm^{-1}$ correspond to a combination of Si-O, Al-O and $M\text{-}O^{-}Li^{+}$ stretching vibrations, perhaps with some contribution from M-O-M bending vibrations. The presence of a small band at 3550 $cm^{-1}$ can be ascribed to isolated νO-H vibrations.[4] In the 900°C sample the broad peak at 3300 $cm^{-1}$ corresponds to νO-H for physisorbed water taken up when the DRIFT sample was made. It is removed in the heat treatments that are made prior to making the ER fluids.

Finally, the presence of small bands in the 1400-1650 $cm^{-1}$ region in the spectra of samples processed below 700°C, can be ascribed to trace amounts of carbonate species.[4] The spectra all looked relatively similar up to 900°C, where there is an abrupt change. The 900°C spectrum contains rather sharp IR absorptions that arise because of the atomic periodicity typical of crystalline materials, indicating crystallization of $Li_2O{\cdot}Al_2O_3{\cdot}2SiO_2$ (see XRD studies below).

Perhaps the most important results of the DRIFT studies are the minimal presence of surface OH groups, as indicated by the absence of strong vibrations in the νO-H 3200-3700 $cm^{-1}$ region. Thus, the surface is essentially anhydrous. Therefore, any ER effects observed are not likely to be due to the presence of physisorbed water.

## XRD Studies

Figure 2 shows the XRD powder patterns for the aluminosilicate at 400°, 700° and 900°C. The powder patterns at 500° and 600°C are identical to the 700°C pattern. All patterns, except the 900°C pattern, indicate an amorphous material. The powder pattern at 900°C is identical to that for $Li_2O{\cdot}Al_2O_3{\cdot}2SiO_2$ (JCPDS File No. 26-839). No extraneous peaks are seen, indicating the absence of a second phase and that the original stoichiometry of the precursor stage is apparently maintained in the ceramic product.

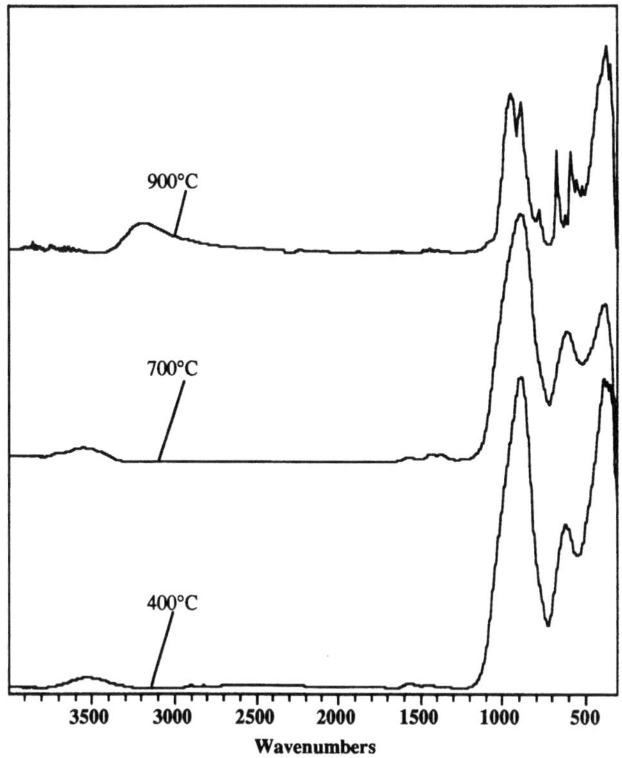


**Figure 1.** DRIFT Spectra for Lithium Aluminosilicate Samples Heated to Selected Temps. Samples ramped at 10°C to temperature in flowing $O_2$ and held at temperature for 2h.

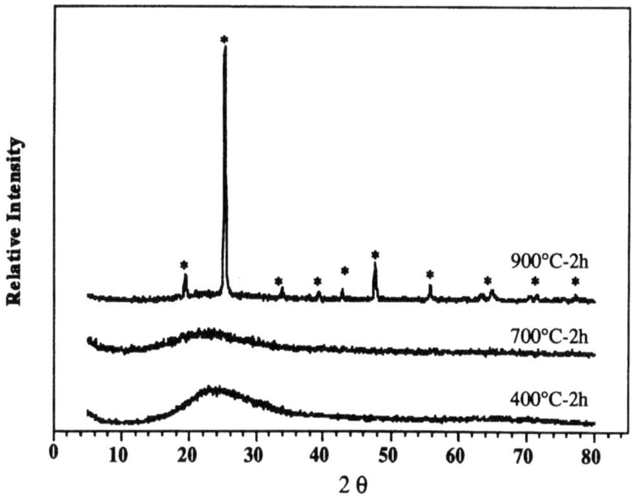


**Figure 2.** XRD Powder Patterns for Lithium Aluminosilicate Samples Heated to Selected Temperatures. Samples ramped at 10°C to temperature in flowing $O_2$ and held at temperature for 2h.

## Porosimetry Studies

The ER activity of a given fluid likely depends on the degree of electrostatic interaction of the suspending medium with the suspended particles. Consequently, a detailed understanding of powder surface area and porosity, especially as a function of processing temperature, should provide a picture of the factors that affect the extent of ER activity. Figure 3 shows a plot of the change in (BET and DFT) surface area as a function of sample processing temperature. Figure 4 shows a corresponding plot of pore volume as a function of processing temperature.

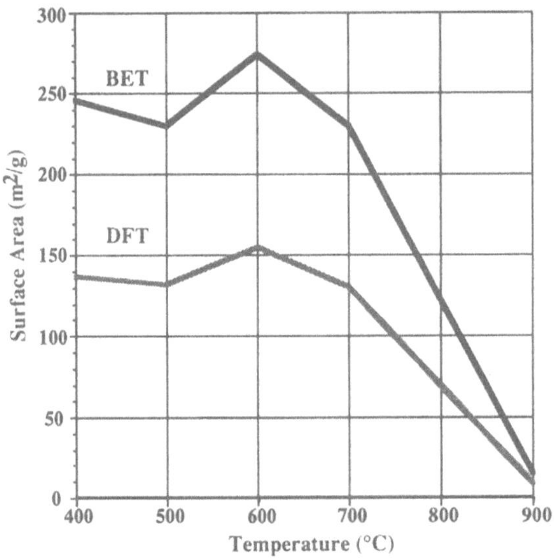


**Figure 3.** Plot of Change in BET/DFT[8-12] Specific Surface Areas as a Function of Processing Temperatures.

Note that the BET method tends to overestimate the total surface area.

The BET method typically overestimates the total surface area; the DFT surface area is deemed more reliable because it analyzes experimental adsorption isotherms using methods based on modern theoretical physical chemistry, not methods dating back to the 1930's. The measured SSAs for the various processing temperatures do not vary significantly over the temperature range of 400° to 700°C ranging from a high of 165 $m^2/g$ at 500°C to a low of 130 $m^2/g$ at 700°C as some densification occurs at this temperature. Likewise, the total pore volume also remains relatively constant, varying between 0.73 cc/g at 600° to 0.6 cc/g at 700°C.

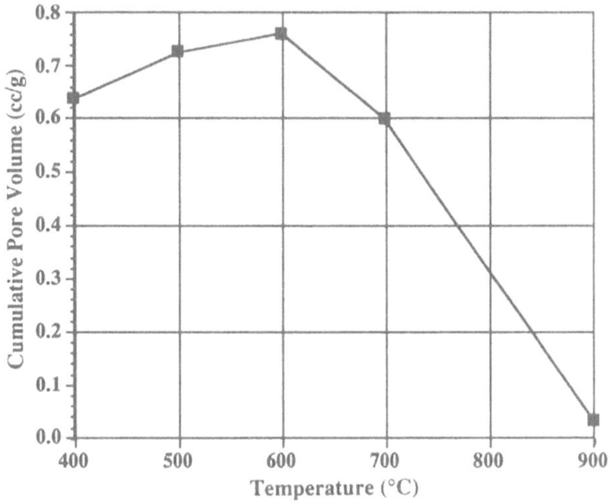


**Figure 4.** Plot of Change in Pore Volume as a Function of Processing Temperatures.

Surprisingly, the pore size distribution (see Figures 5 and 6) does not change much between 400° and 700°C, except for the loss of a slight amount of microporosity. Even on heating to 900°C, the pore size distribution (Figure 7) for those pores not eliminated by crystallization is quite similar to the lower temperatures; although, a small percentage of mesopores now appear in the 150-2500Å region.

However, crystallization of the 900°C sample results in a considerable change in specific surface area (8.5 $m^2/g$) and pore volume (0.03 cc/g) as might be anticipated.

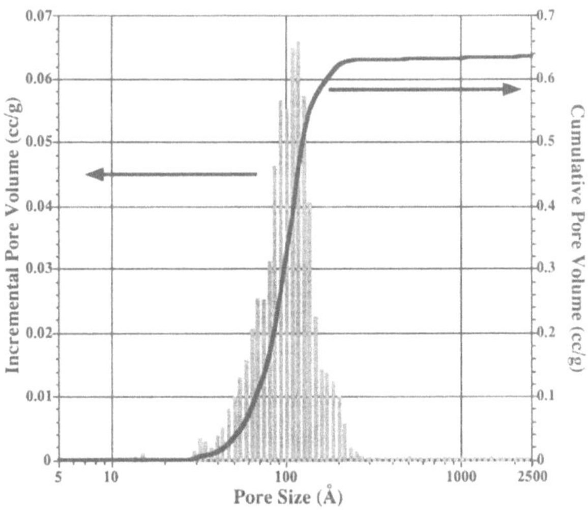


**Figure 5.** Pore Size Distribution of 400°C Lithium Aluminosilicate Sample.

Note the presence of a small quantity of micropores.

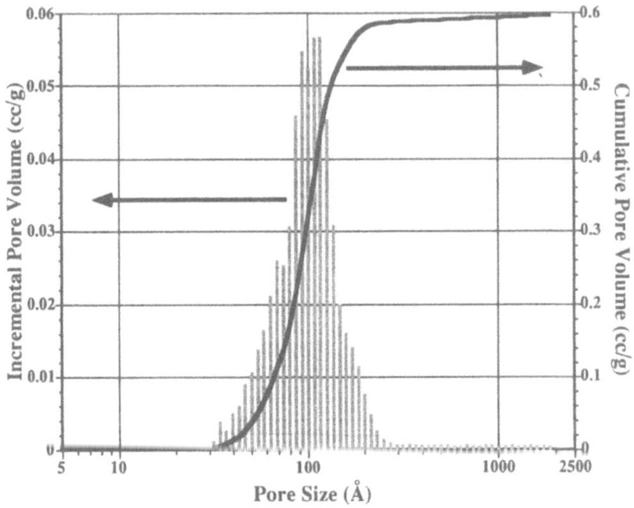


**Figure 6.** Pore Size Distribution of 700°C Lithium Aluminosilicate Sample.

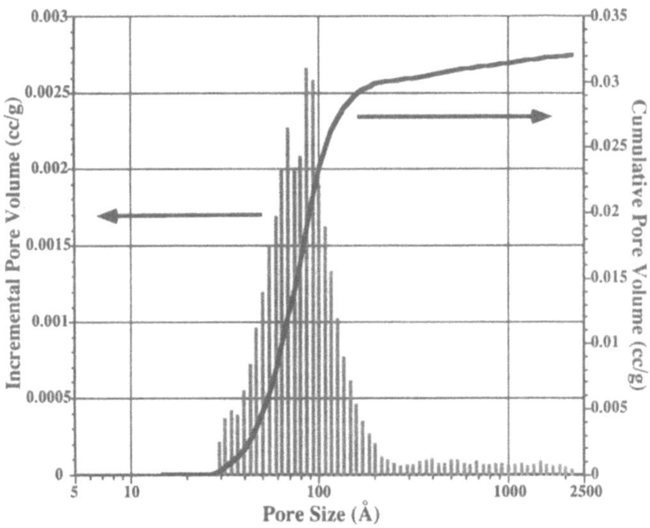


**Figure 7.** Pore Size Distribution of 900°C Lithium Aluminosilicate Sample. Note the presence of mesopores (150-2500 Å).

## ER Behavior

Figure 8 records the ER behavior observed at constant voltage for the various processing temperatures. The values at 5.0 kV/mm were used for the plot because it was observed that samples that were not completely dry or not completely free of carbonaceous material would typically experience dielectric breakdown below this voltage and thus give unreliable data. Only samples that showed ER responses at or above 5.0 kV/mm were deemed reliable.

Although the change in shear is not particularly impressive, it appears to be roughly proportional to the SSAs (or pore volumes) for the given samples. The falloff in ER activity from 75.41 to 15.08 Pa that occurs on going from the 700°C sample's SSA (pore volume) of 130 $m^2$/g(0.6 cc/g) to the 900°C sample's 8.5 $m^2$/g(0.035 cc/g) as the sample crystallizes is

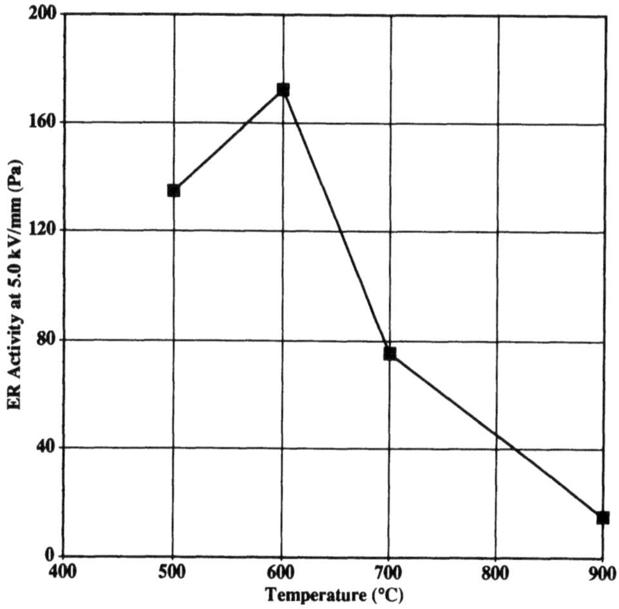


**Figure 8.** Plot of Change in ER Activity as a Function of Processing Temperatures.

not linear. At this early stage, it is not clear whether this result arises because of the phase change or some other phenomena associated with changes that occur at the surface of the crystalline material. Further studies are planned that will examine the effects of changes in group I counterions as well as the effects of different porosities and particle sizes.

## CONCLUSIONS

The "One Pot Oxide Process" provides a simple and general method for preparing a wide variety of porous MAS powders with well defined compositions, SSAs and pore size distributions. The powders studied to date exhibit ER activity that correlates with SSA or pore volume in various heat treated LAS samples. However, the facility with which our synthetic approach permits changes in various individual properties suggests that it will be possible, through a matrix of studies, to identify optimal processing protocols that will provide superior properties and simultaneously permit identification of specific properties that control ER behavior.

## ACKNOWLEDGMENTS

This work was supported by the Army Research Laboratories, the Office of Naval Research, and the Army Research Office. RB thanks National Physical Science Consortium for a fellowship. The authors thank Mr. Adam Schubring and Professor Frank Filisko for extensive help with the ER measurements.

## REFERENCES

1. "Electrorheological (ER) Fluids", A Research Needs Assessment. DOE final report for Contract No. DE-AC02-91ER30172.
2. a. F.E.Filisko, L.H. Radzilowski, "Intrinsic Mechanism for Activity of Alumino-Silicate Based Electrorheological Fluids," *J. Rheol.*, 34:539 (1990). b. D. Gamota, F.E. Filisko, "High Frequency Dynamic Mechanical Study of an Alumino-silicate Electrorheological Material," *J. Rheol.*, 35:1411 (1991) .
3. a. D.R. Gamota, A.S. Wineman, F.E. Filisko," Fourier Transform Analysis: Fundamental Nonlinear Dynamic Response of an Electrorheological Material," *J. Rheol.*, accepted. b. D.R. Gamota, B.L. Mueller, F. Filisko, patent application filed (1994).
4. P. Kansal, R.M. Laine, "Pentacoordinate Silicon Complexes as Precursors to Silicate Glasses and Ceramics," *J. Am. Ceram. Soc.*, 77:875-882 (1994).
5. a. R.M. Laine, B.L. Mueller, T. Hinklin, D. Treadwell, "One Step Synthesis of Neutral Alkoxy Silanes and Alanes from $SiO_2$ and $Al(OH)_3$,"submitted for publication. b. V.E. Shklover, Yu. T. Struchkov, M.G. Voronkov, Z.A. Ovchinnikova, V.P. Baryshok, Dokl. Akad. Nauk SSR. 227:1185 (1984). Chem. Abstracts, 102:37181k (1984).
6. K.Y. Blohowiak, D.R. Treadwell, B.L. Mueller, M.L. Hoppe, S. Jouppi, P. Kansal, K.W. Chew, C.L.S. Scotto, F. Babonneau, J. Kampf, R.M. Laine, "$SiO_2$ as a Starting Material for the Synthesis of Pentacoordinate Silicon Complexes. I.," *Chem. Mater.* in press (1994).
7. C.R. Bickmore, R.M. Laine,"Processing Oxynitride Powders via Fluidized Bed Ammonolysis of Large, Porous, Silica Particles," submitted to *J. Am. Ceram. Soc.*
8. N.A. Seaton, J.P.R.B. Walton, N. Quirke, *Carbon,* 27:853 (1989).
9. J.P.Olivier, W.B. Conklin, presented at 7th Int. Conf. on Surf. and Coll. Sci., Compiegne, France, (1991).
10. R. Evans, *Adv. Phys,* 28:143 (1979).

11. S. Ross, J.P. Olivier, in *On Physical Adsorption*, Chapter 1, Interscience, New York (1964).

12. P. Tarazona, R. Evans, *Mol. Phys.*, 52:847 (1984).

13. A.W. Schubring, F. E. Filisko, "Study of the Variation of the Substitutional Atoms in Amorphous Alumino-Silicates as an Electrorheological Material," *Am. Chem. Soc. Poly. Prpts*, 35:xxx (1994) .

# ELECTRORHEOLOGICAL PROPERTIES OF POLYANILINE DISPERSIONS: EFFECTS OF ACID DOPANT CONCENTRATION

Richard M. Webber

The Lubrizol Corporation
Wickliffe, OH 44092

## INTRODUCTION

A typical anhydrous electrorheological (ER) fluid consists of semiconducting particles dispersed in a low conductivity oil. The electrorheological effect refers to the changes in dispersion rheological properties that occur upon application of an electric field across the fluid; the most commonly identified effect being an increase in the stress required to cause flow.

Many phenomenological aspects of electrorheological response are well described in the context of the electrostatic polarization model. This model relates the interaction force between dielectric particles in a non-conducting oil to the relative particle and fluid dielectric constants and the electric field. These interactions result in particle stranding. Historical development of the electrostatic polarization model and its application to modeling the dynamics of electrorheological response are covered in detail by Zukoski.[1] Of particular note is that the electrostatic polarization model leads to predictions for the electric field strength (E) scaling of static yield stress ($\tau_0$) and the kinetics of structure formation, as well as the effects of shear rate on dynamic yield stress. These predictions are consistent with many experimental observations. However, as Zukoski pointed out, the electrostatic polarization model has a number of short-comings, most notable is our inability to use it to determine appropriate material properties for development of optimal ER fluids. In particular, many practical ER fluids are composed of semiconducting particles, particles that have both finite permittivity and conductivity where the later is explicitly unaccounted for in the electrostatic polarization model.

*Progress in Electrorheology*, Edited by K.O. Havelka
and F.E. Filisko, Plenum Press, New York, 1995

Electrorheological studies of dispersions of polymeric semiconductors in which particle electrical properties were varied have demonstrated that both particle permittivity and conductivity contribute to the electrorheological response of the fluid.[2-4] In these studies, the authors observed increases in yield stress with increasing particle conductivity and permittivity, and in Block *et al.* [3] a maximum in yield stress versus particle conductivity at approximately $10^{-6}$ S/cm. However, the effects of particle electrical properties on current density and on transient rheological response to electric field application, characteristics important to both mechanistic understanding and commercial development of ER fluids, have only been addressed in a limited manner. In this work, the effects of changes in particle permittivity and conductivity of polyaniline dispersions on dielectric and electrorheological properties are studied. The focus of this work is on both steady state and transient electrorheological properties, as well as on the relationship between particle conductivity and suspension conductivity

The electrical properties of polyaniline dispersions were varied by changing the concentration of acid dopant in polyaniline.[5] The effects of degree of protonation of polyaniline on static yield stress, rate of change of rheology upon application of electric field (transient response or response time), current density, and dispersion dielectric properties were measured. Dopant concentration effects on magnitude and rate of polarization from the dielectric spectra of the ER fluids are compared to the effects of dopant concentration on ER properties. A discussion is developed in the context of particle electrical properties effects in a generalized interfacial polarization model.

## EXPERIMENTAL

Polyaniline was synthesized by oxidative polymerization in aqueous acid.[5] Polyaniline was then base washed, subdivided into six samples and re-protonated by equilibration with aqueous HCl solutions. Concentrations of Cl and N were determined; the molar ratio of Cl to N ([Cl]/[N]) for the acid doped polyanilines is shown in Table 1.

Polyaniline samples were dried under vacuum at 423 K for 24 hours and then dispersed at 0.2 weight fraction in 10 cSt Dow Corning 200 polydimethylsiloxane (approximately 0.15 volume fraction, where the density of polyaniline was measured as 1.39 g/cm$^3$). All dispersions contained 0.03 weight fraction of Genesee Polymers EXP-69, a hydroxypropyl functionalized polydimethylsiloxane.

Permittivity ($\varepsilon'$) and loss factor ($\varepsilon''$) of the polyaniline dispersions were measured as a function of electric field frequency ($\omega$) over the range $10^{-1}$ to $10^4$ Hz on a Micromet Instruments Microdielectrometer at approximately 298 K. The amplitude of the electric field (E) is ill-defined due to the comb electrode geometry but was estimated to be ~80 V/mm, which is approximately half an order of magnitude smaller than the minimum electric field in ER experiments.

Rheological measurements were performed on a Carri-Med CSL-500 controlled stress rheometer modified for study of electrorheological fluids. A Trek 610C high voltage power supply was electrically connected to the rheometer through an electrolyte solution contained on the rotating shaft of the rheometer. The rheometer control, high voltage power supply, and electronics were synchronized through Labview® data acquisition software (National

**Table 1.** Degree of protonation ([Cl]/[N]) for the polyaniline samples. $\tau_{\varepsilon}$ and $\alpha$ were determined from fits of eq 2 to dielectric spectra shown in Figure 1.

| sample | [Cl]/[N] | $\tau_{\varepsilon}$ (ms) | $\alpha$ |
|---|---|---|---|
| PANI-1 | 0.002 | 1520 | 0.33 |
| PANI-2 | 0.009 | 260 | 0.39 |
| PANI-3 | 0.025 | 16.0 | 0.47 |
| PANI-4 | 0.028 | 5.6 | 0.53 |
| PANI-5 | 0.033 | 2.5 | 0.51 |
| PANI-6 | 0.037 | 6.3 | 0.54 |

Instruments Corp.) on a Macintosh Quadra 900 equipped with a IEEE-488 controller board, a multifunction I/O board, and a digital I/O board. The Trek 610C was used as a 1000x amplifier where the voltage signal required to generate a particular electric field (E) was supplied from the computer and triggered via a TTL controlled relay. Strain ($\gamma$), E, and current were measured at rates up to 2000 Hz nearly simultaneously; switching delay between channels was ~$10^5$ Hz.

The rheological geometry was a 0.03 m diameter parallel plate set for a gap width of 500 µm. $\gamma$ and shear rate ($\dot{\gamma}$) reported are those at the outer radial edge of the plate. Dispersion viscosities ($\mu$) in absence of electric field were approximately 0.07 Pa·s over the shear rate range $10^2$ to $10^3$ $s^{-1}$. All the rheological experiments reported here were conducted under DC electric fields at a temperature of 298 K.

ER properties of polyaniline dispersions depended on electrorheological history. History effects were minimized by shearing the dispersion at 1000 $s^{-1}$ (E = 0) for 20 s prior to each ER experiment. Static yield stress ($\tau_o$), current density (j), and the rate of change (response time) of effective viscosity upon application of E ($t_R$) were measured as a function of E. All electrorheological measurements reported here were reproducible to better than 5 to 10%.

Static yield stress ($\tau_o$) was measured by applying E across a non-flowing dispersion and then increasing stress ($\tau$) at 0.33 s intervals at a constant rate of 1.7 Pa/s. Static yield stress is dependent on shear and electric field history as well as the rate of increase of stress. Under low electric fields (*e.g.*, less than 0.75 kV/mm), $\tau$ versus $\dot{\gamma}$ curves, specifically the solid to fluid transition, were qualitatively similar to those observed by Lemaire *et al.*[6] on a controlled stress rheometer. At E $\geq$ 1.0 kV/mm, all the polyaniline dispersions (except for PANI-1) exhibited yield stresses greater than the stress required to reach the maximum allowed rotation rate of the rheometer in absence of E, 50 rad/s (1500 $s^{-1}$ for this geometry). A loop in the software was used to monitor strain as a function of the applied stress and to stop the ramp stress experiment when the difference in strain between two consecutive stress increments was greater than 4. $\tau_o$ was defined as the stress required to achieve $\gamma = 2$ which was characteristic of the onset of catastrophic yield for these dispersions. Because by this definition fluids exhibit a yield stress in absence of E, results are presented as the static yield stress due to the ER effect ($\tau_{ER}$), where $\tau_{ER}$ is defined as $\tau_o(E) - \tau_o(E{=}0)$, however typically $\tau_o(E{=}0) \ll \tau_o(E)$.

Current density (j) was measured under static conditions ($\dot{\gamma} = 0$) where E was increased from 0 to 3 kV/mm in discrete steps of 0.33 kV/mm every 2 s. The experiment was repeated three times with each ramp increase of E proceeded by the pre-shear procedure. Data reported as j versus E was the average of three runs.

The rate of change of effective viscosity upon application of E was measured from an initial condition of steady state flow. Steady state flow for rheologically non-time dependent (in absence of E) polyaniline dispersions is a constant stress condition where the applied stress ($\tau_{app}$) and a specified initial shear rate ($\dot{\gamma}_{ini}$) are related by $\mu_{ini} = \tau_{app}/\dot{\gamma}_{ini}$. The experiment consisted of a step change application of E and measurement of the response of shear rate and current density as a function of time (t) (only the time dependence of shear rate is discussed here). For large values of E (>1 kV/mm) where $\tau_{app} < \tau_o(E)$, the rotation rate of the rheometer went to zero with increasing time, indicating structuring of the fluid.

The time interval for the fluid to structure in a rate of change experiment depends on the inertia of the rotating geometry of the rheometer, the electrorheological properties of the fluid, as well as E and $\dot{\gamma}_{ini}$. Instrument inertial effects were corrected through the differential equation describing the torque balance on the rheometer geometry

$$I\frac{d\omega(t)}{dt} = -\frac{\mu(t)}{K}\omega(t) + T_{app} \tag{1}$$

where I is the inertia of the rotating rheometer geometry, ω the measured rotation rate of the rheometer geometry which is a function of time (t), K a geometry specific constant, and $T_{app}$ the applied constant torque.[7] In contrast to standard rheological situations where inertial corrections are typically encountered because of a time dependent torque application, inertial correction is required here because the effective viscosity of the fluid is time dependent ($\mu(t)$) upon application of E. Equation (1) is solved to determine the actual viscosity change of the fluid $\mu(t)$ from the measured apparent viscosity change of the ER fluid $T_{app}/(K\omega(t))$.

Rate of response ($t_R$) was defined as $t_R \equiv 1/\Delta t$ where $\Delta t$ is the elapsed time for an effective viscosity change $\mu(t)/\mu_{ini} = \dot{\gamma}_{ini}/\dot{\gamma}(t) = 50$. Because $\tau_{app} < \tau_o(E)$ and $\dot{\gamma}(t)$ goes to 0, $t_R$ approximates a characteristic transition time to fluid structuring. All values for $t_R$ reported here have been corrected for inertial effects and thus represent the electrorheological response of the fluid. For the PANI ER fluid samples and range of experimental conditions reported here, the inertia correction to $t_R$ ranged from less than 1% to approximately 65% for response times of order 10 s to 4 ms respectively.

## RESULTS AND DISCUSSION

The electrical properties of polyaniline, permittivity and conductivity, depend on dopant concentration which is indicated by the relative concentrations of Cl to N ([Cl]/[N]) in Table 1. Zuo *et al.*[8] studied the dielectric properties of powdered polyaniline samples synthesized by similar methods. They found that over the range of $0.0 \leq [Cl]/[N] \leq 0.08$, the permittivity (ε' measured at 10 Hz) and conductivity (DC value) of polyaniline increased from approximately 10 to 370 and $10^{-9}$ to $10^{-3}$ S/m respectively at 300 K.

The effects of [Cl]/[N] on polyaniline dispersion dielectric properties, both permittivity (ε') and loss factor (ε") versus electric field frequency (ω) at 298 K, are shown in Figure 1 for three polyaniline ER fluids. The other samples exhibited similar behavior. Increasing [Cl]/[N] increased the frequency at which ε' relaxed. The relaxation of ε' and corresponding maximum in ε" with increasing ω in the range of $10^{-1}$ to $10^{4}$ Hz is characteristic of interfacial polarization processes.[9,10]

$\varepsilon'_{DC}$, the limit of ε' as ($\omega \rightarrow 0$), and $\varepsilon'_{\infty}$, the limit as ($\omega \rightarrow \infty$), were estimated by extrapolation of the ε' versus ω curves. Using estimates of $\varepsilon'_{DC}$ and $\varepsilon'_{\infty}$, the dielectric spectra were fit with the empirical circular arc representation of the Debye equation

$$\varepsilon^* = \varepsilon' + i\varepsilon'' = \varepsilon'_{\infty} + \frac{(\varepsilon'_{DC} - \varepsilon'_{\infty})}{1 + (i\omega\tau_{\varepsilon})^{1-\alpha}} \quad (2)$$

where i is $\sqrt{-1}$, $\tau_{\varepsilon}$ is the characteristic relaxation time, and α is a parameter characterizing the broadness of the relaxation process (α = 0 indicates a Debye relaxation process, *i.e.*, a single relaxation time).[9] Fits of eq 2 to dielectric data are shown in Figure 1 as dashed lines and values of $\tau_{\varepsilon}$ and α in Table 1. It is evident that the quality of the fits was better for ε' than for ε".

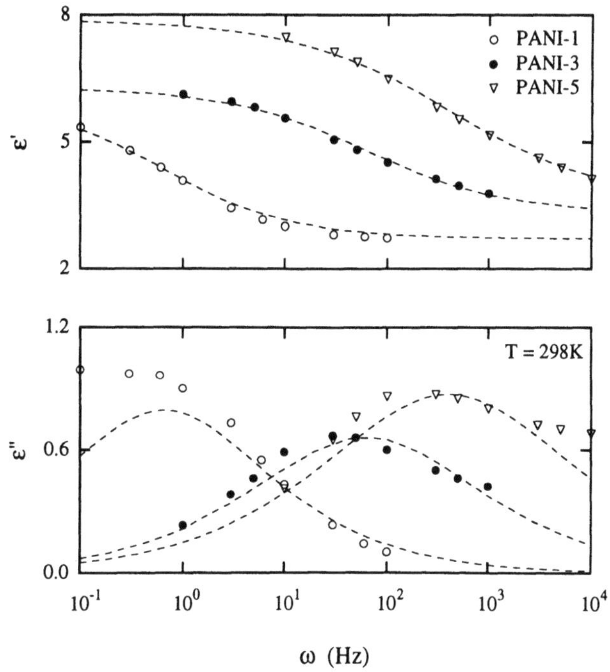


**Figure 1.** Permittivity (ε') and loss factor (ε") as a function of electric field frequency (ω) at 298K for three of the polyaniline ER fluids. The dashed lines are fits of eq 2 to the data, where $\varepsilon'_{DC}$ and $\varepsilon'_{\infty}$ were estimated from the permittivity spectra and the data was fit to determine $\tau_{\varepsilon}$ and α.

The six PANI ER fluids exhibited broad relaxation processes where $\alpha$ ranged from 0.33 to 0.54, *increasing* with increasing dopant concentration as shown in Figure 2. The interfacial polarization process is characterized by polarization within the dispersion on the length scale of particle size; therefore, the breadth of the particle size distribution and heterogeneities due to non-uniform doping are likely to increase the distribution of relaxation times and thus $\alpha$. Particle size distributions of the six sample ER fluids were measured by centrifugation on Horiba CAPA700 instrument and found to be similar and broad (note that all samples were prepared from the same initial batch of undoped PANI). By volume, the majority of the distribution fell in the range 0.1 to 20 µm, but all samples also contained a small fraction of large particles, ≈100 µm. The redoping of polyaniline depends on diffusion of the dopant into the particle core, therefore, a broad particle size distribution may lead to a non-uniform distribution of dopant within particle cores. This heterogeneity may account for tho broadening of the distribution of relaxation times (*i.e.*, increasing $\alpha$) with increasing [Cl]/[N].

Figure 3 shows the magnitude of the interfacial polarization effect ($\varepsilon'_{DC}$ - $\varepsilon'_\infty$) and the rate of polarization characterized by the relaxation time ($\tau_\varepsilon$) as a function of dopant concentration. $\varepsilon'_{DC}$ - $\varepsilon'_\infty$ was independent of dopant concentration while $\tau_\varepsilon$ decreased strongly with increasing [Cl]/[N]. $\tau_\varepsilon$ versus [Cl]/[N] data in Figure 3 scales as approximately $\tau_\varepsilon \sim ([Cl]/[N])^{-2.2}$.

The effects of particle electrical properties on the dielectric properties $\tau_\varepsilon$ and $\varepsilon'_{DC}$ - $\varepsilon'_\infty$ of the polyaniline ER fluid dispersions were characterized with the Maxwell-Wagner model for interfacial polarization.[10] In polarization of a heterogeneous Maxwell-Wagner system, the complex conductivity ($k^*$) of a suspension of a dilute randomly distributed particles is given by

$$k^* = k_c^* \frac{(1+2\phi\beta^*)}{(1-\phi\beta^*)} \tag{3}$$

where

$$\beta^* = \frac{(k_p^* - k_c^*)}{(k_p^* + 2k_c^*)}, \tag{4}$$

$k_p^*$ and $k_c^*$ are the complex conductivities of the particle and continuous phases respectively, and $\phi$ is the particle volume fraction. The complex conductivity of the j^th^ phase is $k_j^* = k_j + i\omega\varepsilon'_o\varepsilon'_j$, where $\varepsilon'_o$ is the permittivity of free space, 8.85x10$^{-12}$ F/m. For a Debye-type relaxation (eq 2 with $\alpha = 0$), $\tau_\varepsilon$ and $\varepsilon'_{DC}$ - $\varepsilon'_\infty$ are given by[10]

$$\tau_\varepsilon = \frac{\varepsilon'_o(2\varepsilon'_c + \varepsilon'_p - \phi(\varepsilon'_p - \varepsilon'_c))}{(2k_c + k_p - \phi(k_p - k_c))}, \tag{5}$$

and

$$\varepsilon'_{DC} - \varepsilon'_{\infty} = \frac{\varepsilon'_c(2k_c + k_p + 2\phi(k_p - k_c))}{(2k_c + k_p - \phi(k_p - k_c))} - \frac{\varepsilon'_c(2\varepsilon'_c + \varepsilon'_p - \phi(\varepsilon'_p - \varepsilon'_c))}{(2\varepsilon'_c + \varepsilon'_p - \phi(\varepsilon'_p - \varepsilon'_c))}$$

$$+3\phi k_c \frac{(2k'_c + k'_p)(\varepsilon'_p - \varepsilon'_c) - (2\varepsilon'_c + \varepsilon'_p)(k'_p - k'_c)}{(2k'_c + k'_p - \phi(k'_p - k'_c))^2}. \quad (6)$$

$k_p$ and $\varepsilon'_p$ as a function of [Cl]/[N] were obtained from fits of the low frequency dielectric data of Zuo *et al.*[8] for polyaniline solids prepared to the low dopant concentrations studied here. The dependence of $\varepsilon'_p$ on [Cl]/[N] was represented by

$$\varepsilon'_p = 10.2 + 1,160\left(\frac{[Cl]}{[N]}\right) + 4,1700\left(\frac{[Cl]}{[N]}\right)^2 \quad (7)$$

and the dependence of $k_p$ on $\varepsilon'_p$ by

$$k_p = 10^{-11.4}(\varepsilon'_p)^{3.4}. \quad (8)$$

Equations 5 and 6 were evaluated using measured values of conductivity and permittivity for silicone oil ($k_c = 2x10^{-12}$ S/m and $\varepsilon'_c = 2.3$) and $\phi = 0.15$. $\tau_\varepsilon$ and $\varepsilon'_{DC} - \varepsilon'_\infty$ as a function of [Cl]/[N] are shown as dashed lines in Figure 3. Calculated values of $\tau_\varepsilon$ and $\varepsilon'_{DC} - \varepsilon'_\infty$ from the Maxwell-Wagner model are significant underestimates to values determined from the measured dielectric spectra (Figure 1).

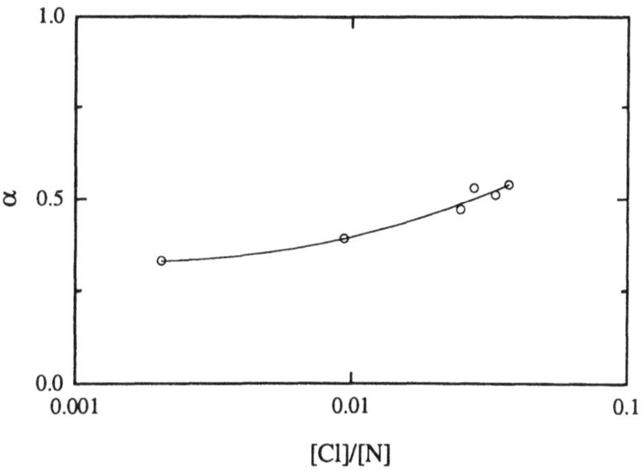


**Figure 2.** Dependence of α from eq 2 on dopant concentration [Cl]/[N]. α values were determined from the fits shown in Figure 1. The line is shown as a guide only.

Application of the Maxwell-Wagner model as presented above requires: 1) that particle and continuous phase complex conductivities ($k_i^*$) are independent of $\omega$ over the range covered by the interfacial polarization process, 2) that electrical properties are independent of the magnitude of the applied field, and 3) that spherical particles are randomly arrayed in the continuous phase. In addition, the assumption is made that our particle electrical properties are well represented by the Zuo *et al.*[8] data for HCl doped polyaniline.

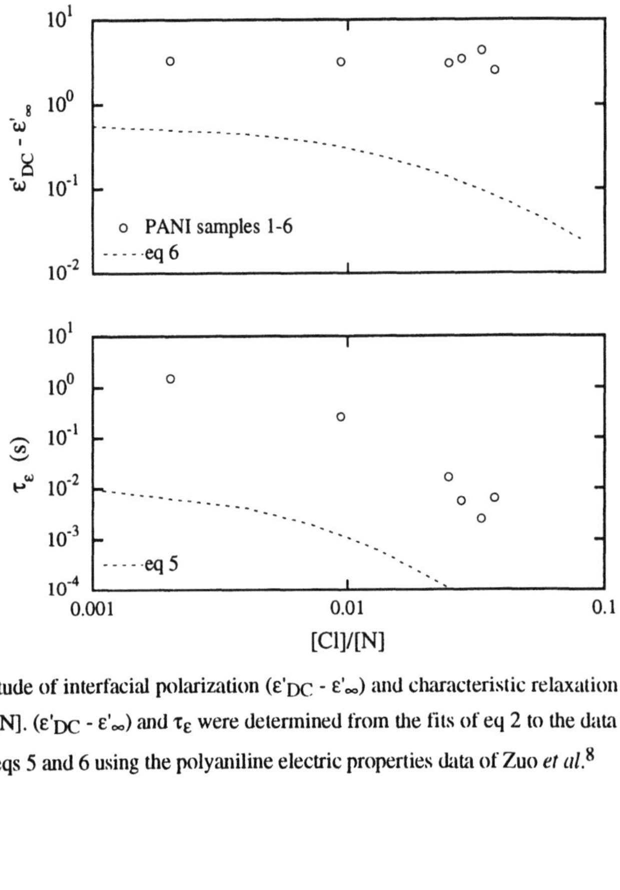


**Figure 3.** Magnitude of interfacial polarization ($\varepsilon'_{DC} - \varepsilon'_{\infty}$) and characteristic relaxation time ($\tau_{\varepsilon}$) and as a function of [Cl]/[N]. ($\varepsilon'_{DC} - \varepsilon'_{\infty}$) and $\tau_{\varepsilon}$ were determined from the fits of eq 2 to the data in Figure 1. The dashed lines are eqs 5 and 6 using the polyaniline electric properties data of Zuo *et al.*[8]

From the Zuo *et al.*[8] data, $k_p$ is approximately independent of $\omega$ up to $10^3$ Hz, while $\varepsilon'_p$ is weakly frequency dependent over the range $10^1 \leq \omega \leq 10^3$ Hz. Relatively small changes in $k_p$ and $\varepsilon'_p$ due to $\omega$ dependencies, however, cannot account for the underestimates of $\tau_{\varepsilon}$ and $\varepsilon'_{DC} - \varepsilon'_{\infty}$ shown in Figure 3. $k_p$ must be decreased by approximately 2 orders of magnitude in order to account for the $\tau_{\varepsilon}$ underestimate. $\varepsilon'_{DC} - \varepsilon'_{\infty}$ is strongly dependent on the random distribution assumption, questionable due to the possibility of electric field induced particulate stranding. The effects of particle stranding on suspension dielectric properties were estimated by accounting for the effects of particle shape on eqs 5 and 6, an analysis presented in van Beek.[10] If particle strands are represented as prolate spheroids, a spheroid with an axis ratio as small as 3 to 4 is enough to account for the underprediction of $\varepsilon'_{DC} - \varepsilon'_{\infty}$ shown in Figure 3. Effects of particle stranding (shape) on $\tau_{\varepsilon}$ are weak with respect to the underestimate shown in Figure 3. None of the explanations investigated above account for the independence of $\varepsilon'_{DC} - \varepsilon'_{\infty}$ from [Cl]/[N] that was observed.

Figure 4 shows the effects of electric field on static yield stress ($\tau_{ER}$), and current density (j) for sample PANI-4 at 293K. Similar behavior was observed for the other samples. Static yield stress scaled as $\tau_{ER} \sim E^{2.1}$ in agreement with polarization model predictions for electric field induced particle-particle interactions in ER fluids.[1] j scales as $j \sim E^{2.3}$, non-ohmic behavior that was typical for polyaniline dispersions over the range of ER electric field strengths, $0.33 \le E \le 3.0$ kV/mm.

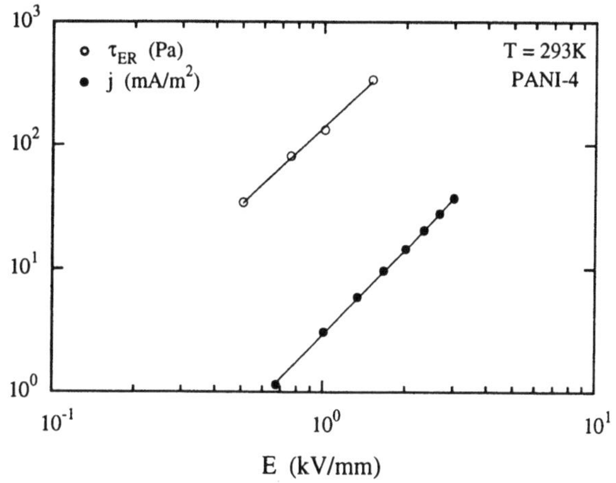


**Figure 4.** Effects of electric field (E) on static yield stress ($\tau_{ER}$) and current density (j) for sample PANI-4. The lines are fits.

Rate of response ($t_R$) depended on both E and the initial shear rate ($\dot{\gamma}_{ini}$) of the fluid at application of E. Figure 5 shows rate of response ($t_R$) as a function of E at constant initial shear rate ($\dot{\gamma}_{ini} = 500\ s^{-1}$) and Figure 6 shows $t_R$ as a function of $\dot{\gamma}_{ini}$ at constant E (3.0 kV/mm). Figures 5 and 6 show that the rheological response time of polyaniline ER fluids ranged over approximately 4 orders of magnitude, from 4 ms to 10 s, depending on E, $\dot{\gamma}_{ini}$, and [Cl]/[N].

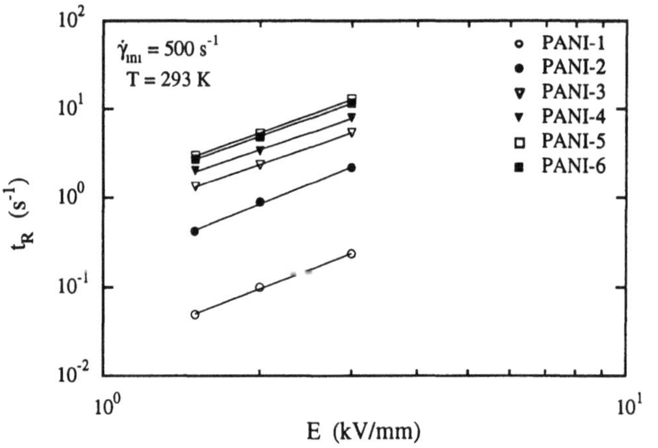


**Figure 5.** Effect of electric field (E) on rate of response ($t_R$) at 293K and initial shear rate ($\dot{\gamma}_{ini}$). The lines are fits.

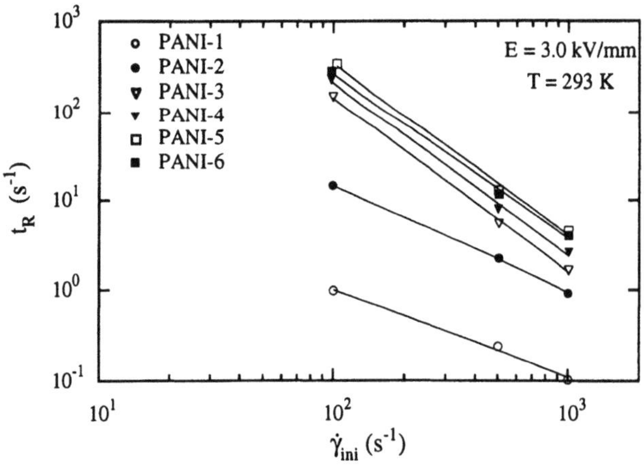


**Figure 6.** Effect of initial shear rate ($\dot{\gamma}_{ini}$) on rate of response ($t_R$) at 293K and constant electric field (E). The lines are fits.

$t_R$ scales with E as $t_R \sim E^{2.0}$, a scaling which was independent of [Cl]/[N]. The observed electric field scaling is consistent with the kinetics of structure formation under quiescent conditions.[11,12] The time scale ($\Delta t$) for structure formation is determined from a balance of particle interactions due to particle polarization and a Stokes drag force on the interacting particles, *i.e.* $\Delta t \sim \mu_c/E^2$ where $\mu_c$ is the continuous phase viscosity. $t_R$ scaling with $\dot{\gamma}_{ini}$ depended on [Cl]/[N], changing from $(\dot{\gamma}_{ini})^{-1.0}$ to $(\dot{\gamma}_{ini})^{-2.0}$ scaling with increasing dopant concentration. A shear rate dependence for $t_R$ is expected based on Mason number arguments presented by Klingenberg and Zukoski.[13] Klingenberg and Zukoski show that under steady state conditions, the ER effect is the result of competition between polarization

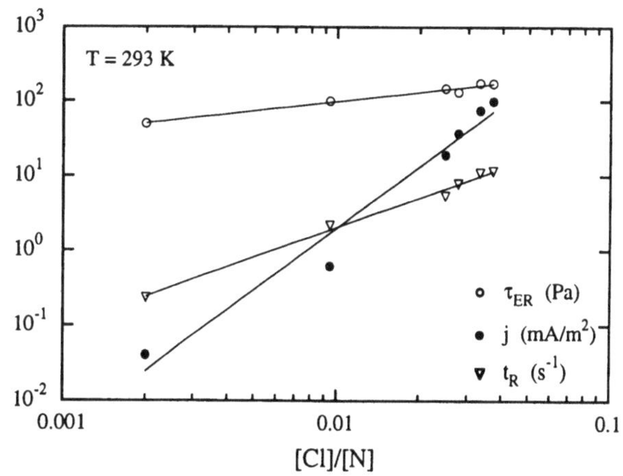


**Figure 7.** Effects of degree of protonation ([Cl]/[N]) on $\tau_{ER}$ (E = 1.0 kV/mm), j (E = 3.0 kV/mm), and $t_R$ (E = 3.0 kV/mm and $\dot{\gamma}_{ini}$ = 500s$^{-1}$) at 293 K. The lines are fits.

forces causing particle agglomeration and alignment and shear forces acting to disrupt this process. Qualitatively at least, this argument applies to the build up of structure under transient conditions in shear flows.

The effects of degree of protonation on $\tau_{ER}$, j and $t_R$ at 293K are shown in Figure 7 (experimental conditions are given in the caption). Dopant concentration scalings of $\tau_{ER} \sim ([Cl]/[N])^{0.4}$, $j \sim ([Cl]/[N])^{2.8}$, and $t_R \sim ([Cl]/[N])^{1.4}$ were observed. The stronger scaling of j with [Cl]/[N] than that for $\tau_{ER}$ or $t_R$ indicates that the efficiency of the ER effect decreases with increasing [Cl]/[N].[14]

The scaling of j with [Cl]/[N] is similar to that for $k_p$ shown in the data of Zuo *et al.*;[8] j versus $k_p$ is approximately linear, $j \sim k_p^{1.1}$, suggesting a correspondence between particle conductivity and current flow through the suspension. j measurements reported here were performed under quiescent conditions, thus the characteristic ER strand structure is expected and is likely to be the mechanism for conduction, *i.e.*, stranded particles acting as conducting wires.

The different scalings of $\tau_{ER}$ and $t_R$ with [Cl]/[N] are qualitatively consistent with material properties effects in the long and short time limits of a generalized polarization model for the ER effect. As noted by Zukoski,[1] the leading order term for the polarization force between two conducting particles in the point dipole approximation scales as

$$F_{pol} \sim |\beta^*|^2 E^2 \tag{9}$$

where $\beta^*$ is defined by eq 4. The static yield stress experiment is a long time experiment, O(10 s) to O($10^2$ s), and therefore $\beta^*$ tends toward its DC limit ($\omega \rightarrow 0$), and becomes a function of conductivities $k_p$ and $k_c$ only.[15] Over the range of [Cl]/[N] in Table 1, $\beta^*$ determined from the $k_p$ data of Zuo *et al.*[8] is constant at the maximum value of 1; therefore, $\tau_{ER}$ is expected to be independent of $k_p$. This observation is unaffected by the possibility that $k_p$ of the polyaniline samples may be significantly less than estimates made from the data of Zuo *et al.*, *i.e.*, in both cases $k_p >> k_c$. If $\tau_{ER}$ is plotted as a function of $k_p$ using the Zuo *et al.* data for $k_p$, a weak scaling relation of $\tau_{ER} \sim k_p^{0.15}$ is observed, qualitatively consistent with this analysis.

The transient rate of response experiment involves application of a DC electric field across the ER fluid, where the measured rheological response of the fluid occurs at short times relative to the time scale of the static yield stress experiment. The applicable frequency range for approaching the high frequency limit of $\beta^*$ is defined with respect to the time scale for interfacial polarization which is the characteristic relaxation time $\tau_\varepsilon$. In Figure 8, plots of $1/t_R$ ($\dot{\gamma}_{ini}$ = 500 $s^{-1}$) and $\tau_\varepsilon$ versus [Cl]/[N] show that the time scale for rheological response is longer than the characteristic time for interfacial polarization at each value of [Cl]/[N]. As $\dot{\gamma}_{ini}$ decreases, these time scales become comparable, see Figure 6. In the limit of short times (or high frequency electric fields, $\omega \rightarrow \infty$), $\beta^*$ depends on permittivities $\varepsilon'_p$ and $\varepsilon'_c$ only.[15] Over the range of [Cl]/[N] in Table 1, $\beta^*$ varies from approximately 0.55 to 0.98 for $\varepsilon'_p$ values determined by Zuo *et al.* Although the transient response clearly does not occur solely in the interfacial polarization defined high frequency limit, this analysis suggests that the transient electrorheological experiment should show a stronger dependence on [Cl]/[N] than the long time experiments; this was observed and is illustrated in Figure 7.

## SUMMARY

The effects of polyaniline electrical properties on dielectric and electrorheological properties were studied by varying the degree of polyaniline protonation, [Cl]/[N]. Literature data for the effects of [Cl]/[N] on polyaniline conductivity and permittivity and the Maxwell-Wagner model for interfacial polarization were used as a basis for interpreting the dielectric spectra as well as the effects of dopant concentration on electrorheology. Under quiescent conditions, current density increased approximately linearly with increasing particle conductivity. This results suggests that the mechanism for current passage is particle/particle contact and stranding. Second, dopant concentration had the strongest effect on current

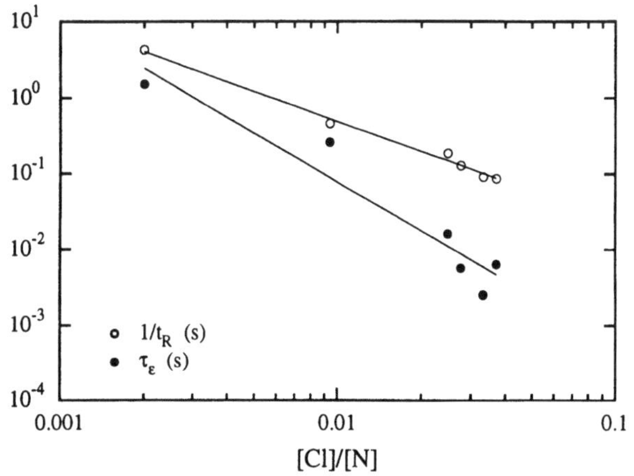


**Figure 8.** Effects of degree of protonation ([Cl]/[N]) on response time ($1/t_R$) and characteristic relaxation time ($\tau_\varepsilon$) ($1/t_R$ from Figure 7 and $\tau_\varepsilon$ and from Figure 3). The lines are fits.

density with respect to its effects on static yield stress and rate of response, an observation that indicates that efficiency of the ER effect decreases as dopant concentration increases. Finally, despite similar $E^2$ scaling, static yield stress and rate of response were observed to scale differently with [Cl]/[N]. The different [Cl]/[N] dependencies for static yield stress and rate of response and the respective time scales for these experiments are qualitatively consistent with the low and high frequency limits of a generalized polarization model. The significance of this results is that different particle material properties may determine ER response in long-time and transient electrorheological experiments.

## ACKNOWLEDGMENTS


I would like to acknowledge D. R. Clark and J. W. Pialet for synthesis and chemical characterization of the polyaniline samples, and B. L. Soukup for dielectric characterization of the polyaniline ER fluids.


## REFERENCES


1. Zukoski, C.F. *Annu. Rev. Mater. Sci.*, **1993**, *23*, 45.
2. Gow, C.J.; Zukoski, C.F. *J. Colloid Interface Sci.*, **1990**, *136*, 175.
3. Block, H.; Kelly, J.P.; Qin, A.; Watson, T. *Langmuir*, **1990**, *6*, 6. Block, H.; Rattray, P.; Watson, T. In *Proc. Int. Conf. Electrorheological Fluids*, R. Tao, Ed.; World Scientific: Singapore, 1992; p.93.
4. Xu, Y.-Z.; Liang, R.; Hao, T.; Zhang, Q.-Z.; Xu, Z.-M.; Chen, Y.-H. In *Proc. Int. Conf. Electrorheological Fluids*, R. Tao, Ed.; World Scientific: Singapore, 1992; p.129.
5. MacDiarmid, A.G.; Chiang, J.C.; Richter, A.F.; Somasiri, N.L.D. In *Conducting Polymers*, L. Alácer, Ed.; Reidel: Dordrecht, 1986; p.105.
6. Lemaire, E.; Bossis, G.; Grasselli, Y. *Langmuir*, **1992**, *8*, 2957.

7. Krieger, I.M. *J. Rheology*, **1990**, *34*, 471.

8. Zuo, F.; Angelopoulos, M; MacDiarmid, A.G.; Epstein, A. J. *Physical Rev. B*, **1989**, *39*, 3570.

9. Smyth, C. P. "Dielectric Behavior and Structure", McGraw-Hill Book Co.: New York, 1955.

10. van Beek, L. K. H. In *Progress in Dielectrics, Volume 7*, J. B. Birks, Ed.; CRC Press: Cleveland, 1967; p. 69.

11. Klingenberg, D. J.; Zukoski, C. F.; Hill, J. C. *J. Appl. Phys.*, **1993**, *73*, 4644.

12. Ginder, J. M.; Elie, L. D. In *Proc. Int. Conf. Electrorheological Fluids*, R. Tao, Ed.; World Scientific: Singapore, 1992; p.23.

13. Klingenberg, D. J.; Zukoski, C. F. *Langmuir*, **1990**, *6*, 15.

14. Boyle, F. P. In *Proc. Int. Conf. Electrorheological Fluids*, R. Tao, Ed.; World Scientific: Singapore, 1992; p.236.

15. Anderson, R. A. In *Proc. Int. Conf. Electrorheological Fluids*, R. Tao, Ed.; World Scientific: Singapore, 1992; p.81.

# ER-FLUIDS BASED ON POLYURETHANE DISPERSIONS: STRUCTURE AND PROPERTIES


Robert Bloodworth and Eckhard Wendt

Central Research and Development
Bayer AG
D-51368 Leverkusen, Germany



## ABSTRACT

This paper describes the structure and properties of a new class of electrorheological fluids based on non-aqueous polyurethane dispersions. The fluids exhibit an attractive combination of properties: low viscosity, high ER effect, and low conductivity. The dispersed phase consists of a specially developed polyurethane elastomer which solvates and stabilizes metal salts. The polymer network density influences the mobility of the dissolved ions, allowing a surprising degree of control over the ER effect. Properties such as the field strength dependence of the ER-effect, switching response, and conductivity of these fluids correlate directly with changes in the polymer structure. Electrorheological measurements in a couette viscometer (shear-mode) and in a model shock absorber (flow-mode) using a commercial polyurethane-based fluid show that the ER effect is also dependent upon the shearing geometry. The implications of these results for the design and application of ER-fluids are discussed.


## INTRODUCTION

The strong electrorheological fluids (ERF) known to date are dispersions of electrically polarizable particles in a non-conducting continuous phase. The oldest and most widely studied of these systems are based on silica gels and zeolites,[1,2] including a number of fluids created in our laboratories.[3] Problems with the abrasiveness and sedimentation of these dispersions led to the development ERF's with polyelectrolyte gels as the dispersed phase.[4] Both systems must be "activated" by the presence of a low molecular weight polar liquid, most frequently water. However, a number of reports suggest that anhydrous polar solvents (even aprotic ones like propylene carbonate) can also be used to activate these systems.[5,6,7,8] Poly(lithium methacrylate) is a well known dispersed phase in ER-fluids. As a dry powder, the lithium ions in the particles are strongly bound by the carboxylate groups. The addition of water or another substance which can solvate the lithium increases the ionic mobility and allows for a surface polarization of the particle in an electric field. We have suggested that the special role of water for the ER-effect lies in its solvating ability, rather than in its inherent polarizability in an electric field.

Filisko[9] has reported ER-effects with some scrupulously dried materials. A number of interesting systems based on anhydrous inorganic conductors such as zirconium doped aluminosilicates have emerged in the patent literature.[10,11,12] More recently, a number of fluids based on semiconducting polymers have been reported.[13] Most if not all of the dispersed phases in these materials contain ionic species (some inadvertently). Unfortunately, the sensitivity of these systems to trace concentrations of the so-called "promoter liquids" often leads to high conductivity and poor reproducibility as well as instability of the ER effect. In an effort to minimize these problems, we have developed new commercial ER fluids using a polymeric solid ionic conductor as the dispersed phase.[14]

Polymers based on polyethers, as well as certain polyesters and polycarbonates can function as solid solvents for metal salts. The high solvating power of polyethylene oxide (PEO) segments for metal atoms stems from the interaction of the electron pairs on oxygen with the metal cation. The powerful cryptand 18-crown-6, an excellent complexing agent for lithium and sodium ions, is based on an advantageous circular arrangement of the oxygen atoms coordinating the metal cation. Similar structures, in which a central metal ion is enwrapped by a helical PEO-chain have been determined for crystalline complexes of ethylene glycol oligomers and alkali metal salts.[15]

Polyethylene glycol oligomers are especially interesting by virtue of their high solvating power and have been used as solid solvents in lithium-polymer batteries.[16] Unfortunately, these systems have poor mechanical properties. Multifunctional polyether polymers can be crosslinked to form elastomers which possess the required dimensional and thermal stability. The ER fluids based on these dispersed phases exhibit an attractive combination of properties: low viscosity, high ER effect, and low conductivity. The chemistry of the polymer can be widely varied to modify the polarity, Tg, and crosslink density, thus affecting the ionic mobility within the solid polymer. In addition to the practical advantage of allowing one to tailor the properties of these ER-fluids to particular applications, this well defined system provides a tool with which to examine the polarization process in the dispersed phase as a function of the ionic concentration and mobility.

A central question of ER phenomenon concerns the underlying mechanism of particle polarization. The purpose of this work was to investigate the effect of variations in fluid formulation such as the volume fraction, polymer structure and ionic strength of the dispersed phase on the ER properties of the fluid.

## EXPERIMENTAL

The polyurethane-silicone oil dispersions were prepared using previously disclosed materials and methods.[17] Measurements of the ER effect in shear mode were made using a modified couette viscometer as described in reference 8. Measurements at constant shear rate were performed by incrementally increasing the field strength and monitoring the shear stress ($\tau$) via torque measurements. The minimum field strength above which the shear stress increases ($E_0$) and the linear rate of increase in the shear stress with increasing field strength, e.g. the electrorheological effect (S) were determined from the intercept and slope respectively of the excess shear stress ($\tau - \tau_{0kV}$) versus field strength curve. Couette measurements were performed at 60°C. Measurements of the ER-effect in Poiseuille flow were made using a pneumatically driven piston assembly.[18] The fluid is forced through an electrified annulus valve by the piston of a modified shock absorber and the shear stress and shear rate calculated from the speed of the piston and pressure drop across the valve. The Poiseuille flow measurements were performed at 40°C.

## RESULTS

### Volume Fraction of the Particulate Phase

The base viscosity of the fluid depends as expected upon the volume fraction of the particulate phase. Figure 1 indicates a limited shear rate dependence of the viscosity at high volume fractions at shear rates below 1000 $s^{-1}$, as well as surprisingly low zero field viscosities at higher volume fractions. A low base viscosity is critical to achieving a large switchable range of shear stresses in an ER device. The zero field viscosity scales approximately linearly with temperature, as would be expected for a silicone oil based fluid.

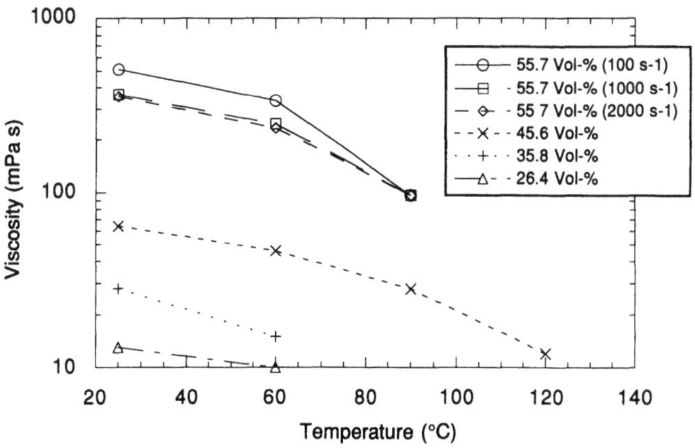


**Figure 1.** Effect of temperature, shear rate and volume fraction solids on the zero field viscosity.

## Concentration of Ions in the Particulate Phase

The main factors contributing to the overall ER-effect are the concentration and mobility of the charge carriers within the particles. Modifying the number of charge carriers should directly influence the magnitude of the ER effect.

An AC voltage source was used to minimize any electrophoresis effects in the following experiments. Figure 2 shows the effect of increasing the concentration of an added salt ($LiNO_3$), which is dissolved in the dispersed phase, on the ER-effect as measured in a couette viscometer. The polyurethane dispersed phase is based on a trifunctional polyethylene oxide polyether (MW 675) crosslinked with 1 equivalent of toluene diisocyanate. The continuous phase is a polydimethylsiloxane oil with a viscosity of 5 mPa·s at 25°C.

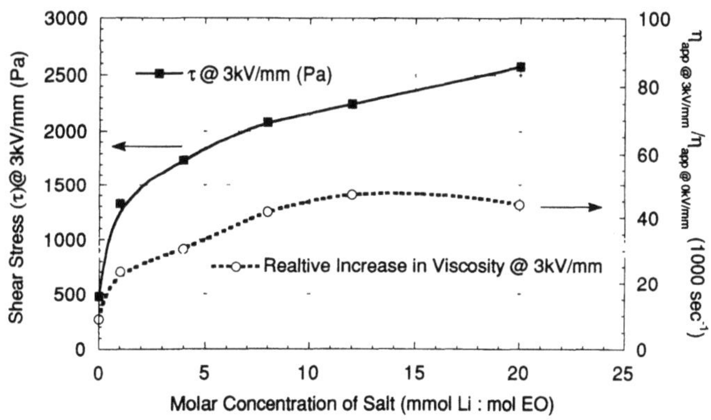


**Figure 2.** Effect of the $Li^+$ concentration on the shear stress (τ) and on the relative increase in the apparent viscosity ($V_{3kV}/V_{0kV}$).

After an initial strong increase in the ER effect, an approximately linear dependence of the shear stress on the concentration of $Li^+$ is observed. The switching factor, the ratio of apparent viscosity measured at 3kV/mm and at 0 kV/mm ($\eta_{3kV}/\eta_{0kV}$), is a measure of the fluids switchable viscosity range and increases only marginally above 10mmol $Li^+$/mol EO.

Examination of Figure 3 yields additional insight into the effect of the ion concentration on the polarization process. The threshold field strength for the onset of ER-activity, $E_0$, is high at low $Li^+$ concentrations but is much less sensitive to higher salt levels. This suggests that a minimum average concentration of ions is necessary for easy polarization of the dispersed phase, but that higher concentrations do not appreciably affect the barrier to polarization. The absolute magnitude of ER-effect, as well as the shear stress at 3kV/mm, appears to depend upon the total number of ions in the particles.

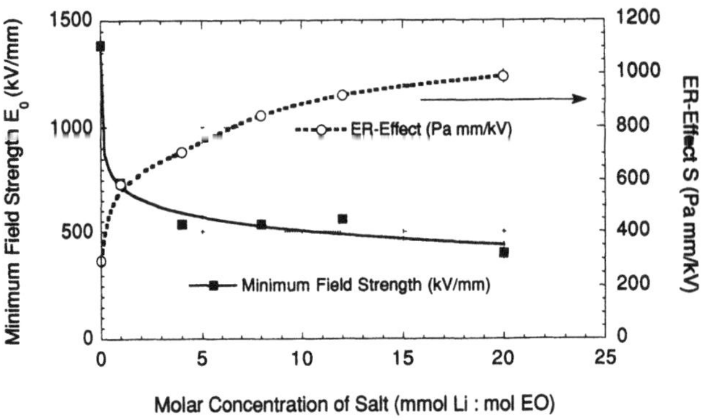


**Figure 3.** Minimum field strength for the ER-effect ($E_0$) and the absolute ER-effect (S) versus lithium ion concentration.

Figure 4 demonstrates the strict linear relationship between the concentration of dissolved ions and the conductivity of the ER fluid. This behavior is expected for solid ionic conductors at dilute concentrations. In the dilute regime, the ions are well solvated and do not agglomerate to form clusters. Although the shear stress and ER-effect continue to rise with the concentration of added salt, a strong increase in the bulk conductivity is associated with high ion concentrations. The concomitant increase in the power requirements of the fluid puts a practical limit on the amount of salt which may be incorporated into the dispersed phase.

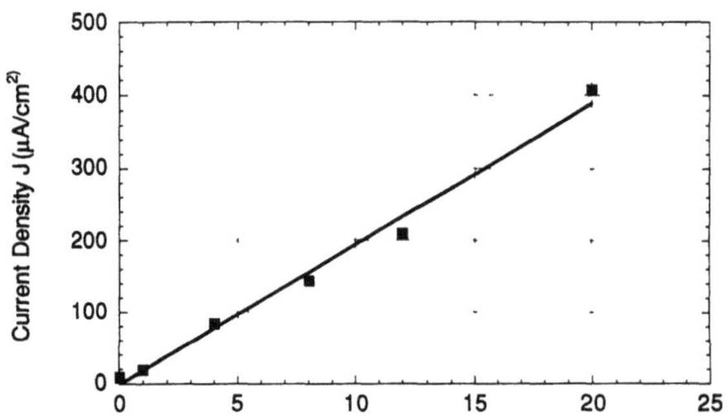


**Figure 4.** Current density (J) as a function of lithium ion concentration.

## Crosslink Density of the Polyurethane

Figure 5 demonstrates the effect of varying the network structure of the dispersed polymer phase. Using mixtures of a trifunctional and a difunctional PEO polyol of the same equivalent weight (ca 225 g/equiv. OH) and crosslinking with toluene diisocyanate, the crosslink density of the polyurethane particles was varied while maintaining a nearly constant concentration of urethane groups within the polymer. The lithium ion concentration was held constant at 10 mmol Li: mole EO.

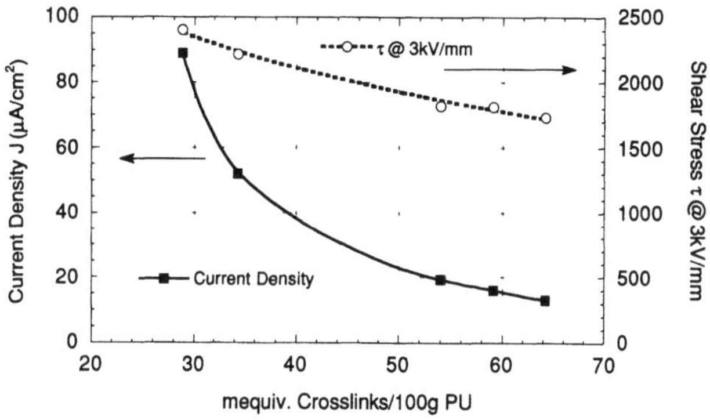


**Figure 5.** Shear stress (τ) and current density (J) versus crosslink density of the PU dispersed phase (ionic concentration of 10mmol Li: mol EO).

Increasing the crosslink density lowers the conductivity of the dispersed phase, indicating a decrease in the long range ion mobility. The magnitude of the shear stress (τ @ 3kV/mm), by comparison, decreases relatively slowly upon increasing the number of crosslinks, suggesting that the particle polarization is less affected.

An examination of Figure 6 reveals that the ER-effect falls approximately linearly with the crosslink density, as does τ. The current density falls off much more rapidly in a tighter network, scaling with the inverse of the crosslink density cubed. More surprisingly, the minimum field strength, $E_0$, remains relatively constant over the entire range of data.

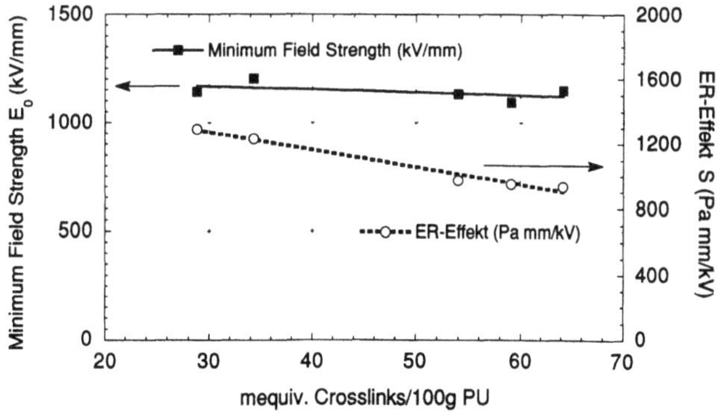


**Figure 6.** Minimum field strength ($E_0$) and ER-effect (S) versus crosslink density of the PU dispersed phase (ionic concentration of 10mmol Li: mol EO).

Additional experiments on related systems have shown that further increases in the crosslink density eventually lead to an increase in $E_0$, but that in a wide range of polymer networks, the polarity of the polymer matrix (as well as the concentration of ions) exerts a much greater influence on $E_0$ than does the crosslink density, or the mobility of the chain segments.

The data supports a model for particle polarization in which short range motions of the ions are responsible for an interfacial polarization[19] of the dispersed phase, and that ion transport *through the dispersed phase* controls the bulk conductivity of the ERF.

One consequence of these results for the design of improved ER-fluids lies in reducing the conductivity of the ERF without adversely affecting the ER response. Apparently, the bulk conductivity of the ERF, observable as the current density in Figure 5, can be reduced by restricting the movement of ions in the polymer matrix, albeit at a small cost to the maximum shear stress and the ER effect. This result suggests that it may be possible to make a strong ERF with a negligible power consumption.

## APPLICATION OF POLYURETHANE BASED ER FLUIDS

### Poiseuille Flow versus Shear Flow

ER fluids behave differently depending upon the velocity profile of the fluid between the electrodes. In the shear mode, the fluid is sheared between electrodes moving relative to each other and the ERF experiences a linear shear gradient. The shear mode is prevalent in clutch applications of ER fluids. In Poiseuille flow (flow mode), the fluid is forced though an electrified annulus which functions as an electrorheological flow control valve. The Bingham character of the fluid in an electric field results in a complex velocity profile. The flow mode is common to a number of commercially attractive ER-applications, including hydraulics and automotive vibration damping.

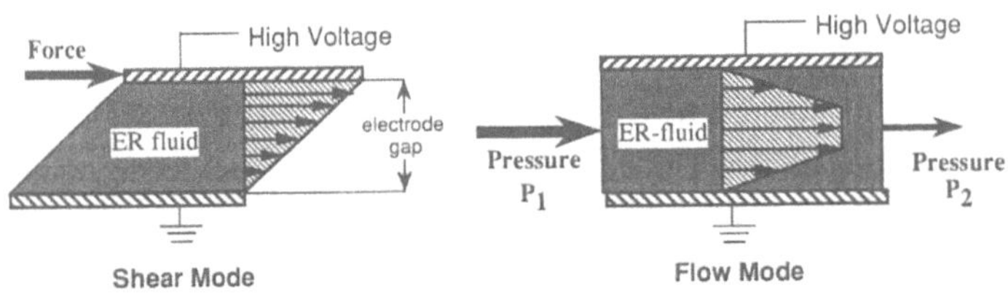


**Figure 7.** Comparison fluid velocity profiles in shear and flow mode applications.

Figure 8 shows the ER response of a commercial polyurethane-based ER fluid (RHEOBAY® VPAI 3565) in a flow mode testing apparatus based on an automotive shock absorber. Corresponding couette viscometer data is included for comparison. The couette viscometer used for the shear mode measurements has a maximum achievable shear stress of app. 2000 $s^{-1}$.The shear stress at 1000 $s^{-1}$ extrapolated from flow mode measurements made at high shear rates lies significantly higher than would be expected from the couette measurement in shear mode. The reason behind this apparent discrepancy is the well known "chaining" phenomenon of the ER dispersed phase in an electric field. The pearl chains moving through the electrified valve do not experience a transverse shear and the fluid acts as a Bingham solid undergoing plug flow. This yields ER effects which are usually much higher at a given field strength than those estimated from standard couette measurements. Shear stresses of well over 10 kPa are possible with the correct selection of valve geometry and fluid.

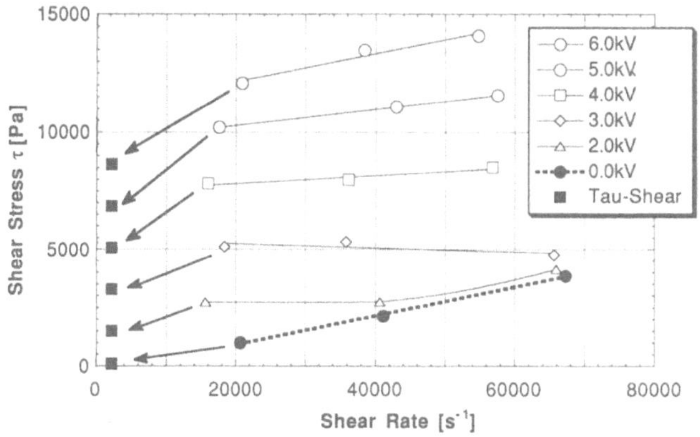


**Figure 8.** Shear stress versus shear rate data measured in flow mode using a modified ER shock absorber as compared to couette measurements (shear mode). Nominal shear rates calculated assuming Newtonian flow.

This complex behavior has implications for the design of ER fluids for flow mode applications. To optimize the plug flow within a valve, the initial polarization to form chains must be fast. However, once the plug has been formed, surface interactions likely dominate the interparticle forces. Maximizing the mechanical interaction of the particles while maintaining colloidal stability remains a difficult task for the designers of ER fluids.

## ER Shock Absorber

The comparison of shear and flow mode data indicates that the polyurethane-based fluids are more effective in flow mode applications. An ER damper was constructed for to demonstrate the application of the polyurethane fluids in an actual device. A standard automotive shock absorber was modified with an external ER valve (Figure 9). The restricted flow of the ER fluid through the external valve controls the damping force generated upon moving the piston.

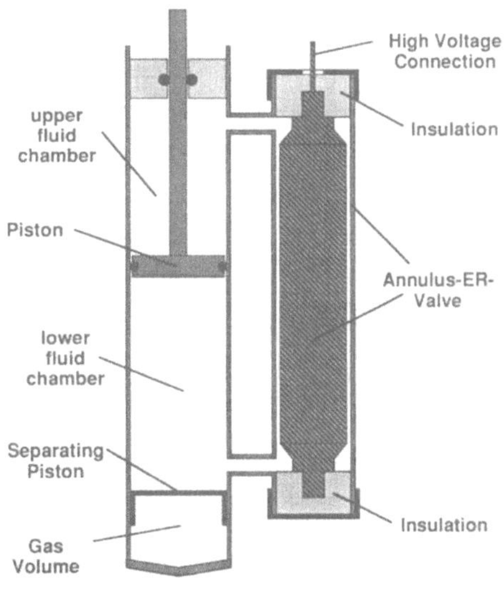


**Figure 9.** ER damper construction.

The damper was filled with a commercial polyurethane-based fluid (RHEOBAY® VPAI 3566, Bayer AG), and the damping force measured as a function of piston speed and electric field strength. The characteristic curves in Figure 10 demonstrate that the system responds symmetrically with a variable damping force between 2kN and 6kN. Forces of this magnitude are adequate for applications in passenger vehicles, thus demonstrating the utility of ER technology using these fluids.

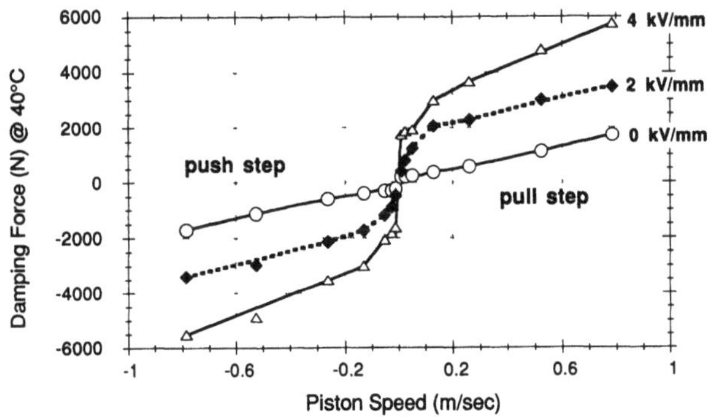


**Figure 10.** Characteristic damping curves of an ER shock absorber containing a commercial polyurethane-based ER fluid (RHEOBAY® VPAI 3566).

A major concern in the application of ER fluids is the long term stability of the ER effect under use. Figure 11 depicts the results of endurance tests on a polyurethane-based ER-fluid using a modified hydraulic apparatus. In this test, the polyurethane-based fluid is continuously recirculated through an electrified annulous valve and the electric field across the valve is varied continually between 0 and 9kV/mm. The curves indicate the change in pressure across the valve as the field strength is varied. The effective shear rate at the wall of the valve (as calculated for a newtonian fluid) was app. 45,000 $s^{-1}$. Over a period of several thousand hours the ER effect remains nearly constant and the fluid properties unchanged. Neither the particles nor the apparatus show any signs of abrasion or degradation.

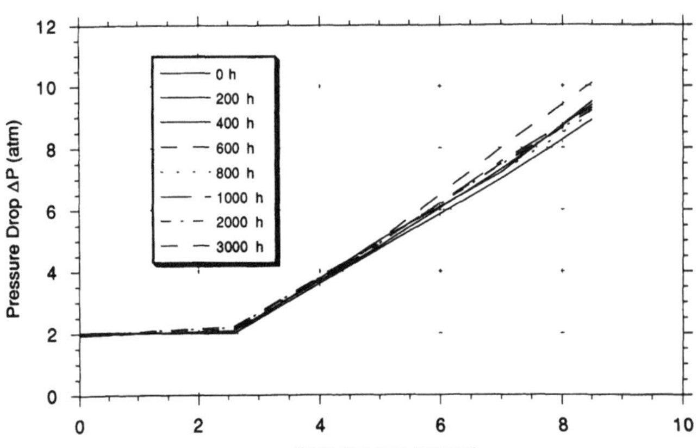


**Figure 11.** Endurance test of a polyurethane-based ERF in an hydraulic apparatus at a temperature of 50°C and a nominal shear rate of 45,000 $s^{-1}$.

## CONCLUSIONS

New water free ER-fluids based on polyurethane dispersions exhibit a combination of properties: low viscosity, high ER effect, and low conductivity, which makes them attractive for commercial applications. The dispersed phase consists of solid solutions of metal salts dissolved in a crosslinked polyurethane matrix. The conductivity and polarizability of the particles can be modified by varying the crosslink density and polarity of the polyurethane as well as the concentration of added salt, thus allowing a considerable degree of control over the ER properties of the fluid. Electrorheological measurements in a couette viscometer (shear mode) and in a model shock absorber (flow mode) using a commercial polyurethane-based fluid show that the magnitude of the ER effect is dependent upon the velocity profile of the fluid between the electrodes. This result underscores the importance of carefully matching fluid properties and device design in ER applications. A continuously variable damping forces of 2-6 kN could be generated using a standard automotive damper modified with an ER valve, thus demonstrating the practical utility of these ER fluids. The fluids exhibit a stable ER effect which remain virtually unchanged even after prolonged use at elevated temperatures and high shear rates.

## REFERENCES


1. Klass, L. D., Martinek, Th. W., *J. Appl. Phys.*, 38(1):67(1966).
2. Winslow, W. M., *J. Appl. Phys.*, 20(0):1137(1949).
3. DE 3427499, DE 3517281, DE 3536934
4. UK 1570234
5. JP 1253110
6. EP 374525
7. JP 3074494
8. EP 509571
9. US 4,744,914
10. US 5,139,690
11. US 5,139,691
12. US 5,139,692
13. Block, H., Kelly, J. P., Watson, T., *Spec. Publ. - R. Soc. Chem.* 87:51(1991).
14. Bloodworth, R., Electrorheological fluids based on polyurethane dispersions, *in:* Electrorheological Fluids, Mechanisms, Properties, Technology and Applications, R. Tao and G.D. Roy, ed., World Scientific, Singapore (1994)..
15. S. Okamura and Y. Chatini, *Polymer Preprints Japan*, 34:2089 (1985).
16. Wright, Perter V., *J. Makromol. Sci. Chem.*, A26(2-3):519(1989)
17. DE 4026881, US 5,268,118
18. Wendt, E. and Büsing, K.W., *Proc. Actuator '94*, Bremen, Germany, in press.
19. Block, H., Kelly, J.P., Quin, A., Watson, T., *Langmuir*, 6(1):6(1990).

# THE IMPACT OF CRYSTALLINE PHASE MORPHOLOGY ON THE WATER-PROMOTED ELECTRORHEOLOGICAL EFFECT OF POLYSACCHARIDES

Arthur J. Stipanovic and Jeffrey P. Schoonmaker

Lubricants Research Division
Texaco Research and Development Department
P.O. Box 509, Beacon, New York 12508, U.S.A.

## INTRODUCTION

Polysaccharides represent the world's largest class of naturally occurring organic macromolecules and, as a result, potentially provide an excellent source of "particulate" materials for the design of electrorheological (ER) fluids. The ability of these polymers to exhibit an ER effect when dispersed in mineral, vegetable, or silicon oils has been well established[1-3] although they generally require adsorbed water to promote activity which could limit their thermal range of useful application. As shown in Figure 1, typical polysaccharides contain glucose sugar monomeric residues which can interact readily with water via hydrogen bonding through a number of hydroxyl groups and ether oxygen atoms on the polymer.

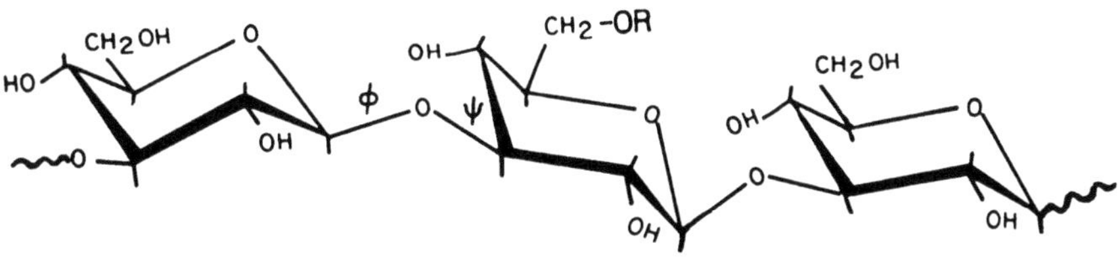


**Figure 1.** Chemical Structure of Curdlan [(1-3)-Beta-D-Glucan]. Extracellular polysaccharide from *Alcaligenes faecalis var. myxogenes* bacterium. R = hydrogen.

In the solid state, water / polysaccharide interactions can occur in a non-specific fashion within "amorphous" domains or on crystallite surfaces, while stronger, site-specific hydrogen bonds typically exist inside the polysaccharide crystal matrix. From the perspective of ER fluids, it would be beneficial if the water bound by polysaccharides was sufficiently mobile to promote polarization in an electrical field while remaining fixed in the crystal lattice thereby minimizing dehydration or the development of current in the fluid. Previous work by Uejima[1] and Conrad[2] using dispersions of the

*Progress in Electrorheology*, Edited by K.O. Havelka
and F.E. Filisko, Plenum Press, New York, 1995

polysaccharides cellulose and corn starch, respectively, has illustrated that adsorbed water promotes ER activity (viscosity, yield strength) up to a level of 8-10 wt % water above which activity is diminished due to electrical conductivity and field strength breakdown associated with excessive water mobility. X-ray crystal structure determinations on these polysaccharides have shown that the crystal lattice of cellulose, in several polymorphic forms, does not contain water[4] while corn starch (so-called "A" amylose) contains less than one water molecule per glucose residue[5]. As a result, water adsorbed to cellulose and starch is relatively loosely bound with conductivity at high moisture levels (> 10%) being a logical consequence for ER fluids formulated with these polymers.

In this study, the water-promoted ER activity of two glucose-containing homopolymers, pustulan and curdlan (Figure 1), has been evaluated. These polymers exhibit crystalline polymorphic forms containing site-bound water in their lattice structures[6,7] in contrast to cellulose and starch. In addition, we have investigated the ER response of cyclodextrin oligosaccharides, containing 6-8 glucose residues per molecule, arranged in a cyclic geometry. These molecules are known to "complex" water and other ligands in the cavity of their cyclic structure. A key objective for both systems was to determine, empirically, if water-activated ER fluids could display the thermal and conductivity properties typical of "anhydrous" systems if water molecules are effectively "trapped" in the particulate phase of the fluid.

## MATERIALS AND EXPERIMENTAL METHODS

Samples of curdlan [([1-3)-Beta-D-Glucan] and pustulan [(1-6)-Beta-D-Glucan] were obtained from Wako Chemical and Calbiochem-Behring, respectively. *Alpha, beta* and *gamma* cyclodextrins were provided from American Maize Products Company (Amaizo). Water content levels were controlled by exposure to high humidity environments for periods of 12 hours to several days. ER fluids were prepared by dispersing the polysaccharide or cyclodextrin in a paraffinic mineral oil at 5 - 20% volume fraction. This mineral oil exhibited a viscosity of 4.4 cSt at 100°C, a dielectric constant of 2.19, and a conductivity of 1.25 ($10E^{-12}$ S/m). In all cases, the ER fluids were not stabilized against phase separation and some settling occurred with time Rheological measurements were made at the University of Connecticut - Institute of Materials Science on a Rheometrics System 4 rheometer retrofit for application of high voltages or at North Carolina State University on a Physica ER Fluid Rheometer. In addition, estimates of the "relative" yield strength for several ER fluids evaluated in this study were made using a simple parallel plate device (2 mm gap) described previously[8]. In this apparatus, a PTFE block was supported on the surface of an ER fluid "solidified" by the application of an electrical field (1500 V / mm; fluid surface cross section: 2mm x 20 mm). The critical voltage at which the block was no longer supported by the ER fluid surface, recorded with decreasing voltage, was converted into a "relative" yield strength by assuming that stress scales directly with voltage squared as demonstrated in Figure 2 for a polysaccharide-in-oil system.

Cross Polarization / Magic Angle Spinning (CP/MAS) $^{13}C$-NMR was used to characterize the polymorphic states of solid and crystalline curdlan and pustulan samples according to methods outlined elsewhere [6,7]. This technique is complimentary with x-ray diffraction methods and is quite diagnostic for identifying specific crystalline polymorphic structures for polysaccharides and other biopolymers. Transformations between the various hydrated forms of these polysaccharides, summarized below in Tables 1 and 2 , have also been detailed in other sources[8].

Table 1. Crystalline Polymorphic Forms of Curdlan

| *Polymorph:* | *Anhydrous* | *Low Temp. Hydrate* | *High Temp. Hydrate* |
|---|---|---|---|
| Anneal Temp. | 20 -25 °C | < 50 °C | 180 °C |
| Water Content | 0 - 20 % | > 25% | 18 % |
| Water Binding Strength | Weak Adsorption | Loosely Held in Gel-Like Matrix | Hydrogen Bonded in Lattice Sites |

## RESULTS AND DISCUSSION

### Curdlan-Based ER Fluids

As shown in Table 1, curdlan exists in at least three crystalline polymorphic forms, distinguishable by CP/MAS $^{13}$C-NMR, that are obtained at different conditions of moisture and annealing. Although the "anhydrous" form contains no water in its crystal lattice, amorphous domains and crystal surfaces may adsorb moisture. The relationship of applied voltage and the yield stress for an ER fluid containing "anhydrous" curdlan (5% volume fraction) at 11% water content is shown in Figure 2. This system displays the "voltage squared" dependence typical of many polymer-based ER fluids[1,10] and conductivity was relatively low for this fluid.

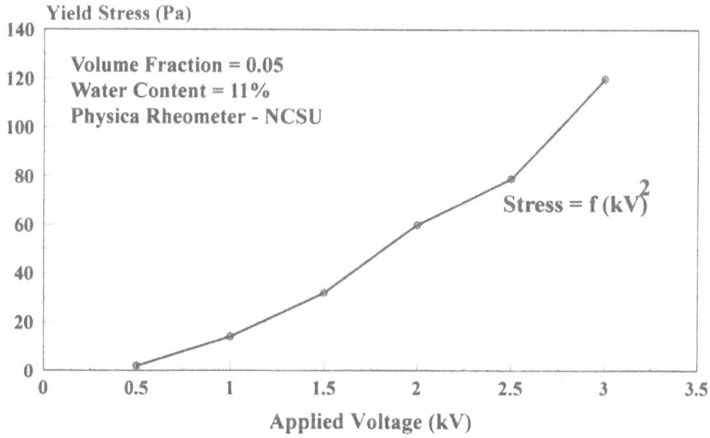


**Figure 2.** ER fluid "Yield Stress" for "anhydrous" curdlan in paraffinic mineral oil.

With the 11% moisture curdlan sample described above, an anomalous time or strain-induced increase in shear stress was observed with each successive test run of the sample, as illustrated in Figure 3. For this fluid, it is possible that the electric field induces a time-dependent electrophoretic movement of water molecules and / or ionic species through the bulk particle resulting in an accumulation at the particle poles and an increase in apparent ER activity with successive test runs. This phenomenon may be similar to the time-dependent increase in ER fluid viscosity and current density reported by Asako and co-workers[11] for systems containing sulfonated polystyrene particles.

The influence of moisture content on the "relative" yield stress of ER fluids prepared from "anhydrous" curdlan is given in Figure 4. Although water molecules in this system are not crystallographically bound, maximum yield stresses are observed at moisture contents significantly above those observed for cellulose and starch (20% vs. 8-10%; Refs. 1,2) suggesting that curdlan binds water more strongly, even in an amorphous matrix.

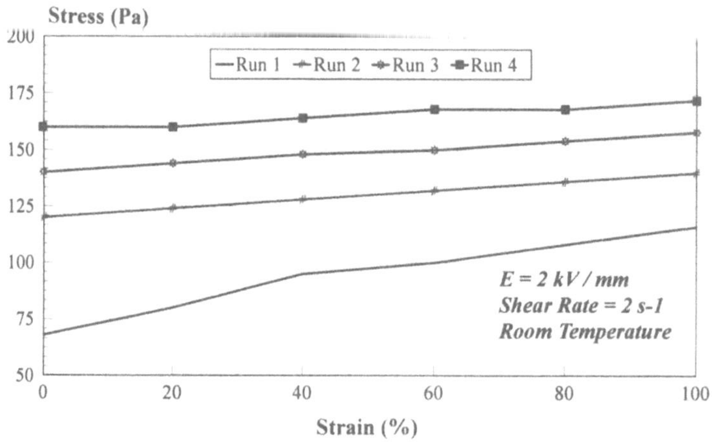


**Figure 3.** "Aging" Effect of 11% moisture "anhydrous" curdlan in paraffin oil.

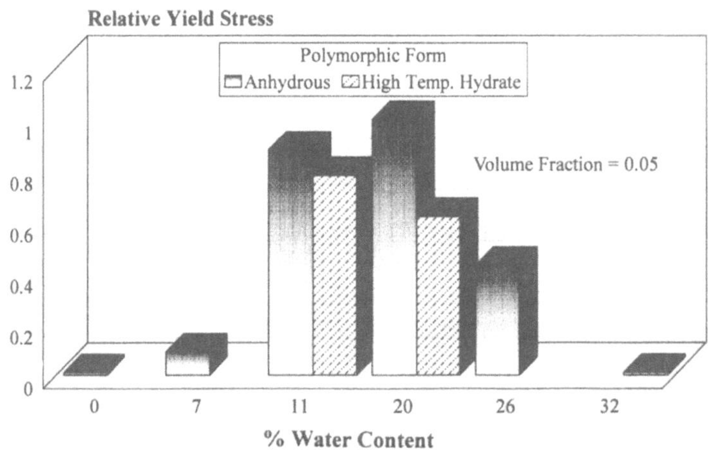


**Figure 4.** Influence of moisture content on the ER activity of curdlan polymorphs.

A crystal lattice transformation from the "anhydrous" form to a "low temperature hydrate" morphology occurs for curdlan at approximately 30% water content as evidenced by CP/MAS $^{13}C$ NMR. At this moisture level, ER fluids containing curdlan hydrate show little ER activity and increase significantly in conductivity (data not shown). Under these circumstances the "swollen" character[7] of curdlan hydrate appears to provide little inhibition to water migration and current densities are very high (> 1 mA / $cm^2$ at 1 kV / mm, 25°C). On the other hand, the "high temperature hydrate" form of curdlan, generated by autoclaving curdlan in the presence of moisture at 180° C, contains tightly bound water hydrogen bonded to the polysaccharide triple helices as illustrated in the crystal structure reported by Chuah and co-workers[12]. In this study, conversion to the "hydrate" crystal structure was confirmed by CP/MAS $^{13}C$-NMR spectra. When formulated into an ER fluid, this polymorph exhibited a reduction in yield stress compared to the "anhydrous" polymorph at comparable water concentrations suggesting that tightly bound water may be unavailable for promotion of the ER response, as shown in Figure 4. However, the high temperature "hydrate" retained a significantly larger fraction of its ER activity after heating for 16 hours at 105 °C, as shown in Figure 5, suggesting that water mobility may be restricted in the curdlan crystal lattice and the resulting ER fluid, in effect, behaves as if it were anhydrous. Such speculation needs to be confirmed experimentally in future work.

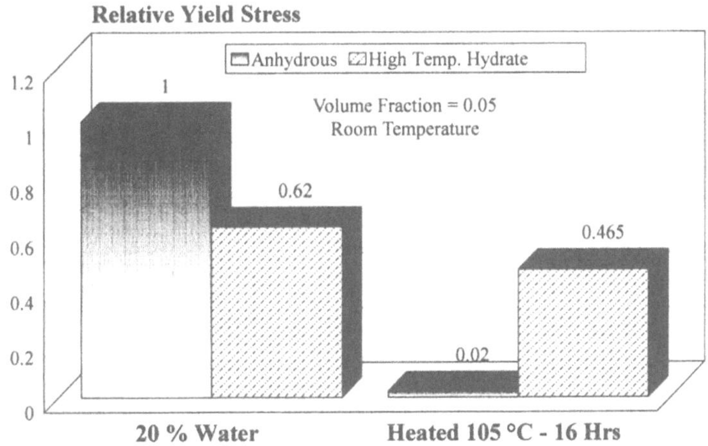


**Figure 5.** Impact of heating on the ER activity of curdlan polymorphs.

## ER Activity of Pustulan

The polysaccharide pustulan is a gel-forming, water-soluble extract from a foliose lichen named _Umbilicaria pustulata._ For this material, a transformation from a non-crystalline amorphous structure to a crystalline hydrate morphology occurs with the adsorption of water, at room temperature, in the range of 20-25% moisture content[6,8]. This conversion can be monitored by x-ray diffraction and / or by $^{13}C$-NMR in the solid state, as described above. Table 2 indicates that the "amorphous" polymer displays an optimum ER fluid yield stress at a water content level (10-12%) similar to cellulose and starch[1,2]. At higher levels of water, the resulting "hydrate" polymorph does not hold water sufficiently tightly, however, to prevent the development of current in ER fluids at modest field strengths (1000 kV / mm). This observation emphasizes that the behavior of high temperature curdlan "hydrate" is quite unique among glucose-containing polysaccharides. The dehydrated "amorphous" and crystalline "anhydrous" states of pustulan are not active as ER fluid particulate phases.

Table 2. Properties of ER fluids based on pustulan polysaccharide.

| *Polymorph* | *% Water* | *Relative Yield Stress* |
|---|---|---|
| Amorphous | 0 | 0 |
| | 12 | 1.0 |
| | 17 | < 0.1 |
| Hydrate | 25 | 0, Current |
| | 150 | 0, Current |
| Anhydrous | 0 | 0 |

## Cyclodextrin Particulate Phases for ER Fluids

**Water-Activated Systems.** The commercial introduction of a unique family of glucose-containing oligosaccharides, called cyclodextrins, has provided an opportunity for designing novel particulate phases for application in ER fluids. Cyclodextrins are produced by a "highly selective enzymatic synthesis" of 6-,7- and 8-membered ring structures referred to as *alpha*, *beta* and *gamma* cyclodextrins, respectively[13]. These molecules are generated from a starch (amylose) feed stock by the action of the bacterial microorganism *Bacillus macerans* and adopt a "donut" structure as shown in Figure 6 with cavity dimensions outlined in Table 3 below[14]:

Table 3. Geometry of cyclodextrin (CD) structures.

| | *alpha-CD* | *beta-CD* | *gamma-CD* |
|---|---|---|---|
| | $\alpha$ | $\beta$ | $\gamma$ |
| Glucose / Ring | 6 | 7 | 8 |
| Waters / Ring | 6 (10%) | 12 (16%) | 17 (19%) |
| *Diameter (Å)* | | | |
| Inner | 5.7 | 7.8 | 9.5 |
| Outer | 13.7 | 15.3 | 16.9 |

Although the outer rims of the cyclodextrin structure (both upper and lower) contain hydroxyl groups which can readily hydrogen bond to water molecules, the inner cavity is hydrophobic in character and a broad spectrum of small organic molecules (aromatics, alcohols, fatty acids, esters, etc.) can be bound. This complexation or "molecular encapsulation"[13] has enabled the use of cyclodextrins in a variety of pharmaceutical, agricultural and food applications where "controlled release" properties are desired. The impact of cyclodextrin ring size on "guest molecule" binding characteristics is an important parameter in selecting commercially useful systems and has been reviewed previously[14,15]. As shown in Table 3, crystal structure analysis has determined that *alpha*, *beta* and *gamma* cyclodextrins contain 6, 12, and 17 water molecules per ring, respectively, corresponding to moisture contents of 10%, 16% and 19% (by weight) for fully saturated crystal lattices[14].

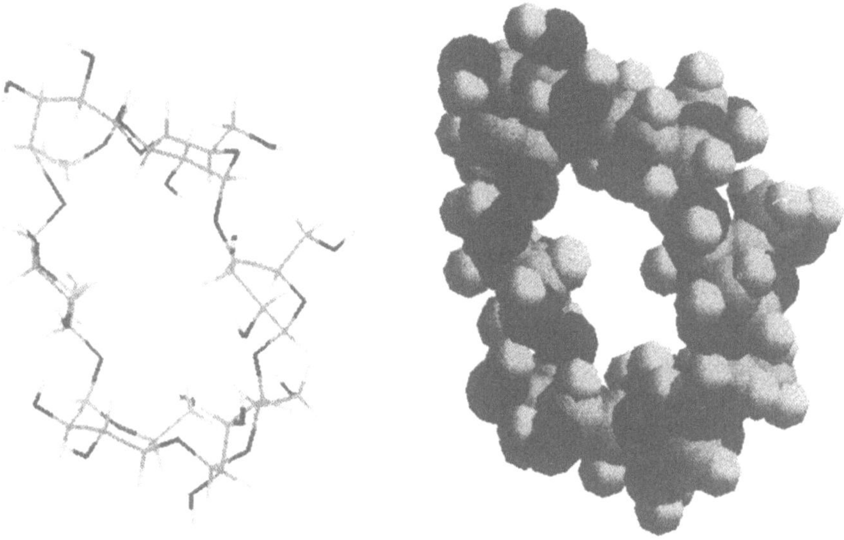

**Figure 6.** Molecular model of alpha cyclodextrin (6 glucose monomers / ring)

Figure 7 illustrates that cyclodextrins dispersed in paraffinic mineral oil display significant ER activity that depends on both moisture content and ring size. A large viscosity increase is observed, over a range of applied voltages, when ring size is increased from 6 to 7 glucose units (alpha to beta cyclodextrin), as additional water molecules are held within the structure. As ring size is further increased from 7 to 8, a similar increase in viscosity is evident although both the beta and gamma samples contained a comparable level of bound water (13%; below total lattice saturation). This observation suggests that increasing ring size in itself may provide improved performance, perhaps by allowing water molecules more mobility compared to a smaller ring. This hypothesis, unfortunately, is not supported by the current density data contained in Figure 8 for these ER fluids. It is evident that the three cyclodextrin-based ER fluids display very low, but quite comparable, current densities which would not be expected if water molecules were significantly more mobile in the case of the gamma structure. For all cyclodextrins, dehydrated particles yielded no significant ER activity.

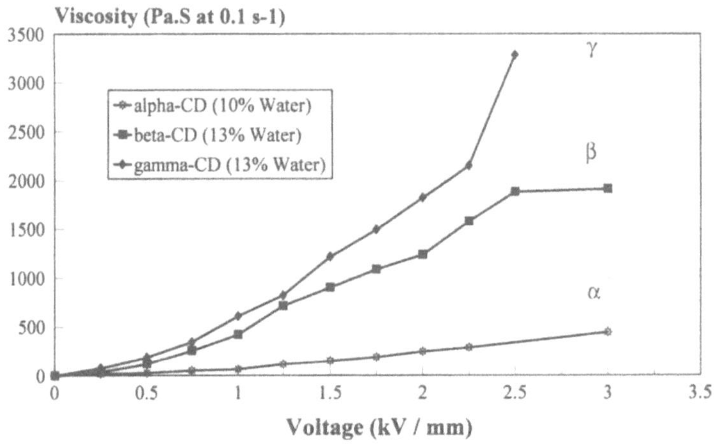


**Figure 7.** Impact of ring size and moisture content on the ER activity of cyclodextrins in paraffinic mineral oil.

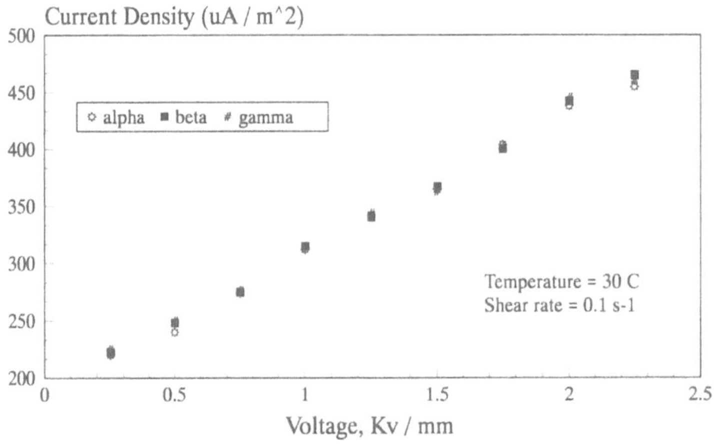


**Figure 8.** Current density of cyclodextrin ER fluids.

**"Anhydrous" Cyclodextrin-Based ER Fluids.** The ability of cyclodextrins to incorporate guest molecules in their cavity[14] , other than water, provides a novel means to synthesize inclusion complexes which display ER activity in the "anhydrous" state. Although work was still in progress at the time of this writing, molecules such a sodium dodecyl sulfate and sodium p-nitrophenolate[16] have been successfully complexed to cyclodextrin and these derivatives were found to display ER activity at levels of adsorbed water which do not promote the ER effect in the absence of complexation (1-2%). Further, sustained heating of these complexes dispersed in paraffinic mineral oil, at temperatures in the range of 100-150 °C, did not result in a loss of effectiveness. As in the case of the sulfonated polystyrene-based ER fluids reported previously[11], it is likely that the sodium (or other) counterion is responsible for ionic mobility and particle polarizability in these systems. However, for the cyclodextrin complexes it is possible that "encapsulation" of the ionic species may further inhibit their ability to migrate into the liquid phase of the ER fluid thereby reducing the potential for current development. Studies to better characterize these mechanisms are in progress.

## CONCLUSIONS

Based on the empirical investigations detailed above, it may be generally concluded that the potentially negative impact of "mobile" water on the performance of water-activated particulates in ER fluids may be mediated through appropriate complexation and / or inclusion within a crystal lattice. Further, electrorheologically active "anhydrous" systems can be fabricated using ionic species, such as sodium salts of organic acids and alcohols, which results in thermally stable ER fluids. Specific conclusions are provided below:

1. Hydration is a key element in promoting the ER activity of glucose-containing homopolysaccharides such as curdlan and pustulan. In general, heating above 100 °C results in deactivation of these particulates.

2. Crystalline "Anhydrous" curdlan containing 11% adsorbed moisture displayed a curious "aging" effect in an applied electric field whereby viscosity was found to increase with time, possibly due to electrophoretic mobility of water molecules within the ER fluid particulate phase.

3. Water molecules bound in crystalline phases may be less mobile and more resistant to dehydration than those adsorbed in "amorphous" domains, although this depends on the specific crystalline polymorph being evaluated.

4. Tightly bound waters in "high temperature" curdlan hydrate appear less capable of providing an ER enhancing effect relative to "free" adsorbed water in "anhydrous" curdlan but activity is maintained even after heating above 105 °C.

5. Cyclodextrins display a strong moisture-activated ER response which can be moderated by water content and / or ring size.

6. Anhydrous ER fluids can be prepared from cyclodextrin complexes containing sodium salts of certain organic molecules.

## ACKNOWLEDGMENTS

The authors would like to thank American Maize Products Company (AMAIZO) for providing cyclodextrin samples and Dr. Mani Subramanian of the Texaco R&D Computer Services Section for his molecular modeling work on these compounds. The experimental support and consultations of Professor Hans Conrad and Dr. Ying Chen of

North Carolina State University and Professor Montgomery Shaw and Mr. Rex Kanu of the University of Connecticut are gratefully appreciated.

## REFERENCES AND NOTES


1. H. Uejima, "Dielectric Mechanism and Rheological Properties of Electro-Fluids", *Japanese J. Appl. Phys.*, 11 (3), 319 (1972).

2. Y. Chen and H. Conrad, "Effects of Water Content on the Electrorheology of Corn Starch / Corn Oil Dispersions", Proceedings ASME Annual Meeting, New Orleans, November 28-December 3, 1993.

3. H. Block and J.P. Kelly, "Electrorheology", *J. Phys. D.*,21, 1661 (1988).

4. A. Sarko and R. Muggli, "Crystal Structure of Native Cellulose I From Valonia", *Macromolecules*, 7, 486 (1974).

5. H. Wu, Ph.D. Dissertation, SUNY College of Environmental Science and Forestry, Syracuse, N.Y. 1977 (Subsequent papers in *Carbohydrate Res.*)

6. A.J. Stipanovic, P.J. Giammatteo and S.B. Robie, "Cross-Polarization / Magic Angle Spinning 13C-NMR of (1-6)-Beta-D-Glucan (Pustulan): Mechanism of Gelation", *Biopolymers*, 24, 2333 (1985).

7. A.J. Stipanovic and P.J. Giammatteo, "Curdlan and Scleroglucan - NMR Characterization of Solution and Gel Properties", *ACS Advances in Chemistry Series 223*, 74 (1989).

8. A.J. Stipanovic and J.P. Schoonmaker, " The Impact of Crystalline Phase Morphology on the Water-Promoted Electrorheological Effect of Polysaccharides", *Polymer Preprints*, Vol. 35 (2), 365 (1994).

9. A.J. Stipanovic and P.J. Giammatteo, "CP/MAS 13C-NMR Studies on Aqueous Polysaccharide Gels", *Progress in Biotech.* , 3, 281 (1987).

10. A.P. Gast and C.F. Zukoski, "Electrorheological Fluids as Colloidal Suspensions", *Advances in Colloid and Interface Science*, 30, 153 (1989).

11. Y. Asako, S. Ono, R. Aizawa and T. Kawakami,"Properties of Electrorheological Fluids Containing Numerously Sulfonated Polymer Particles", *Polymer Preprints*, 35 (2), 352 (1994).

12. C.T. Chuah, A. Sarko, Y. Deslandes and R.H. Marchessault, "Triple-Helical Crystalline Structure of Curdlan and Paramylon Hydrates", *Macromolecules*, 16, 1375 (1983).

13. "Cyclodextrin" Marketing Brief, American Maize Company (AMAIZO).

14. R.J. Begeron, "Cycloamylose-Substrate Binding", *Inclusion Compounds - Part 3*, Academic Press, London, 391 (1984).

15. W. Saenger, "Structural Aspects of Cyclodextrins and Their Inclusion Complexes", *Inclusion Compounds Part 2*, Academic Press, London, 231 (1984).

16. R.J. Bergeron, M.A. Channing, G.J. Gibeily and D.M. Pillor, "Disposition Requirements for Binding in Aqueous Solution of Polar Substrates in the Cyclohexaamylose Cavity, *J. Amer. Chem. Soc.*, 99 (15), 5146 (1977).

# ER FLUID INKS: A RAINBOW OF POSSIBILITIES

K. D. Weiss, D. A. Nixon, J. D. Carlson, and A. J. Margida

Lord Corporation
Cary, North Carolina 27511

## INTRODUCTION

The concept of using electrorheological (ER) fluids as inks originated with the design of several electroviscous recording devices some 25 years ago.[1,2] The colored, ER active pigments used in these early inks included carbon black, copper phthalocyanine derivatives (Skyline Blue & Monstral Green B), flavanthrone (Indofast yellow toner), quinacridone (Quindo Magenta), 3,4,9,10-bis(n,n'-(p-methoxyphenyl)-imido)perylene (Indofast Brilliant Scarlet toner), and 1,2,5,6-di-(c,c'-diphenyl)thiazole anthraquinone. In addition to utilizing colored pigments in an ink, the possibility of coloring conventional ER fluids (i.e., those prepared by W. Winslow[3], etc.) through the addition of dyes also was suggested.[1,4] Examples of the oil soluble dyes used to colorize known ER fluids included Oil Yellow AB, Oil Red, Oil Blue G Extra and Oil Black.

Two decades ago the lack of a mechanistic understanding regarding the ER phenomenon hindered the development of ER fluids that could exhibit the yield strength and response time necessary to be successfully used as controllable inks. This paper discusses the effects various types of pigments and dyes have on the mechanism of particle polarization in an ER fluid ink. The outcome of this investigation is the identification of a class of pigments that improves the yield strength and response time exhibited by ER fluid inks. A description of the various limitations encountered using different types of dyes to colorize conventional ER fluids also is provided.

## EXPERIMENTAL

### Preparation of ER Fluid Inks

ER fluids were prepared using colored pigments both in an "as received" and a "dried" condition. In the "as received" condition the particles contained moisture physically adsorbed to their surface. This adsorbed moisture was removed in the "dried" condition by heating the pigments in a convection oven at 130°C for 48 hours. Both the "as received" and "dried" ER fluids contained 30% pigment, 3% polymeric surfactant (Solsperse® 17000, ICI Americas Inc.), and 77% fluorinated polyether (Galden® D-10, Montedison USA Inc.) oil by weight. All pigments were obtained from Aldrich Chemical Company. In each formulation the colored pigment was dispersed in the oil/dispersant mixture using a conventional paint shaker.

*Progress in Electrorheology*, Edited by K.O. Havelka
and F.E. Filisko, Plenum Press, New York, 1995

The conventional ER fluid to which small amounts of various colorants were added was a transparent formulation prepared according to a previously disclosed procedure.[5] This transparent ER fluid contained 12.7% precipiated silica (Union Carbide Co.) as the ER active particle, 0.4% water as an activator, 2.5% glycerol monooloeate (Witco Chemical Co.), and 84.4% light mineral oil (Alfa Products) by weight. All colorants added to the base ER formulation were obtained from Aldrich Chemical Company.

### Measurement of ER Fluid Properties

A description of the properties associated with the ER effect (i.e., static yield stress, dynamic yield stress, current density, plastic viscosity, etc.) and the techniques used to measure these properties have previously been published.[6] The properties of the ER fluids used in this investigation were measured using a stress controlled rheometer equipped with a concentric cylinder cell having a 0.5 mm electrode gap.

The experimental procedures used to estimate the response time associated with particle polarization and fibril formation in ER fluids have previously been published.[7] A dielectric spectroscopic technique was used to measure the time necessary for individual particles to respond to a change in electric field. A channel flow technique was used to measure the time required for particles to form a macroscopic structure.

## RESULTS & DISCUSSION

The classification of colored organic molecules as either ER active pigments or colorant additives is necessary to differentiate between two very different types of ER fluid inks. Inks containing colored, ER active pigments may be essentially anhydrous ("dried") or contain a small amount of ambient moisture ("as received"). ER fluid inks that are colorized by the use of an additive are further categorized based upon the solubility of this colorant in the carrier oil and activator phases present in the ER fluid.

### ER Active Pigments

The utilization of ionic pigments, such as the sulfonated salts of conventional organic dyes[8], as ER active particles has been found to improve both the response time and yield strength exhibited by a controllable ink. Due to their ionic structure and ability to adsorb moisture these colored pigments exhibit improved polarizability when subjected to an electric field (D.C. voltage). Organic pigments that are ER active typically undergo electric field induced polarization through an ionic mediated mechanism. This type of polarization is predominately controlled by the migration of ions within a moisture layer adsorbed onto the surface of the particles.

The polarizability of a particle in an oil is known to be the key factor in determining if a fluid will exhibit electrorheological activity.[6,9,10] This polarizability, $\beta$, is given by the equation, $\beta = [(\varepsilon_2-\varepsilon_1)/(\varepsilon_2+2\varepsilon_1)]$, where $\varepsilon_1$ and $\varepsilon_2$ are the complex permittivities of the carrier oil and particle, respectively. In order for a dispersion of particles to polarize and respond as an ER fluid, the particles and the carrier oil must have different complex permittivities so that $\varepsilon_2 - \varepsilon_1 \neq 0$. By definition, the complex permittivity of an oil or particle is dependent upon both the dielectric constant and conductivity of the oil or particle. Thus the dielectric properties of the oil and particle affect both the response time and yield stress exhibited by an ER fluid.

ER fluids that utilize an ionic mediated polarization mechanism are known to undergo a Debye-like relaxation process.[7] According to Debye theory, the time dependent dielectric properties of an ER fluid can be used to calculate the relaxation time associated with the polarization decay process. This measurement provides a lower limit for the time interval in which an ER fluid response can be expected. In other words, an ER fluid can never exhibit a response to an electric field faster than indicated by this measurement. The relaxation times measured for ER fluid inks containing ionic pigments in an "as received" condition are typically on the low end of the millisecond time frame. The relaxation times measured for analogous ER fluid inks containing conventional nonionic pigments are typically greater than 50 milliseconds.

For a Debye-like relaxation process a measure of the Cole-Cole distribution in relaxation times, $\alpha$, exhibited by an ER fluid can be obtained by fitting the measured dielectric properties with a semicircular arc.[7] This Cole-Cole diagram is basically a plot of the measured dielectric loss data as a function of relative permittivity as shown in Figure 1 for an ER fluid ink containing an "as received" ionic pigment (Acid Green 25 - Anthraquinone classified dye). Values for the material's static permittivity, $\varepsilon_s$, and the relative permittivity at infinite frequency, $\varepsilon_\infty$, are obtained from this curve fit as the x-axis intercepts of the semicircle. The deviation of the dielectric loss data at high values of relative permittivity is an indication of loss due to conduction at low frequencies. The relaxation time, $\tau$, associated with this ER fluid ink is calculated using the equation $(2\pi f_{max})\tau = 1$, where $f_{max}$ is the frequency at which the dielectric loss data goes through a maximum in the semicircular curve fit. The relaxation time associated with the ER fluid ink described in Figure 1 was found to be on the order of 0.006 milliseconds. The validity of this analysis was substantiated by modeling the measured time dependent dielectric data with the Cole-Cole derivation for complex permittivity, as well as by applying a linearity test as previously described in the literature.[7]

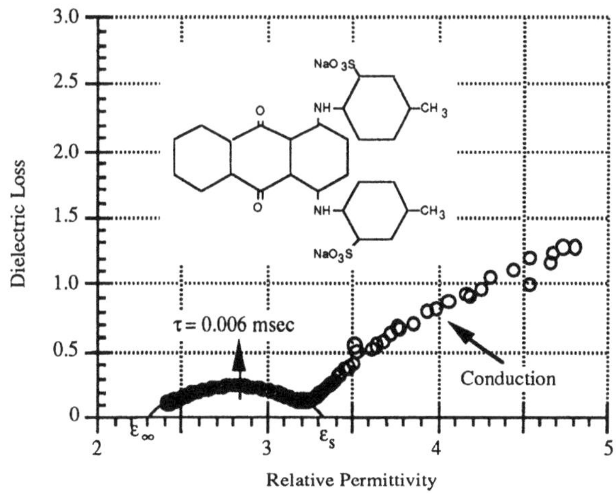


**Figure 1.** Cole-Cole Analysis of Dielectric Data Obtained For ER Fluid Ink Containing Ionic Particles (Acid Green 25).

ER fluid inks containing conventional nonionic pigments exhibit relaxation times that are considerably slower than those measured for inks containing ionic pigments. An example of a Cole-Cole arc diagram for an ER fluid ink containing an "as received" nonionic pigment (Solvent Green 3 - Anthraquinone classified dye) is shown in Figure 2. Even though the measured dielectric data allows only a portion of the semicircular arc to be visible, values for $\tau$, $\varepsilon_\infty$, $\varepsilon_s$, and $\alpha$ can easily be obtained from this analysis. For the ER fluid ink described in Figure 2, the relaxation time was found to be on the order of 62 milliseconds. The validity of this analysis was substantiated by modeling the measured time dependent dielectric data with the Cole-Cole derivation for complex permittivity and by applying the test of linearity as described above.

Reduction in the ambient moisture adsorbed onto the surface of the pigments prior to preparing an ER fluid ink was found to increase the measured relaxation time. However, ER fluid inks containing "dried" ionic pigments were still observed to respond to an electric field faster than inks containing their "dried" or "as received" nonionic pigment counterparts. The relaxation times for ER fluid inks containing "dried" Acid Green 25 or Solvent Green 3 particles were measured to be 0.12 and 430 milliseconds, respectively.

The relaxation time measured for ER fluid inks that contain nonionic pigments can be decreased by incorporating additional activator into the formulation. The addition of 0.2% water by weight into an ink containing the "as received" nonionic pigment (Solvent Green 3) was found to decrease the measured relaxation time from 62 to about 2 milliseconds. However, the high current density level encountered upon applying an electric field to this nonionic pigment based ink makes this approach impractical.

When placed in channel flow, ER fluid inks containing ionic pigments also are found to be capable of going from a voltage-off baseline pressure to a final, voltage-on equilibrium pressure in less time than nonionic pigment based inks. For example, the response times measured for the "dried" ER fluid inks containing the Acid Green 25 and Solvent Green 3 pigments were 3.0 and 1000 milliseconds, respectively. While these response times are related to the previously described relaxation times, they also include the additional time necessary for macroscopic structure formation in the fluid. Thus ER fluid inks containing ionic pigments have been found to vastly out perform their nonionic pigment based counterparts with respect to response time.

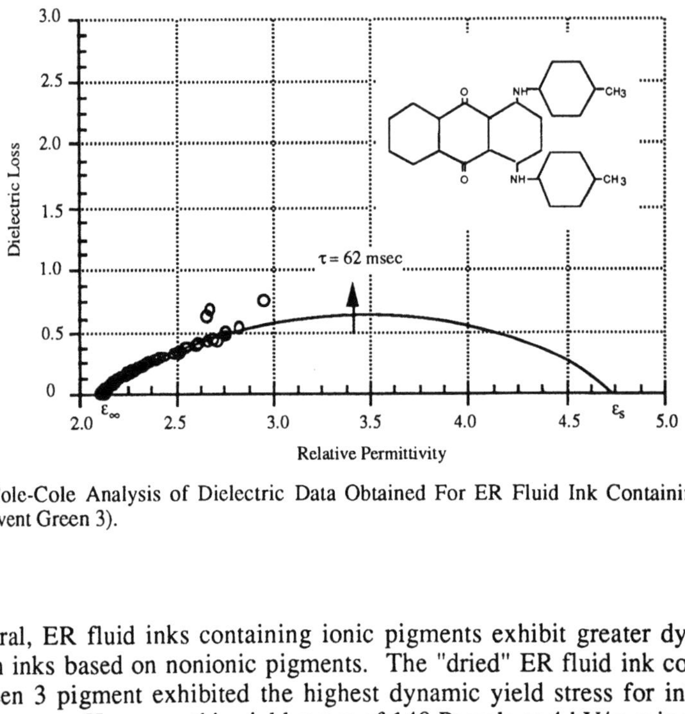


**Figure 2.** Cole-Cole Analysis of Dielectric Data Obtained For ER Fluid Ink Containing Nonionic Particles (Solvent Green 3).

In general, ER fluid inks containing ionic pigments exhibit greater dynamic yield stresses than inks based on nonionic pigments. The "dried" ER fluid ink containing the Solvent Green 3 pigment exhibited the highest dynamic yield stress for inks based on nonionic pigments. However, this yield stress of 140 Pascals at 4 kV/mm is considerably smaller than the 700 Pascals yield stress measured for a "dried" ink containing the Acid Green 25 pigment. The higher yield stress values exhibited by ER fluid inks containing ionic pigments can be attributed to improved polarizability of the particle component. An attempt to increase the polarizability of a nonionic pigment in an ink through the addition of 0.2% water by weight to the formulation was unsuccessful in improving the properties exhibited by the controllable ink. Thus the polarizability of the ionic pigments in an ER fluid is far superior to the polarizability of their nonionic counterparts even in the presence of additional activator.

The current densities measured for "dried" ER fluids containing the Acid Green 25 and Solvent Green 3 pigments at 4.0 kV/mm were 19.7 and 0.9 $\mu A/cm^2$, respectively. The addition of water into the ER fluid containing the Solvent Green 3 pigment increased the measured current density for this formulation to an unacceptable level as demonstrated by electric breakdown (arcing) at electric field strengths in excess of 1.0 kV/mm. The modest current density levels exhibited by ER fluid inks containing ionic pigments is satisfactory for utilization in many applications.

The unique ability of copper phthalocyanine pigments to undergo electronic polarization, as well as ionic polarization is acknowledged.[11] However, "anhydrous" ER fluids based on copper phthalocyanine particles are limited to the application of an A.C. electric field in order to minimize electrophoretic movement of the particles. The motivation in developing "anhydrous" ER fluids is to broaden the operating temperature range available for these formulations. This same motivation does not exist in the development of ER fluid inks because the temperature range associated with applications using these inks is typically quite mild (i.e., 25-90°C). Electrophoretic movement of the "as received" or "dried" ionic pigments present in the ER fluid inks used in this investigation has not been observed upon the application of a D.C. electric field.

## Colorant Additives

The addition of dyes to conventional ER fluids can have adverse affects on the performance of the ER fluid depending upon the solubility of the dye in either the carrier fluid or activator. If the dye is insoluble in both the carrier fluid and activator used in the ER fluid, then the dye is more accurately defined as a pigment. If this colored pigment is electrorheologically active, the addition of a small amount of it to colorize an existing ER fluid may actually increase the dynamic yield stress exhibited by the formulation. However, since most conventional ER fluids are already formulated with a high volume fraction of particles in order to maximize the ER effect, the addition of a colored pigment to these formulations can also increase viscosity, as well as lead to the development of "stiction" related behavior. For these reasons the addition of less than 5% by weight of an insoluble pigment is recommended for the coloration of most conventional ER fluids.

The major affects that a dye, which is soluble in the carrier fluid, may have on the performance of an ER fluid is a decrease in the polarizability of the ER active particles and an increase in current density. These affects are attributed to the dye inadvertently increasing the conductivity of the dielectric oil upon dissolving. In order to minimize any detrimental affect a dye may have on the properties exhibited by a conventional ER fluid, the dye should be added in an amount that is less than 5% by weight.

The type of colorant additive that has the most pronounced affect on the performance of an ER fluid is one that dissolves in the activator phase. The interaction of a "soluble" pigment with the surface of an ER active particle may either increase or decrease the particle's overall conductivity. "Soluble" pigments that increase the particles' conductivity have been found to have minimal affect on the properties exhibited by an ER fluid ink when present in an amount less than about 3-5% by weight. On the other hand, in order for "soluble" pigments that decrease the particles' conductivity to have minimal impact on the performance of an ER fluid ink, they must be present in an amount less than about 0.1-0.5% by weight.

The determination as to whether a "soluble" pigment increases or decreases the conductivity exhibited by ER active particles is made using Maxwell-Wagner theory in accordance with the previously described dielectric analysis.[7] For example, both Acid Green 25 and Acridine Orange Salt were found to be soluble in the water activator used in a conventional ER fluid. Analysis of the dielectric data measured for the base ER fluid and inks containing these "soluble" pigments according to Maxwell-Wagner theory[7] demonstrated that the particle conductivity in the base ER fluid of 0.0042 $\mu S/m$ increased to 0.0096 μS/m and decreased to 0.0022 μS/m with the addition of 0.1% by weight of Acid Green 25 and Acridine Orange salt, respectively.

The addition of Acid Green 25 to the base ER fluid was found to have little detrimental affect on the electrical or mechanical properties of the fluid as shown in Table 1 until the colorant amount exceeded 3.0 % by weight. In fact, a slight improvement in the relaxation time and the yield stresses (static and dynamic) exhibited by the base ER fluid was observed upon the addition of about 3% or less of this "soluble" pigment.

**Table 1.** The Properties Exhibited by a Conventional ER fluid Containing Various Amounts of a "Soluble" Pigment (Acid Green 25).

| | Weight % of Acid Green 25 Added to ER Fluid | | | |
|---|---|---|---|---|
| Description of the Measured Property | 0.0 | 1.0 | 3.0 | 5.0 |
| Viscosity (mPa-s) | 160 | 141 | 157 | 149 |
| Zero-Field Yield Stress (Pa) | 11 | 13 | 14 | 11 |
| Dynamic Yield Stress (Pa) at 3.0 kV/mm | 291 | 315 | 346 | arc |
| Static Yield Stress (Pa) at 3.0 kV/mm | 325 | 343 | 365 | arc |
| Current Density ($\mu A/cm^2$) at 3.0 kV/mm | 0.82 | 1.50 | 1.70 | arc |
| Relaxation Time (ms) | 22.6 | 11.7 | 12.6 | 17.5 |

The addition of Acridine Orange Salt to the base ER fluid was found to decrease the observed electric field-induced yield stresses (static and dynamic) and increase the relaxation time. These detrimental affects are minimized as shown in Table 2 by limiting the addition of this "soluble" pigment to the base ER fluid at about 0.1% by weight. When a larger amount of this "soluble" pigment is added to the ER fluid, the resulting large change in properties that occurs renders the fluid unusable for most applications.

**Table 2.** The Properties Exhibited by a Conventional ER fluid Containing Various Amounts of a "Soluble" Pigment (Acridine Orange Salt).

| | Weight % of Acridine Orange Salt added to ER Fluid | | | |
|---|---|---|---|---|
| Description of the Measured Property | 0.0 | 0.1 | 0.5 | 1.0 |
| Viscosity (mPa-s) | 160 | 162 | 149 | 150 |
| Zero-Field Yield Stress (Pa) | 11 | 8 | 1 | 2 |
| Dynamic Yield Stress (Pa) at 3.0 kV/mm | 291 | 252 | 156 | 100 |
| Static Yield Stress (Pa) at 3.0 kV/mm | 325 | 295 | 186 | 75 |
| Current Density ($\mu A/cm^2$) at 3.0 kV/mm | 0.82 | 0.48 | 0.64 | 0.51 |
| Relaxation Time (ms) | 22.6 | 39.8 | 331.0 | ***** |

***** = high dielectric loss inhibited accuracy of relaxation time measurement

## SUMMARY & CONCLUSION

ER fluid inks based on colored, ionic pigments were found to provide faster response times and higher yield stress values than corresponding ER fluids containing colored, nonionic pigments. In addition, these ionic pigment based ER fluid inks were found to respond to a D.C. voltage without significant electrophoresis.

The incorporation of a colorant additive into a conventional ER fluid can be achieved without seriously affecting the performance of the fluid provided certain guidelines are followed. These guidelines include the classification of the colorant additive as a dye, pigment, or "soluble" pigment. A "soluble" pigment that decreases the conductivity of ER active particles dispersed in an oil has the most severe impact on the electrical and mechanical performance of an ER fluid ink.

## REFERENCES

[1]Clark, H. US Patent 3,484,162 (1969).
[2]Carreira, L.; Mihajlov, V. US Patent 3,553,708 (1971).
[3]Winslow, W. US Patent 3,047,507 (1962).
[4]Kokai Tokkyo Koho (A) 158926 (1979).
[5]Carlson, D. US Patent 5,075,021 (1991).
[6]Weiss, K.; Coulter, J.; Carlson, D. J. Intel. Mat. Syst. & Struct. (1993), 4(1), 13-34.
[7]Weiss, K.; Carlson, D. Int. J. Mod. Phys.: B (1992), 6(15-16), 2609-2623
[8]Kamakura, T.; Yokohama, T.; Kawasaki, K.; Ichikawa, E. US Patent 4,554,555 (1985).
[9]Block, H.; Kelly, J.; Qin, A.; Watson, T. Langmuir (1990), 6, 6-14.
[10]Gast, A.; Zukoski, C. Advances in Colloidal and Interface Science (1989), 30, 153-202.
[11]Block, H.; Kelly, J. US Patent 4,687,589 (1987).

# EFFECT OF CATION CONCENTRATION ON ELECTRORHEOLOGICAL ACTIVITY OF AMORPHOUS ALUMINO-SILICATES

Adam W. Schubring and Frank E. Filisko

Department of Materials Science and Engineering
The University of Michigan
Ann Arbor, Michigan, 48109-2136

## INTRODUCTION

Many materials have been shown to demonstrate electrorheological (ER) activity, defined specifically as materials (most commonly fluids) which undergo a dramatic increase in resistance to flow (i.e. apparent viscosity) with applied electric fields. Until recently, the most common ER fluids consisted of suspensions of nonconducting particles in nonconducting liquids with significant amounts (5-20%) of an activator (the most common and effective being water) adsorbed onto the particles.[1] Such materials became known as "wet" or hydrous ER fluids. Because of the possible instabilities of the water layer, these hydrous fluids were severely limited for any large scale applications.[2] Essentially any particle with sufficient amounts of water adsorbed onto it could form an ER fluid, suggesting that the ER activity was not a primary function of the particle chemistry. The function of the water in creating ER activity is still unresolved.

More recently, certain particle suspensions have been shown to form ER active fluids without the need for adsorbed water or any other activator.[3,4] By virtue of their intrinsic character (chemical, physical, and/or morphological) these particles probably in conjunction with an appropriate non-conducting fluid, result in ER active materials which are not limited by the many restrictions associated with the adsorbed water for the hydrous systems. ER fluids which have been demonstrated to be intrinsically active include suspensions of various ionic conductors (the most common being the crystalline and amorphous alumino-silicates), suspensions of various polyelectrolytes,[5] suspensions of certain polymeric semiconductors (specifically polyacene quinones and polyanilines),[1] and solutions of poly (hexyl isocyanates)[6] and poly γ-benzyl L-glutamates. As with the uncertainty of the function of the

*Progress in Electrorheology*, Edited by K.O. Havelka
and F.E. Filisko, Plenum Press, New York, 1995 

adsorbed water in the hydrous systems, the mechanisms responsible for the ER activity of the "anhydrous" or intrinsic systems is still more obscure.

The most popular explanations for the ER activity of the particulate-based ER fluids center around interfacial polarization as first described by Maxwell[7] in the Maxwell-Wagner-Sillars (MWS) polarization model[8] and also around the character and behavior of the electrical double layer (EDL). In the case of solutions, the most probable explanation involves the rigidity and high polarizability of the individual molecules in solution, or the liquid crystalline nature of both of the reported systems, since both the PBLG and poly (hexyl isocyanates) form liquid crystals in solution.

For suspensions, the MWS interfacial polarization explanation seems reasonable because of the presence of an "anomalous" dispersion which seems to be present in all intrinsically active (as well as hydrous) systems; however, the better ER active fluids do not readily fall within the MWS explanation. For instance, the magnitude of the MWS dispersion is directly related to the magnitude of the differences between the permittivity and/or conductivity of the dispersed and liquid phases, as shown in many studies with dispersed conductive phases.[9] Despite possessing the greatest extreme for permittivity and conductivity differences, the dispersed metal particles show little or no ER activity (when properly insulated) as shown by many ER researchers.[10] Likewise, particles of low conductivity but high permittivity, such as anhydrous $TiO_2$ dispersions, represent a high permittivity difference but also demonstrate no ER activity.[2] On the other hand, many reported intrinsic systems, specifically the alumino-silicates and polyelectrolytes, have very low conductivities and also low permittivities, yet show very significant ER activity.

The basis for the MWS interfacial polarization is that mobile charge carriers associated with the particles, either as ions or electrons, will move to the interface in response to an imposed electric field, thus establishing continuity of the field equations. The carriers accumulate at the interface because they are prevented from moving through the barrier provided by the nonconducting liquid. As opposed to usual ion relaxational frequencies which occur in the range of $10^8$ Hz, these carriers show characteristic frequencies in the range of $10^2$ - $10^3$ Hz at room temperature. These mobile carriers which are associated with the particles differ from the fixed charges (the Stern layer) on the surface of the particles and are involved in formation of the electrical double layer. However, the diffuse layer of the EDL does contain mobile ions which may give rise to a similar dielectric dispersion.

Explanations involving the EDL in relation to the ER activity are also speculative. The basis for the EDL is well developed in electrolytic suspensions,[11] but they are poorly understood for suspensions in nonconductive liquids such as ER fluids.[12] Polarization of the diffuse layer of the EDL has been studied[13,14] and large low frequency dielectric dispersions have been observed and related to polarization of the diffuse double layer.[13,15] These studies were performed in electrolytic media and making a direct extrapolation was difficult. In addition, it was reported that particles with large surface charge or Zeta potential produce poor ER fluids.[16] The poor ER response could be explained by the fact that permanent fixed surface charges cause particles to be attracted to one or both electrodes which create a lubricating layer of oil (devoid of particles) either adjacent to an electrode or in the center of the fluid (if particles are drawn to both electrodes). Lastly, it was shown that the stronger ER fluids were produced with particles having the lowest electrophoretic mobility.

In this study, we are investigating the relationship between mobile ion charge concentration on the particles, characteristics of the anomalous dispersion, and the ER activity of the corresponding ER fluids. The amorphous alumino-silicates (AAS) which were synthesized in our laboratories were particularly suited for this study as a model system. The cation concentration is controlled by the extent of substitution of Al for Si and can be varied over a wide range without apparent change of the morphology, i.e. AAS demonstrating no x-ray crystallinity. The AAS differ greatly from the crystalline materials (Zeolites) because in the latter, changes of more than a few % in Al/Si ratio will result in morphological changes as well. Additionally, since water is a constant concern, the AAS particles are stable to temperatures as high as 900°C before crystallizing, thus allowing more critical drying. Finally, the AAS have a much lower water affinity than the Zeolites, further reducing the concern for water adsorption during a measurement.

## EXPERIMENTAL

The particulate phase of the ER materials investigated consist of three amorphous aluminosilicate powders, each with different Al/Si ratios, dispersed in anhydrous paraffin oil. The three powders studied were the sodium aluminosilicates $Na_{0.5}Al_{0.5}SiO_3$, $NaAlSiO_4$, and $Na_2Al_2SiO_6$. Studies used to characterize the powders included electrorheological, dielectric, surface area/pore volume, X-ray powder diffraction (XRD), and environmental scanning electron microscopy (ESEM).

### Sample Synthesis

The aluminosilicates were synthesized according to procedures similar to those of Kansal and Laine.[17] The following is a general description of the synthesis of the aluminosilicate $NaAlSiO_4$. The difference in the preparation methods for the $Na_{0.5}Al_{0.5}SiO_3$ and the $Na_2Al_2SiO_6$ are in the molar quantities of reactants.

The preparation of the aluminosilicates can be divided into 3 sections: preparation of the sol-gel precursor, sol-gelation, and drying.

**Preparation of the sol-gel precursor.** Equimolar ratios of aluminum hydroxide powder and triethanolamine were combined with ethylene glycol. The mixture was refluxed and distilled under dry nitrogen at 200°C to remove residual water until the aluminum hydroxide went into solution. After the aluminum hydroxide fully dissolved, the solution was sealed and allowed to cool.

In the second step, a mixture of ethylene glycol and equimolar amounts of fumed silica and sodium hydroxide pellets were mixed, heated, and refluxed under dry nitrogen until the two solids dissolved. This solution was then added to the previous aluminum hydroxide/ triethanolamine complex to form the precursor.

The precursor was then mixed and refluxed again under dry nitrogen (to assure a water-free homogeneous solution), sealed and allowed to cool.

**Sol-gel method.** Distilled de-ionized water was stirred and heated to a temperature of 85-90°C. When the temperature of the precursor decreased to the temperature of the water, the precursor was added to the water, continuously stirring. The gel formed almost immediately, and it was allowed to mature for a period of 24 hours.

**Drying of the gel.** Once the gel had matured, it was heated at 100°C for 24-48 hours until most of the water had evaporated. The temperature was then increased to 300°C to burn off the organic components (ethylene glycol and triethanolamine). The remaining carbonaceous agglomerates were then pyrolized at 600°C where the carbon was oxidized. This step could be accelerated by flowing a continuous stream of $O_2$ through the sample. The white granules were then ball-milled in anhydrous methanol until a particle size range of 1-50 μm was obtained.

**ER Fluid Preparation**

For the electrorheological and dielectric studies, the preparation of the aluminosilicate powders required high temperature drying to remove any absorbed water on the powder's surface. The method used was described by Filisko and Radzilowski.[4] The samples were slowly heated to 600°C then placed under vacuum with a liquid nitrogen trap for four hours and allowed to cool. When the samples cooled to room temperature, anhydrous mineral oil was added without breaking the vacuum seal. All the ER fluids used in this study contained 14g of powder to 30 ml mineral oil.

**Electrorheology.** Data was collected at room temperature using a Weissenberg rheogoniometer modified for ER measurements. The samples were loaded into a couette cell on the rheogoniometer which was connected to a high voltage, low wattage DC supply (12kV - 5mA maximum). The dimensions of the couette cell are the following: the bob - 24.77 mm diameter, 24.13 mm in length and the cup - 27.30 mm inside diameter. The cup was rotated at a constant RPM ranging from 0.00142 to 142 $s^{-1}$. For each RPM, a torque was measured on the upper bob. Torque deflection readings were taken at every 500 V/mm until the material experienced dielectric breakdown.

**Dielectric testing.** Measurement of the dissipation factor (tan δ) as a function of temperature (-150°C to 150°C) and frequency (13 Hz to 200 kHz) were collected on a computer receiving data through a GenRad 1693 RLC Digibridge connected to a Hewlett

Packard 34401A multimeter. Activation energies were calculated from the resulting dispersions.

**Surface area/Pore volume.** Surface area was calculated with the Micromeritics 2000 ASAP analyzer by application of the Brunauer, Emmett and Teller (BET) theory.

**X-ray powder diffraction.** The degree of crystallinity was determined with the use of a Rigaku Rotaflex X-ray diffractometer. Data was taken between 5 and 85 degrees 2θ.

**Particle size.** The particle size was measured with an ElectroScan environmental scanning electron microscope.

## Powder Characteristics

The x-ray diffraction (XRD) plots of Figures 1 and 2 compare the non-crystalline sodium aluminosilicate, $NaAlSiO_4$, to a crystalline aluminosilicate, 4Å. XRD's were taken of the other non-crystalline samples, $Na_{0.5}Al_{0.5}SiO_3$ and $Na_2Al_2SiO_6$, and each demonstrated similar diffuse diffraction patterns shown in Figure 1.

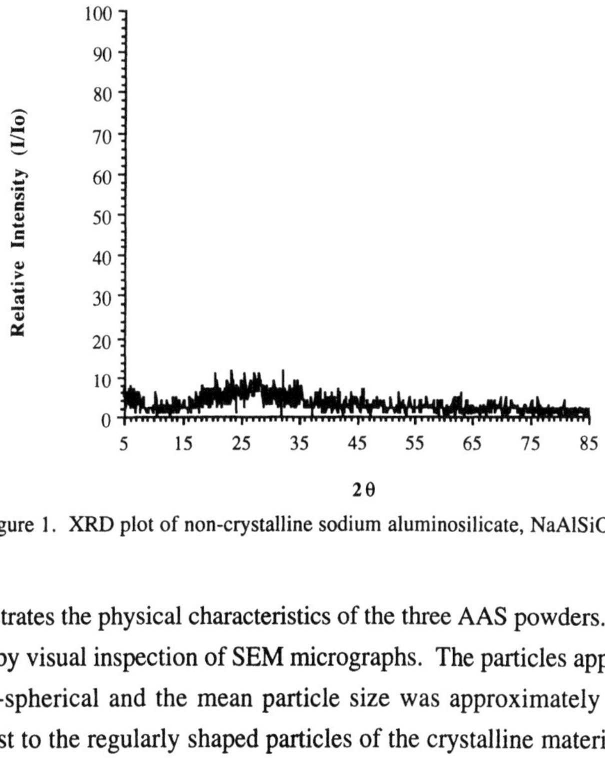


Figure 1. XRD plot of non-crystalline sodium aluminosilicate, $NaAlSiO_4$.

Table 1 illustrates the physical characteristics of the three AAS powders. The particle size was determined by visual inspection of SEM micrographs. The particles appeared irregularly shaped and non-spherical and the mean particle size was approximately 10 μm. This is clearly in contrast to the regularly shaped particles of the crystalline materials. The surface area values were taken from BET measurements and show a decreasing value as the ratio of aluminum to silicon increases from 0.5 to 2. A description of the surface area for non-crystalline aluminosilicates are not as well defined as the crystalline systems.

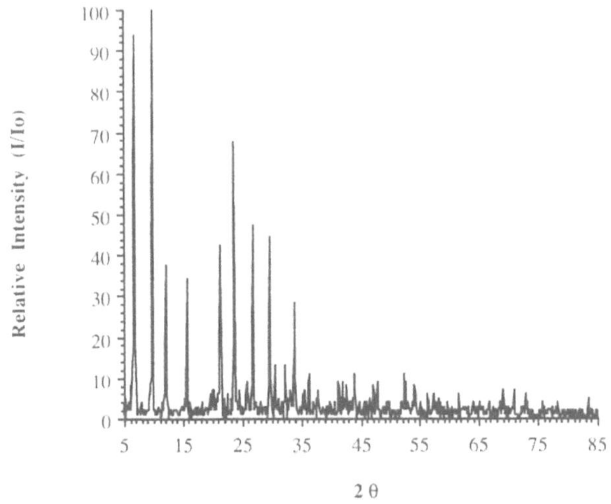


Figure 2. XRD plot of crystalline aluminosilicate, 4Å.

Table 1. Particle size range and surface area of three non-crystalline aluminosilicates.

| Sample | Particle Size | Surface Area |
|---|---|---|
| $Na_{0.5}Al_{0.5}SiO_3$ | 1-50 μm | 236 $m^2/g$ |
| $NaAlSiO_4$ | 1-50 μm | 166 $m^2/g$ |
| $Na_2Al_2SiO_6$ | 1-50 μm | 125 $m^2/g$ |

Figures 3 - 5 illustrate incremental and cumulative pore area distributions as a function of pore widths for the three non-crystalline samples. As the ratio of aluminum to silicon

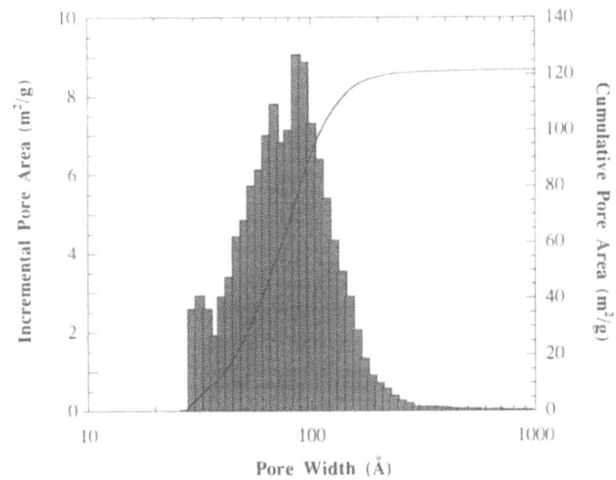


Figure 3. Pore area distribution of $Na_{0.5}Al_{0.5}SiO_3$.

increases from 0.5 to 2, the peak of the **Incremental Pore Area** distribution flattens and the development of small pores (~12 Å) occurs with the $NaAlSiO_4$ and the $Na_2Al_2SiO_6$ samples. The 12 Å pores constitute approximately 23% of the total pore area in the $NaAlSiO_4$ sample, and almost 25% in the $Na_2Al_2SiO_6$ sample. Another trend that occurs with the increase in the Al/Si ratio is the decrease in the surface area (Table 1). As the ratio increases from 0.5 to 2, the surface area decreases from 236 $m^2/g$ to 125 $m^2/g$. While the effects of surface area is not clear on the effects of ER, it is generally accepted that it is an important parameter to consider.

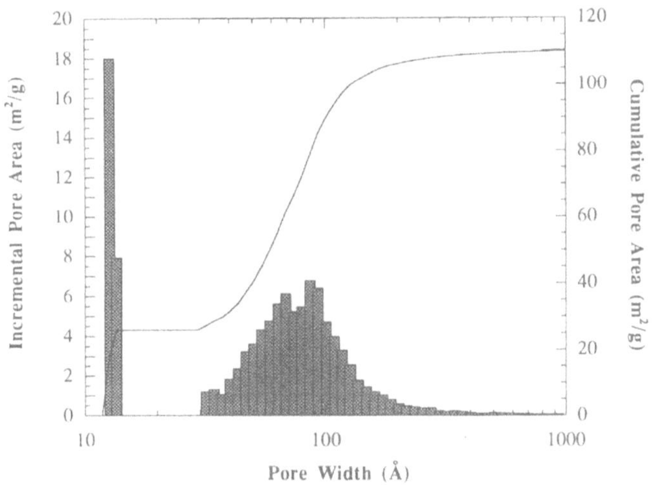


Figure 4. Pore area distribution of $NaAlSiO_4$.

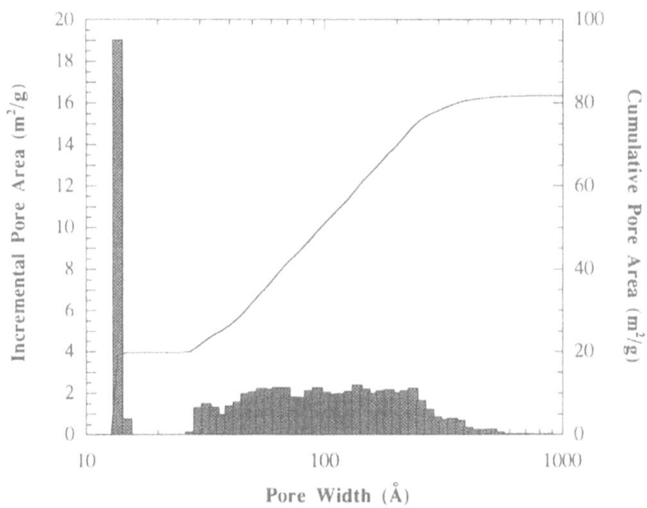


Figure 5. Pore area distribution of $Na_2Al_2SiO_6$.

## RESULTS

Figures 6-8 show electrorheological responses for the three samples: $Na_{0.5}Al_{0.5}SiO_3$, $NaAlSiO_4$, and $Na_2Al_2SiO_6$. For four separate shear rates, the shear stress is monitored as a function of electric field.

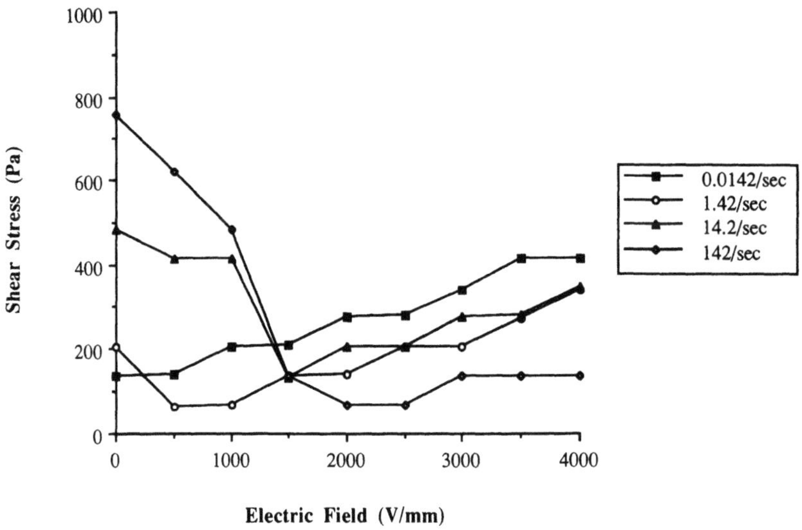


Figure 6. ER activity for $Na_{0.5}Al_{0.5}SiO_3$ at 4 RPM's.

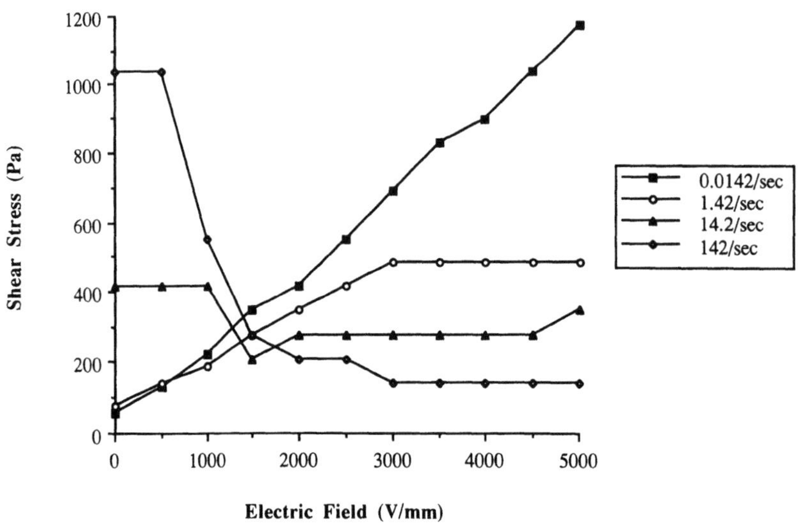


Figure 7. ER activity for $NaAlSiO_4$ at 4 RPM's.

It is evident that in each system, as the electric field increases up to 1500 V/mm, the high shear rate (>1.42 $sec^{-1}$) plots show a rapid decrease in the shear stress. At fields beyond

1500 V/mm, the responses appear constant. Conversely, at low shear rates (<1.42 $sec^{-1}$), the ER response increases with electric field at a constant rate. The mechanism responsible

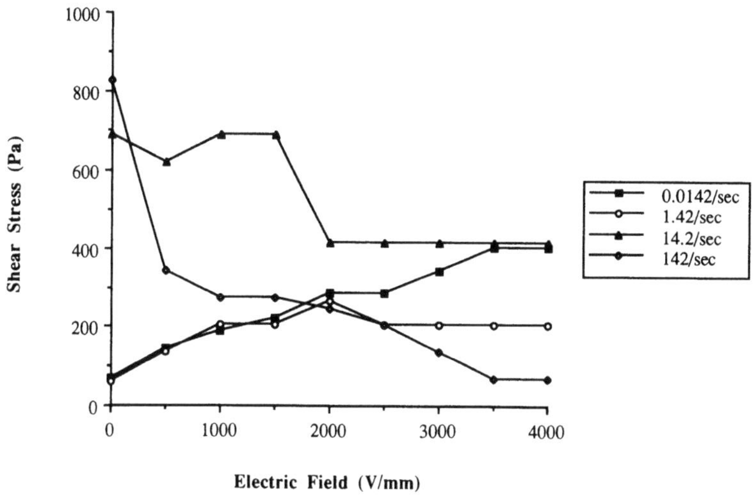


Figure 8. ER activity for $Na_2Al_2SiO_6$ at 4 RPM's.

for this transition of shear stresses with decreasing shear rate is unknown, but a considerable trend does occur. The most evident transition can be seen in Figure 6. At the lowest shear rate, the ER response is nearly linear. As the shear rate increases, the shear stress decreases to a minimum then is followed by a linear increase. At each increasing increment in shear rate, the minimum value shifts down with electric field. The shifting continues up to the maximum shear rate when the material no longer shows ER activity. For instance, at 1.42 $sec^{-1}$, the minimum occurs at approximately 1500 V/mm, while at 142 $sec^{-1}$, the minimum is shifted down to approximately 2250 V/mm. Figures 7 and 8 also suggest that a minimum occurs, but they are not as evident as the $Na_{0.5}Al_{0.5}SiO_3$ system.

To compare the ER response of the three samples, the shear stress values taken at the lowest shear rate appear most valid. At the shear rate of 0.0142 $sec^{-1}$, Figures 6 and 8 have very similar stresses with respect to electrical field, but the $NaAlSiO_4$ sample has a lower zero-field stress as well as a higher maximum shear stress. A comparison of the zero-field stresses is shown in Figure 9. All three samples appear to behave similarly as a function of shear rate.

At zero-field, Figure 9 illustrates that increases in shear rate result in increases in shear stress. In Figures 6 - 8, however, increases in shear rate coupled with increases in electric field result in decreases in shear stress. Figure 10 re-illustrates Figure 7 ($NaAlSiO_4$) for shear stress as a function of shear rate for electric fields of 0 up to 5000 V/mm. At approximately 1-10 $sec^{-1}$, the shear stress curves appear to converge for all electrical fields. To avoid redundancy, similar trends are seen in the $Na_{0.5}Al_{0.5}SiO_3$ and $Na_2Al_2SiO_6$ samples.

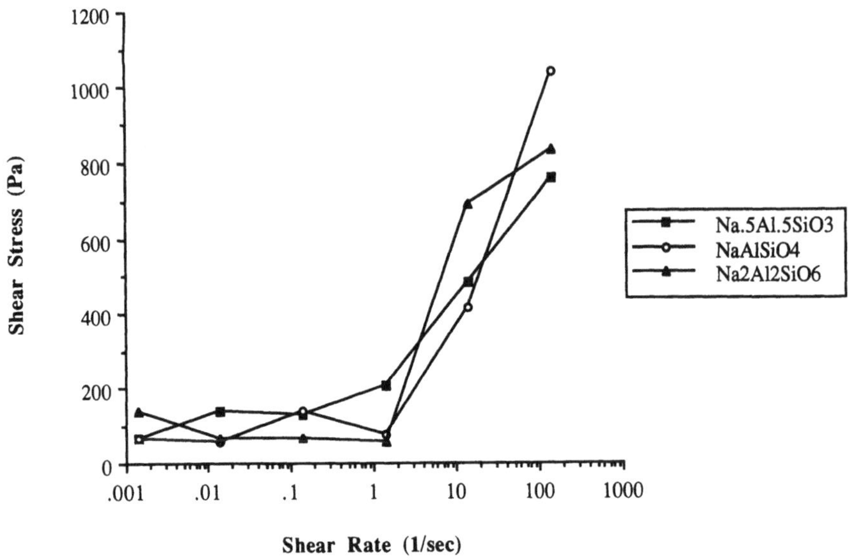


Figure 9. Rheological behavior for $Na_{0.5}Al_{0.5}SiO_3$, $NaAlSiO_4$, $Na_2Al_2SiO_6$ at zero electrical field.

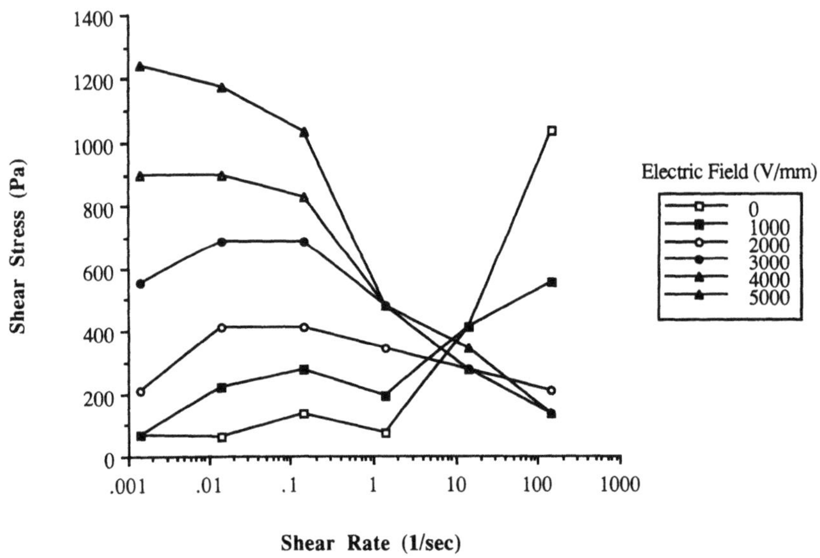


Figure 10. ER activity for $NaAlSiO_4$ for electric fields of 0 - 5000 V/mm.

A dielectric study was performed on all three samples to determine if anomalous dispersions were present. It has been suggested that the dispersions may be linked to ER activity. As shown in Figure 11 for the $NaAlSiO_4$ sample, the dissipation factor as a function of frequency is recorded at temperatures from 50 to 150°C. For each temperature curve, a characteristic frequency (indicated by the peak) could be found. Again, to avoid redundancy, similar dispersions over similar temperature ranges were present for the other amorphous samples.

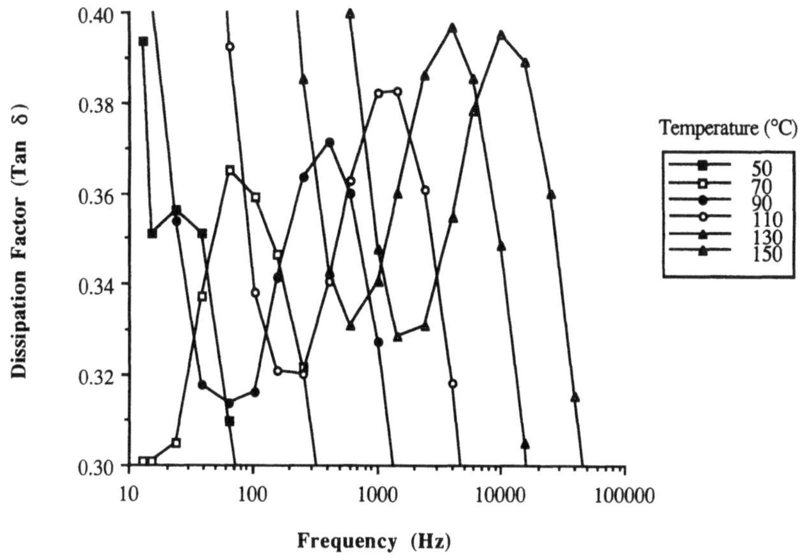


Figure 11. Dielectric study of $NaAlSiO_4$, showing Tan δ as a function of frequency.

From the peaks of the dispersions of Figure 11 and from the subsequent plots of the other samples, further plots were constructed to determine if the molecular mechanisms responsible for the dispersions followed an Arrhenius type relationship. Shown in Figure 12 are plots of each of the materials illustrating frequency as a function of 1/T.

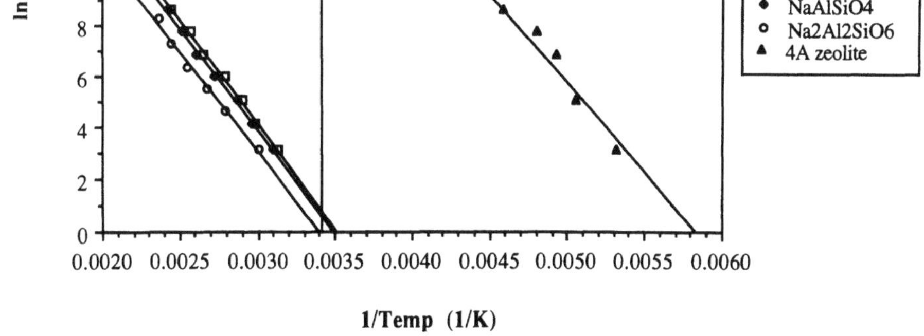


Figure 12. Activation energy plot comparing non-crystalline and crystalline alumino-silicates.

The straight lines indicate that the molecular mechanisms are well characterized by an Arrhenius approach. Because the materials follow an Arrhenius relationship, an activation energy can be approximated by multiplying the slope by Boltzman's constant. Table 2 illustrates the comparisons between the measured activation energies.

Table 2. Comparison of activation energies and Na+ concentration between samples.

| Sample | Activation Energy (kcal/mol) | $Na^+$/Tetrahedra concentration |
|---|---|---|
| $Na_{0.5}Al_{0.5}SiO_3$ | 16.22 | 0.33 |
| $NaAlSiO_4$ | 16.0 | 0.5 |
| $Na_2Al_2SiO_6$ | 15.6 | 0.67 |
| 4Å | 13.7 | - |

From Figure 12 and the data of Table 2, comparisons can be drawn between the samples. First, as the ratio of aluminum to silicon increases from 0.5:1 to 2:1, the curves shift to the left, indicating that the high aluminum ratio materials are activated by a higher temperature at any given frequency. Conversely at any given temperature, the highest ion concentration material has the lowest characteristic frequency. In comparison to the crystalline zeolite system, the non-crystalline materials, as a class, have a much lower characteristic frequency. While the reason for this is unknown, the characteristic frequency for the non-crystalline system at room temperature is approximately 1 $sec^{-1}$. The second observed trend is the decrease in activation energy as the ratio increases from 0.5 to 2. While the significance of the differences in activation energies are not established, the materials do exhibit a notable trend.

## DISCUSSION

The structure of a crystalline alumino-silicate can be characterized as consisting of a labyrinth of channels and cavities of well defined dimensions,[18] whereas the structure of AAS has not been and can not be that well characterized. Figure 1 illustrates quite clearly the absence of long range translational order, i.e. x-ray detectable crystallinity, in these materials. Furthermore, in crystalline materials the pore distribution is quite narrow and the pore size is of near crystallographic dimensions, while the vast majority of pores in the AAS are very large and cover a tremendous range of sizes (Figures 3 - 5). This suggests that the microstructure of the AAS may consist of microscopic glassy beads, randomly packed together with a substantial amount of the porosity existing as unfilled space between these beads. This would also explain the relatively high amount of surface porosity for the AAS as opposed to the characteristic 3% for the Zeolites.[19] From these types of measurements, the surface is distinguished from internal porosity by rate of desorption of the adsorbed species, in this case $N_2$. The cations are able to reside on the surfaces of these glassy beads as well as in the internal pores (pores within the beads) which were created by the irregular packing of the tetrahedra. It might be expected that the cations on the surface of the glassy beads would possibly create a substantial bulk current as well as be fairly mobile over the surface of the particles. In contrast, the internal cations would be expected to have considerably greater difficulty moving through a dense and irregular maize of tetrahedra. However, there is no significant increase of bulk conductivity observed with fluids of these materials, suggesting that the bulk of the cations are internal to the glassy beads or the glassy bead concept of the structure is in error. We are currently attempting to resolve this by looking for these beads with electron optical methods and small angle x-ray diffraction. Our view of the micro-structure seems the most plausible considering the size and distribution of the pore sizes.

The ER behavior of all three samples is presented in figures 6, 7, and 8. The obvious strange behavior of these fluids at higher shear rates, i.e. shear stress decreasing dramatically with increasing electric field may at first appear unusual but is a behavior which we have consistently observed. At first it was convenient to explain this (apparent negative ER effect) by proposing that fixed surface charges of the particles (Stern layer) caused migration of the particles toward an electrode when under an electric field. This would create a lubricating layer of oil (devoid of particles) adjacent to the opposite electrode. If this was true, then an effect opposite to that observed here would be expected, i.e. lower shear rates would encourage more pronounced particle migration. At the present time we have no definitive explanation for this behavior other than it appears consistently in many systems and shows a consistent systematic behavior with field and shear rate. We suspect it is related to a complex coupling of the shear rate and particle dynamics with the characteristic times of the molecular (relaxational) mechanism associated with the intrinsic ER activity of the various particles. This explanation is further supported by plotting the shear stress vs. ln shear rate at various fields as in figure 10. This shows that the field clearly has a significant part in this behavior, i.e. the behavior of the fluid at higher fields "dropping off" more dramatically then for the lower fields. The fact that the curves at different fields converge at about 10/sec suggests that this shear rate may correspond to a characteristic frequency of some relevant internal relaxational mechanism. More will be said about this shortly.

In terms of ER activity, a standard comparison between the three fluids is only reasonable at the very low shear rates, specifically 0.0142/sec. At these shear rates, the AAS with an Al/Si of 1, which corresponds to the intermediate cation concentration of 0.5 (Table 1), shows an increase of yield stress from about 75 - 1240 Pa. The powders with Al/Si of 0.5 and 2 both show similar but much smaller increases of from 125 - 400 Pa, which correspond to cation concentrations (Table 1) of 0.33 and 0.67 respectively. Clearly, there is no direct correlation between ER activity and cation concentration in the context of this study, suggesting that additional factors must be important.

For all 3 samples, very distinct anomalous dispersions are evident and the corresponding Arrhenius plots are shown in Figure 12. The most obvious feature of these is the tremendous difference in the characteristic frequencies of this mechanism between the crystalline and AAS. At room temperature, the AAS have characteristic frequencies in the range of 1 - 5 /sec which are very slow in contrast to the crystalline material which is in the range of $10^8$/sec. On the other hand the activation energies, obtained from the slopes, as shown in Table 2, show a definitive decrease with increasing cation concentration from

16.22 - 15.6 kcal/mol. These are still significantly higher than the activation energy for the crystalline material. As mentioned previously, the room temperature frequencies of this molecular mechanism in the AAS is in the vicinity of 1-5 Hz which corresponds with the crossover in Figure 10. Whether or not this behavior is meaningful remains to be confirmed but clearly the relatively sluggish frequencies of the mechanisms responsible for this dielectric dispersion in the AAS could easily be affected by shear rates in this range.

As mentioned, the most obvious difference between the AAS and the crystalline materials is the tremendous difference in characteristic frequencies. In a classical view, i.e. a simple harmonic oscillator, it is suggested that the corresponding apparent strengths associated with the cation bonding is much greater for the crystalline than the amorphous materials. It would be further suggested that cations trapped within the internal pores of the crystalline materials are more strongly held then the cations within the pores of the amorphous materials. This is consistent with observations that the pores in the AAS are substantially larger then for pores in the crystalline materials. It is not unlikely that the morphology of the pores, which are dramatically different for the two, is also of considerable importance in their ER activity.

## CONCLUSIONS

The ER activity of AAS in suspension do not correlate directly with the charge concentrations nor with the other particulate parameters normally associated with ER activity, i.e. surface area. The AAS do show anomalous dispersions with activation energies similar to that of the crystalline system. However, major factors which are unaccounted for in these studies are the characteristic frequencies of the relevant molecular mechanisms (cation mobility) which are tremendously different between the crystalline and amorphous alumino silicates, and the morphology of the pores.

## ACKNOWLEDGEMENTS

This work was supported by NSF grant #BCS-9201787.

## REFERENCES

1. Block, H. and J. P. Kelly, "Electrorheology", J. Phys. D: Appl. Phys. 21:1661-1677 (1988).

2. DOE report, "Electrorheological (ER) Fluids, A Research Need Assessment - Final Report", Contract No. DE-AC02-91ER30172 (1993).

3. Block, H., J. P. Kelly, A. Qin, and T. Watson, "Materials and mechanisms in electrorheology", Langmuir 6(1):6-14 (1990).

4. Filisko, F. E. and L. H. Radzilowski, "Intrinsic mechanism for activity of alumino-silicate based electrorheological fluids", J. Rheol. 34(4): 539-552 (1990).

5. Treasurer, U., L. H. Radzilowski and F.E. Filisko, "Polyelectrolytes as inclusions in water free electrorheological materials: chemical characteristics," J. Rheol. 35(4) (1991).

6. Yang,I-K and A. D. Shine, "Electrorheology of a nematic poly (n-hexyl isocyanate) solution", J. Rheol. 36(6),1079-1104 (1992).

7. Maxwell, J.C., Electricity and Magnetism, Vol.1. 452, Claredon Press, Oxford (1892).

8. VonHippel, A. R. Dielectric and Waves, pp. 228-234, John Wiley & Sons, Inc., New York (1954),.

9. Dukhin, S. S. and V. N. Shilov, "Dielectric phenomena and the double layer in disperse systems and polyelectrolyte", translated from Russian by D. Lederman, Keter Publishing House Jerusalem Ltd (1974).

10. Inoue, A., "Study of new ER fluid" in "Electrorheological Fluids: Proceedings of 2nd International Conference on ER fluids", eds. J. D. Carlson, A. F. Sprecher and H. Conrad (Technomic Publ., Lancaster), 176-183 (1990).

11. Deinega, Y. F. and G. V. Vinogradov, "Electric fields in the rheology of disperse systems," Rheol. Acta 23:636-651 (1984).

12. Kitahara, A., "Nonaqueous systems", in Electrical Phenomena at Interfaces: Fundamentals, Measurements and Applications, eds. A. Kitahara and A Watanabe, pp. 119-143, Marcel Dekker, Inc., New York (1984).

13. Schwan, H. P., Schwarz, G., Maczuk, J., and Pauly, H., “On the Low-Frequency Dielectric Dispersion of Colloidal Particles in Electrolyte Solution”, J. Phys. Chem., 66, 2626 (1962).

14. Fixman, M., “Thin double layer approximation for electrophoresis and dielectric response”, J. Chem. Phys., 78, 1483 (1983).

15. Schwarz, G., “A theory of the low-frequency dielectric dispersion of colloidal particles in electrolyte solution”, J. Phys. Chem., 66, 2636 (1962).

16. Filisko, F. E. and Gamota, D. R., "Electrorheological Materials: Mechanisms and Mechanical Properties", ASME: AMD, Vol. 153, FED, Vol. 141, 75-87 (1993).

17. Kansal, P. and Laine, R. M., “Pentacoordinate silicon complexes as precursors to silicate glasses and ceramics” Journal of the American Ceramic Society 77:875-82 (1994).

18. Breck, S.W. Zeolite Molecular Sieves, John Wiley & Sons, Inc., New York (1974),.

19. Barrier, R.M., Brit. Chem. Eng. 5:1 (1959).

# HIGH FIELD CONDUCTION OF LIQUIDS IN CONTACT WITH POLYMERIC MATERIAL WITH REFERENCE TO ELECTRORHEOLOGICAL FLUIDS

P. Atten, J.N. Foulc and H. Benqassmi

Laboratoire d'Electrostatique et de Matériaux Diélectriques
CNRS and Joseph Fourier University
BP 166 – 38042 Grenoble-Cedex 9 (France)

## ABSTRACT

The conduction model of electrorheological (ER) effect appears to account qualitatively for the dependence of yield stress on electric field intensity in some ER fluids under DC applied voltage. This model is based on the assumption of the electrical conductivity of the suspending liquid varying exponentially as a function of the square root of the electric field. The field dependence of the liquid conduction in conditions typical of those of working ER fluids is investigated experimentally by examining the conduction properties of a layer of mineral oil lying between sheets of a slightly conducting polymer covering the electrodes. The results presented show that the current in the liquid arises mainly not from the field enhanced dissociation of electrolytic species dissolved in the liquid bulk but from injections of ions occurring at the solid/liquid interfaces.

## 1 - INTRODUCTION

It has been suggested recently that conduction properties of the solid and liquid constituents of the electrorheological (ER) fluids play the major role in promoting the ER effect under D.C. and low frequency A.C. fields [1,2,3]. Indeed, in these conditions the electric field distribution is controlled by the conductivities of the different media and the ER effect can only occur in the case of the conductivity of the solid particles being greater than that of the suspending liquid [3]. A conduction model of the ER effect has been developed which gives expressions of the attraction force between two identical spheres depending on the electric field E and the ratio $\Gamma$ of conductivities of solid and liquid phases [4,5]. This model has been tested using i) a system metallic plate-sphere of slightly conducting polymer [4] and ii) two identical spheres of slightly conducting polymer [5,6]. Both experimental tests show that the model fairly well predicts the attraction force and its change from quadratic to quasi linear dependence on E at high field strengths. Moreover, the same change of variation law has been obtained for the static yield stress of an ER fluid [7].

*Progress in Electrorheology*, Edited by K.O. Havelka
and F.E. Filisko, Plenum Press, New York, 1995

In the conduction model, the quasi linear dependence of the interaction force between spheres on E arises from the field enhanced conductivity of the liquid [4,5,6]. The retained variation with E of the effective liquid conduction is plausible in the case of spheres of centimeter size. But it is questionable in the case of ER fluids where the distances between particles interfaces are on the order of micrometers because of the very limited amount of dissociable species in such very small liquid volumes. More generally we can remark that the proposed conduction model [4,5,6] which appears to be sound from the electrical point of view disregards the physico-chemical processes responsible for the charge transfer at the solid/liquid interfaces and should be considered as a first approximation. A knowledge of the electrochemical characteristics of the interfaces is a prerequisite for developing more realistic models of the attraction between particles.

We present here the results of an experimental investigation on a particular system making it possible to draw some conclusion concerning the basic conduction processes. The study is performed on a planar layer of liquid sandwiched between two sheets of slightly conducting polymer, both liquid and solid materials being identical to the ones used in the test experiment on attraction between spheres of centimeter size [4,5,6]. Before presenting the study it is worthwhile to recall the basic phenomena involved in the electrical conduction of the so-called "insulating" dielectric liquids.

## 2 - RECALL ON CONDUCTION IN LIQUIDS

The first basic fact to point out is that, except for prebreakdown phenomena occurring at very high fields (E > 300 kV/mm) in highly purified liquids, there are no free electrons in dielectric liquids. The current is associated only to ions (and maybe very small particles) which, under the action of the electric field E, migrate at a velocity $v_i = KE$ proportional to the field (K is the charge carrier mobility). The second basic fact is the existence of two different processes of creation of the ions which give rise to the current [8] : i) dissociation of neutral species in the bulk and ii) creation of ions of the polarity of the electrode at the metal/liquid interface, a process referred to in the following as "injection".

Generally the dissociation of neutral species is accompanied by recombination of ions of opposite signs and the conductivity of the liquid characterizes the thermodynamic equilibrium of both processes. If the rate of dissociation is not very small (as can be the case in strongly polar liquids), the current due to bulk conduction (and also the conductivity) will decrease rapidly due to the ionic purification. For very small dissociation rates (case of non-polar liquids), conversely, the current approximately keeps constant for hours or days.

In "insulating" liquids, under not too high electric fields, charge injection is generally of electrochemical nature and most often occurs at one electrode only, either anode or cathode [8]. In dielectric liquids of high enough resistivity, the current is non-ohmic and the injected current often strongly dominates over the residual conduction current [9]. We examine now the two processes of ions creation with some detail, mainly in the case of non polar or weakly polar liquids.

### 2-a Bulk conduction

The simplest model for describing pure bulk conduction is to assume univalent ions $A^+$ and $B^-$ to be created from dissociation in the bulk of a unique dissociable species (formally denoted by (AB)) and to recombine into (AB) [10] :

$$(AB) \underset{k_D}{\overset{k_R}{\rightleftarrows}} A^+ + B^-$$

$k_D$ and $k_R = 2K/\varepsilon_L$ ($\varepsilon_L$ : liquid permittivity) are the dissociation and recombination kinetic rate constants respectively (we show below that $k_D$ depends on the electric field). At thermodynamic equilibrium the charge density $n_o$ of both positive and negative ions is related to the concentration c of neutral species through :

$$k_D\, c = k_R\, n_o^2 \tag{1}$$

which leads to a $c^{1/2}$ dependence for the liquid conductivity $\sigma_L = 2Kn_o$ (in this section we make the approximation of anions and cations having the same mobility).

Applying an electric field perturbs this equilibrium. The analysis developed for conduction of gases [11] is easily transposed to the case of liquids. When the transit time $t_T = d^2/KV$ (V : applied voltage) required for the ions to migrate between two plane parallel electrodes, a distance d apart, is much greater than the relaxation time $\tau = \varepsilon_L/\sigma_L$ , the equilibrium is only slightly perturbed; there appear two heterocharge layers near the electrodes the typical thickness of which is $\lambda = KE\tau$ [11] ($\lambda$ is the distance drifted by the ions during the relaxation time $\tau$). In these conditions the current is ohmic :

$$j = \sigma_L E \tag{2}$$

j being the current density. When $t_T \ll \tau$ conversely, the system is fully out of equilibrium and the charge density of ions takes values lower or much lower than $n_o$. At steady state the recombination is negligible and the amount of electric charges reaching the electrodes per time unit is equal to the rate of production of ions in the bulk [11]. The saturation current density $j_s$ is then :

$$j_s = k_D c d \tag{3}$$

When the dissociation rate is constant, $j_s$ is independent of V and can be expressed in terms of the conductivity $\sigma_L$ :

$$j_s = \sigma_L^2 d/(2K\varepsilon_L) \tag{4}$$

The saturation field $E_s$ for which the ohmic current is equal to the saturation current is then :

$$E_s = \sigma_L d/(2K\varepsilon_L) \tag{5}$$

In non-polar or weakly polar liquids the electrostatic interaction between ions plays the prime role and the dissolved salts give rise to ions pairs ($A^+B^-$) which have an electrostatic interaction energy on the order of 1 eV [12,13]. This explains the extremely low rate of dissociation of these ions pairs into free ions. The electric field favors the dissociation of the ions pairs by slightly lowering the potential barrier and increasing the probability of escape (this is a Schottky like effect). Onsager [14] gave a treatment of the problem and showed that the field enhanced dissociation is such that :

$$k_D(E) = k_D(0)\, F(E/E_c) \tag{6}$$

where the function F is expressed in terms of a modified Bessel function :

$$F(x) = 2\, I_1(x^{1/2}) / x^{1/2} \tag{7}$$

and where the characteristic field $E_c$ is :

$$E_c = \pi\varepsilon U^2/e, \qquad U = kT/e \tag{8}$$

kT being the thermal energy and e the electron charge. For a non-polar liquid with $\varepsilon = 2.2\ \varepsilon_0$ we have at room temperature $E_c \cong 0.24$ kV/mm which implies a strong dependence of $k_D(E)$ on the field for E greater than about 10 kV/mm. Onsager's theory leads to very good prediction of conduction characteristics in non-polar and weakly polar liquids in the absence of injection from the electrodes [15,16,17,18]. Fig. 1 (taken from Denat's thesis [19]) gives a

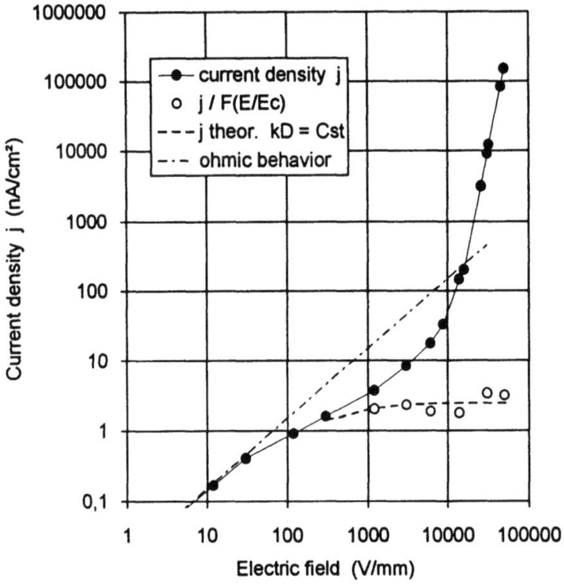


**Figure 1**. Current density as a function of electric field in cyclohexane (d = 0.35 mm) with added AOT at concentration c = $10^{-4}$ mole/liter. The solution lies between metallic electrodes under a hydrostatic pressure of 10 MPa. The lower curve is the current divided by the function $F(E/E_c)$ given by relation (7) and characterizing the enhancement of dissociation by the field (*after Denat* [19]).

typical current-voltage characteristic in a non-polar liquid into which an ionic surfactant has been added. There is a clear tendency toward saturation of the current at medium fields and then a steep increase for E > 10 kV/mm. Fig. 1 also shows that Onsager expression (7) fairly well accounts for the field enhanced dissociation.

## 2-b Injection phenomena

In strongly polar liquids the creation of charge carriers at the metal/liquid interface is of electrochemical nature and results from oxidation and reduction reactions [20]. This was established unambiguously in nitrobenzene where the creation of the nitrobenzene anion radical was detected and was shown to follow the same kinetics independently of the liquid resistivity [21]. In some cases the reactions concern the solvent itself but most often, even in carefully deionized polar liquids, the injections are promoted by minute traces of electroactive species (water for example). This makes in practice the injections to be unpredictable.

In non polar liquids, conversely, the injection process is somewhat better understood and often the injected currents follow pretty well defined laws. As for the bulk conduction, this can be ascribed to the dominance of electrostatic interactions. For a salt dissolved in a non-polar liquid, there are strong interactions with the metallic electrodes. Due to image force, ions (and also ions pairs) are "associated" with the metal and the electrostatic "association" energy of ions is on the order of 1eV. We therefore deal with a compact layer of ions of both polarities characterized by very high charge densities compared with the bulk ionic densities. An applied electric field tends to extract ions (of the relevant polarity) from this reservoir of charges, again through a Schottky like mechanism often referred to as a one-dimensional Onsager effect [22,23]. The theoretical treatment predicts an injected charge density $q_i$ varying as [22] :

$$q_i = q_o\, G(E/E_c^{'}) \tag{9}$$

where the function G is expressed in terms of the modified Hankel function :

$$G(x) = 1 / [x^{1/2}\, K_1(x^{1/2})] \tag{10}$$

and where the critical field $E_c^{'}$ is :

$$E_c^{'} = 4\pi\varepsilon U^2/e = 4\, E_c \tag{11}$$

The constant $q_o$ is related to the charge density in the reservoir. It depends on the nature of the liquid and on the ions mainly through their radius a (the association energy between an ion and the metal is $W_{i,el} = -(1/4\pi\varepsilon_L)\,(e^2/4a)$). In practice, the difference in the radii of positive and negative ions and the effect of specific adsorption result in the occurrence of a unipolar injection, i.e. of injection of one ionic species (positive or negative) much stronger than the other [9,24].

The law (9) very well accounts for the variations versus the applied field (up to field values about 20 kV/mm) of the injected charge densities obtained in non-polar liquids when adding quaternary ammonium salts giving ions of small size (for $E > 20$ kV/mm, the injection is stronger) [25]. Fig. 2 gives an example of the good prediction of the variation of $q_i$ with E and also shows that the current curve increases more than linearly and clearly differs from the one relative to bulk conduction alone (Fig. 1). Note that at high fields the current increase is very strong and similar to that corresponding to field enhanced dissociation of a salt at saturation (§ 2-a) because of similar exponential behavior as a function of the square root of the applied field. In practice it is therefore not possible to distinguish between the bulk and the electrode origin of the ions from the high field behavior alone.

In the experiments reported in [25] the charge density $q_o$ is found to vary proportionally to the bulk conductivity and the injected current usually dominates over that arising from dissociation in the bulk [9,25]. With big ions ($a > 12$ Å) the ion/electrode association energy is lower and corresponds to much lower $q_o$ values and presumably very low injection levels which are not detectable by electric measurements [26]. In this case the conduction phenomena are controlled by the dissociation/recombination processes.

The liquids purified by the usual techniques have a weak residual conductivity even in the absence of added species. This is presumably due to very small amounts of impurities which can hardly be identified. Under the action of the electric field the current often exhibits an increase which is steeper than with added salts [24]. This current most often arises from injections.

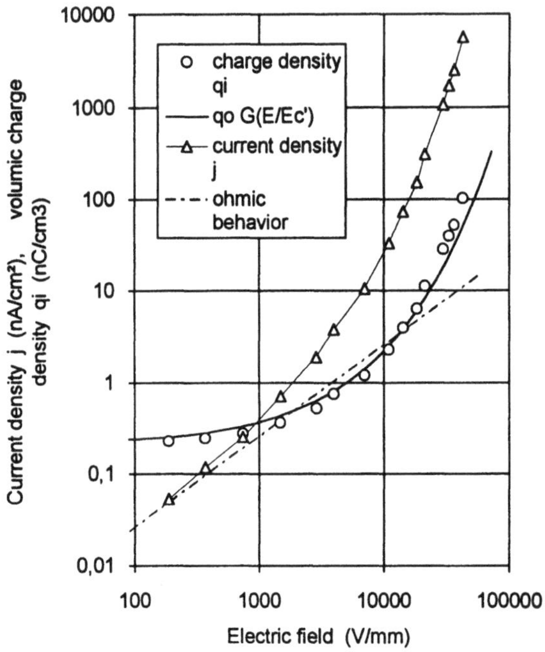


**Figure 2.** Current density as a function of electric field in cyclohexane (d = 0.28 mm) with added TiAP (triisoamylammonium picrate) at concentration c = 5 × $10^{-4}$ mole/liter. The solution lies between metallic electrodes under a hydrostatic pressure of 10 MPa. The full line curve corresponds to the function G(E/$E'_c$) given by relation (10) and characterizes the field dependence of the injection strength (*after Denat* [19]).

## 3 - EXPERIMENTAL SET-UP

In order to investigate the conduction properties of the liquid lying between polymer spheres (or more generally between particles), we designed a test cell (Fig. 3) where a planar layer of liquid (mineral oil) of thickness d lies between two circular disks of a slightly conducting polymer coating the electrodes ($\phi$ = 3 cm,). Teflon spacers allow us to work with different liquid thicknesses d. The cell is immersed in mineral oil in a teflon vessel in order to avoid discharges that could be induced by the applied D.C. high voltage.

The polymer disks of about 0.5 mm in thickness have been obtained by moulding in order to have good planar surfaces. The polymer used, polyamid Hüls 2124, is a polyamid 12 containing about 13% of plasticizer and some additives to promote its weak electrical conductivity. The current-voltage characteristics of the polyamid disks show only a slight departure from ohmic behavior and give $\sigma_S \approx 1.5 \times 10^{-9}$ S/m. As the polymer disks are much more conducting than the liquid, the major part of the overall voltage is applied to the liquid (for all presented results the drop of potential in the polymer disks was taken into account in determining the mean electric field in the liquid). In order to have the best possible electric contact between the electrode and the polymer, the polymer disks were moulded directly on the electrodes. One of these was made with a guard ring (see Fig. 3) in order to obtain the current density with a good confidence.

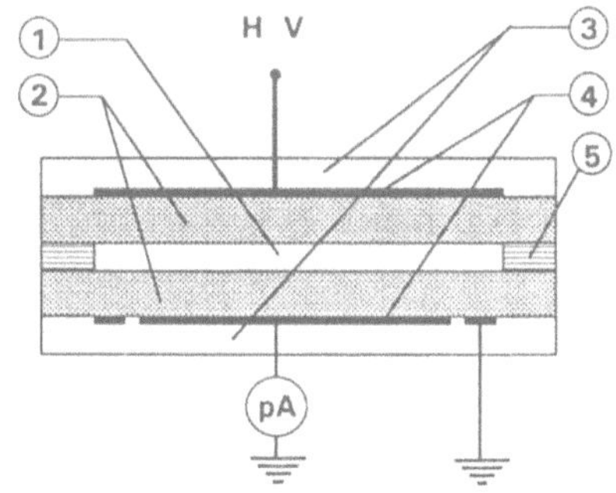


**Figure 3**. Schematic cross section of the test cell. (1) : liquid; (2) : disks of polyamid; (3) : insulating epoxy resin; (4) : electrodes (the lower one with a guard ring); (5) : spacer.

The tested liquid was the mineral oil Elf TF50 (from Elf-Atochem) with initial conductivity $\sigma_L \sim 10^{-12}$ S/m. In order to vary the liquid conductivity we used the petroleum antistatic additive aerosol OT (or AOT) : sodium di-2-ethylhexylsulfo-succinate which is a ionic surfactant [17]. We observed that the AOT added in mineral oil tends to move into the polymer disks whose conductivity could noticeably increase (after applying the field during one day the conductivity could become up to 3 times greater).

The I(V) characteristics relative to the liquid have been obtained by taking the quasi steady current values at times < 1 minute after change of applied voltage (we used a high voltage DC supply SPELLMAN 20 kV and a digital electrometer KEITHLEY model 616). Due to the complex structure of the system we did not examine the transient response to a voltage step. The conductivity of the oil (with the AOT when added) was measured before filling the cell using a high sensitivity resistivimeter (IRLAB model LDTRP-2 - see [27]).

## 4 - RESULTS AND DISCUSSION

The conduction properties of the mineral oil sandwiched between polyamid sheets have been investigated varying the liquid thickness (from 0.1 to 1 mm), using applied fields up to about 50 kV/mm, adding in some cases the ionic surfactant AOT at different concentrations and looking at the long term behavior. The reference field $E_c$ given by (8) takes for the oil we used the value $E_c \cong 0.24$ kV/mm. From the results shown below, it appears that the pure conduction processes associated with dissociable species dissolved in the liquid bulk cannot account for the observed liquid conduction. Injection processes play a role which is often dominant. We have to point out, however, that the amount of charge carriers injected at the polymer/liquid interface(s) is very different from the one characterizing the metal/liquid interfaces.

The difference in injection laws is clear when we compare the currents in the same liquid with different "electrodes". For mineral oil (d = 0.2 mm) lying between i) two tin electrodes, ii) one tin and one polymer coated electrodes and iii) two polymer electrodes, the current versus field curves have grossly the same shape (Fig. 4). Above the ohmic range (E less than a few V/mm) the current exhibits some sub-ohmic behavior, but the increase as a function of the field is more marked than expected from field enhanced dissociation effect

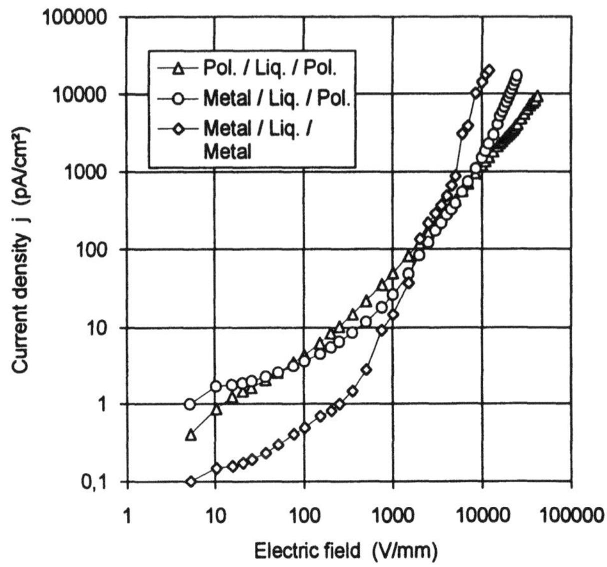


**Figure 4.** Current density-field characteristics in mineral oil (layer thickness d = 0.2 mm). 1) between two tin electrodes; 2) with one electrode coated with polyamid; 3) with both electrodes coated with polyamid.

alone [14-18], especially in the case of two tin electrodes. For fields above about 1 kV/mm there is a rather strong increase that we ascribe to an "injection" process.

At low field values the current with one or two polymer electrodes is higher than with two metallic electrodes (Fig. 4) and this arises either from the passage into the liquid of some electrolytic species present in the polymer or from an injection from the polymer/liquid interface or from both processes. At high enough field between metallic electrodes, the strong current increase is very likely due to injection(s) at the metal/liquid interface(s) as described in §2-b. Covering one electrode with polymer suppresses one strong injection (Fig. 4). For the liquid lying between two polymer coated electrodes, the field dependence of the current density is even weaker (Fig. 4). Taking for granted that the current is controlled by "injections" from the polymer, it is clear that the injection law is then much smoother than from a metallic electrode.

The existence of injection from the polymer/liquid interface is not at all obvious from the relevant curve of Fig. 4. We came to this conclusion through several consistent observations in the polymer/liquid/polymer arrangement. First it is interesting to work in conditions such that we would have a pure conduction between metallic electrodes. This can be achieved by adding the ionic surfactant AOT which is known to increase the liquid conductivity without promoting any detectable injection from the metallic electrodes [17]. The j(E) curve obtained just after filling the cell (surfactant concentration $c = 10^{-4}$ M/l) exhibits three parts, ohmic, saturation-like and of strong increase (Fig. 5). The full line curve of Fig. 5 is the theoretical prediction based on the measured liquid conductivity with the ohmic part up to the saturation field $E_S$ and the field enhanced saturation current [14,17,18] for $E > E_S$. The measured current is clearly greater than the predicted one, by a factor of

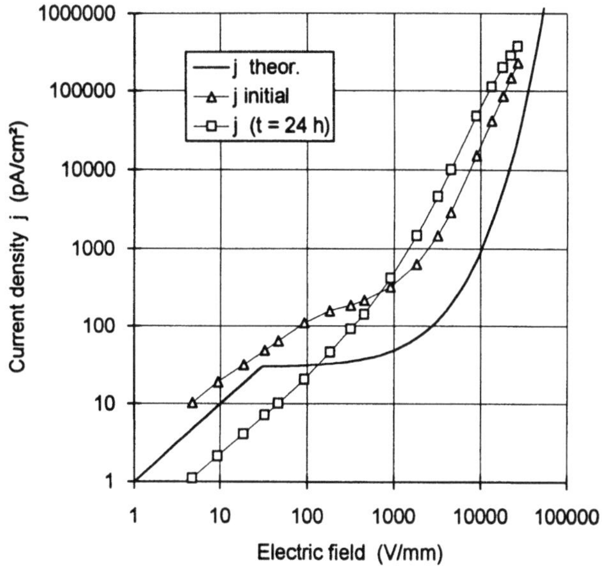


**Figure 5**. Current density-field characteristics in mineral oil (d = 0.2 mm) with added AOT of concentration c = $10^{-4}$ mole/liter (electrodes coated with polyamid sheets). 1) initial current ; 2) : after applying a voltage V = 1 kV during 24 hours and then waiting 40 hours without field. The full line curve is the theoretical pure conduction current calculated from the liquid conductivity measured independently.

about 2 in the ohmic part and between 10 and 20 in the high field part (E ≥ 1 kV/mm). The difference can hardly be accounted for by an assumed adjunction of a dissociable (electrically neutral) species coming from the polymer and it appears plausible to infer some injection of ions from the polymer into the liquid.

This hypothesis is strongly supported by the results obtained after subjecting the cell to a field of 5 kV/mm during 24 hours and then leaving it without field during 40 hours. The shape of the j(E) curve then obtained is reminiscent of the ones relative to dominant injections from metallic electrodes (see Fig. 2). The action of the field during 24 hours promotes an ionic purification of the liquid [8]. Assuming that the current associated with the drift of charge carriers generated by dissociation of AOT is 2 nA/cm$^2$ under the field E = 5 kV/mm leads to the result that the added AOT can be fully removed in one day. This explains why the low field current can be one order of magnitude lower than at the beginning of the experiment. The unexpected phenomenon, however, is that the high field current takes higher values (Fig. 5). This can only arise from injections which are enhanced.

Similar results are obtained when working with "pure" oil, i.e. oil without additive. Fig. 6 shows that the initial characteristic has the same shape than the corresponding curve of Fig. 5, the current values being about 10 times lower. After ionic purification (at a lower field : 1 kV/mm during 24 h), the j(E) curve has a regular shape without tendency towards a plateau but the high field current in this case is lower than the initial one.

One important test to discriminate between pure conduction or injection origin of the current is to examine the influence of liquid thickness d. In the ohmic regime the current

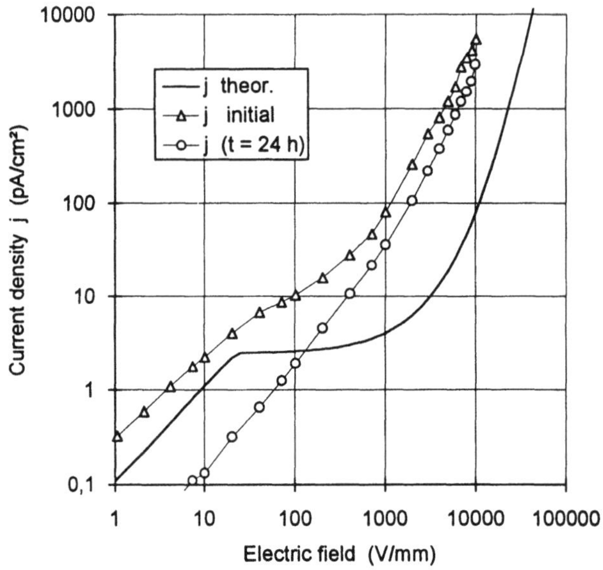


**Figure 6.** Current density-field characteristics in mineral oil (d = 1 mm) without additive (electrodes coated with polyamid sheets). 1) initial current ; 2) after applying a voltage V = 1 kV during 24 hours. The full line curve is the theoretical pure conduction current calculated from the liquid conductivity measured independently.

density is proportional to E and does not depend on d. In the saturation regime of pure conduction, the measured current should be proportional to d (see relation (3)). The injection current, conversely, should depend on E only and not on d (provided that the polymer and its interface have the same properties).

The test on the influence of liquid thickness d was performed on oil without additive. We see on Fig. 7 that, to a first approximation, j does not depend on the thickness for $E > 200$ V/mm (the slight dependence is inverse to what is expected for pure conduction) and this strongly suggests that the injection component of the current is the dominant one. At low field ($E < 100$ V/mm) the variations are qualitatively consistent with the assumption of the current having two components, a dominant one due to injection and a smaller one associated with the weak conduction in the bulk. After subjecting the cell to a field E = 1kV/mm during 24 hours, the results (Fig. 8) reasonably confirm that there is no noticeable influence of d on the current for $E > 1$ kV/mm (note the measurements done for d = 0.1 mm up to the field value $E \cong 75$ kV/mm).

The conclusion that arises from this study is that the passage of current in the liquid is mainly due to ions which are "injected" at the polymer/liquid interfaces. It is clear that the process of dissociation of electrolytic species initially present in the liquid cannot account for the current intensity and for its independence on the gap d. Moreover, application of the voltage for long times should eventually decrease the contribution of bulk conduction by eliminating these species. The injection process very likely is fully different from the one prevailing with metallic electrodes. The dependence of the injected current on the electric field is much smoother than that given by relation (9) and we can hardly imagine the

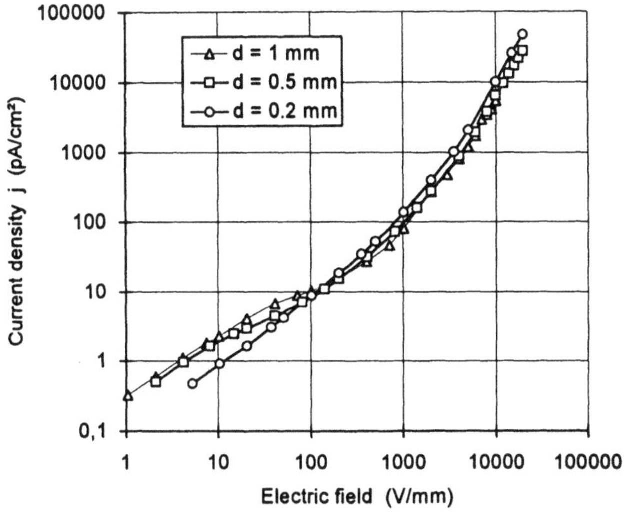


**Figure 7.** Current density-field characteristics in mineral oil without additive between electrodes coated with polyamid sheets. Initial current for various liquid thicknesses.

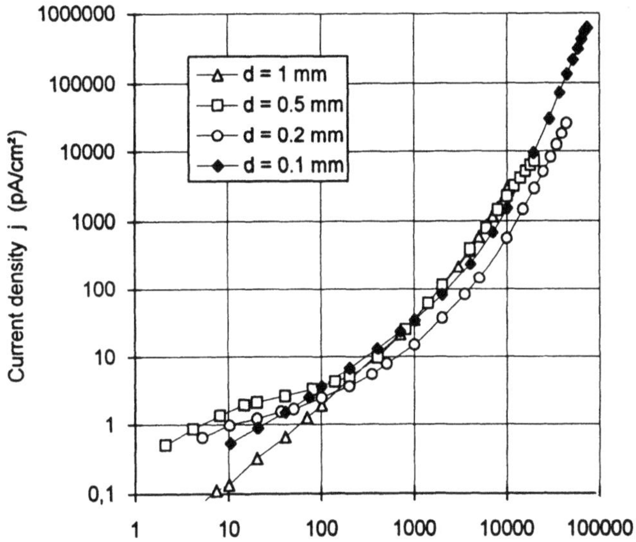


**Figure 8.** Current density-field characteristics in pure mineral oil between electrodes coated with polyamid sheets. Current for various liquid thicknesses measured after application of a field E = 5 kV/mm during 24 hours.

existence, at the polymer/liquid interface, of a compact layer with very high charge densities as conjectured on metals.

For the particular material tested here, a plasticized polymer, there are presumably exchanges of various species until the system polymer + liquid be in thermodynamic equilibrium. There are fluxes of both liquids (oil and plasticizer), of electrolytic species and even of ions across the polymer/liquid interface. In both media the charge carriers presumably are only ions and, under the action of the electric field, the ions can pass from one medium to the other as they do with permselective membranes. Applying the field during a long time can induce a ionic purification of the liquid, the ions entering the polymer sheets, which induces a gradual change of the nature and concentration of ions in the polymer very close to the interface. This, in turn, can modify the amount of injected ions, i.e. the ions coming out of the polymer and penetrating into the liquid layer. We guess that such a mechanism might explain the observation of an increase of injection strength at high enough field after the elimination from the liquid of the AOT initially added (Fig. 5).

## 5 - CONCLUDING REMARKS

The current through a layer of mineral oil lying between two electrodes coated with sheets of a slightly conducting polyamid exhibits a strongly non ohmic character. For liquid thicknesses ranging from 0.1 to 1 mm, the passage of current is not due to ions created by dissociation of electrolytic species in the liquid bulk. It is mainly determined by the amount of charge injected by the polymeric material into the liquid. As a consequence, the model of field enhanced conductivity of the liquid phase used in a previous work [4-6] does not provide a strictly correct description of the phenomena (even though the law retained for the variation of the current density as a function of the field qualitatively accounts for the injected current and therefore leads to plausible predictions for the attraction force between spheres).

In the case of ER fluids involving particles of small enough size ($\phi$ < 100 µm), the typical gap between particles in a fiber is in the range of micrometers and the corresponding liquid volume is so small that its bulk conduction plays but a negligible role. The passage of current from a particle to a neighboring one is determined by the injections at the interfaces. We then deal with a rather complex electrochemical problem and different cases must be distinguished. For polymers characterized by a weak ionic conduction as, for example, impregnated polymers with added salts (the case examined in this work) or polyelectrolytes, the injection process very likely depends on the ability of the suspending liquid to penetrate the polymer and of the ions present in the polymer to escape into the liquid. Using semi-conducting polymers for which the electric conduction is of electronic nature leads to a fully different problem, that of electrochemical reactions involving an electron transfer across the solid/liquid interface. Another category concerns the inorganic materials having a ionic conduction for which impregnation by the liquid is unlikely; the injection process at the interface might be due to the transfer of electrons from ions to neutral molecules or vice versa. Much work has to be done in order to identify the physico-chemical and electrochemical processes occurring at the interfaces and to determine the injection laws.

## ACKNOWLEDGMENTS

We wish to thank the management of Elf-Atochem for permission to publish this paper and the Centre National de la Recherche Scientifique (CNRS) for partially supporting the study developed in collaboration (contract CNRS/Société Elf-Aquitaine No 50.8016). We are indebted to A. Denat, B. Gosse, N. Felici and C. Boissy for fruitful discussions.

## REFERENCES


[1] R.A. Anderson, Proceed. 3rd Intern. Conf. Electrorheological Fluids, R. Tao (Ed.), *World Scientific*, Singapore, 1992, pp. 81-90.

[2] L.C. Davis, *J. Appl. Phys.*, **72**, p. 1334 (1992).

[3] J.-N. Foulc, N. Félici and P. Atten, *C. R. Acad. Sci. Paris*, **314**, Ser. II, pp. 1279-1283 (1992).

[4] J.-N. Foulc, P. Atten and N. Félici, *J. Electrostatics*, **33**, pp. 103-112 (1994).

[5] J.-N. Foulc, P. Atten and N. Félici, *C. R. Acad. Sci. Paris*, **317**, Ser. II, p. 5-11 (1993).

[6] P. Atten, J.-N. Foulc and N. Félici, *Int. J. Modern Physics B*, **8**, pp. 2731-2745 (1994).

[7] J.-N. Foulc and P. Atten, Proceed. 4th Intern. Conf. Electrorheological Fluids, R. Tao (Ed.), *World Scientific*, Singapore, 1994, pp 358-371.

[8] N. Felici, *Dir. Cur.*, **2**, pp 90-99 (1972).

[9] A. Denat, B. Gosse and J.-P. Gosse, *J. Electrostatics*, **7**, pp 205-225 (1979).

[10] G. Briere and F. Gaspard, *J. Chimie Physique*, **64**, pp. 1071-1084 (1967).

[11] J.J. Thomson and G.P. Thomson, "*Conduction of electricity through gases*", Cambridge Univ. Press, London, 1928.

[12] N. Bjerrum, Kgl. Danske Vid. Selskab, Math - fys medd., 7, p. 9 (1926).

[13] R.M. Fuoss, *J. Amer. Chem. Soc.*, **82**, pp. 1013 (1960).

[14] L. Onsager, *J. Chem. Phys.*, **2**, pp 599-615 (1934).

[15] A. Persoons, *J. Phys. Chem.*, **78**, pp. 1210 (1974).

[16] F. Nauwelaers, L. Hellemans and A. Persoons, *J. Phys. Chem.*, **80**, pp 767-775 (1976).

[17] A. Denat, B. Gosse and J.-P. Gosse, *J. Electrostatics*, **12**, pp 197-205 (1982).

[18] Z. Randriamalala, A. Denat, J.P. Gosse and B. Gosse, *IEEE Trans. Electr. Insul.*, **EI-20**, pp 167-176 (1985).

[19] A. Denat, *Doctoral Thesis*, Grenoble University, 1982.

[20] G. Briere, G. Cauquis, B. Gosse and D. Serve, *J. Chimie Phys.*, **66**, pp 44-53 (1969).

[21] B. Gosse, *Electroanal. Chem. & Interfacial Electrochem.*, **61**, pp 265-279 (1975).

[22] D.F. Blossey, *Phys. Rev. B*, **9**, pp. 5183-5187 (1974).

[23] .K.P. Charle and F. Willig, *Chem. Phys. Lett.*, **57**, pp 253-258 (1978).

[24] A. Alj, A. Denat, J.P. Gosse, B. Gosse and I. Nakamura, *IEEE Trans. Electr. Insul.*, **EI-20**, pp 221-231 (1985).

[25] A. Denat, B. Gosse and J.-P. Gosse, *J. Electrostatics*, **11**, pp 179-194 (1982).

[26] A. Alj, J.P. Gosse, B. Gosse, A. Denat and M. Nemamcha, *Revue Phys. Appl.*, **22**, pp 1043-1053 (1987).

[27] M. Hilaire, C. Marteau and R. Tobazeon, *IEEE Trans. Electr. Insul.*, **EI-23**, pp 779-787 (1988).

## STRUCTURE EVOLUTION OF FERROFLUID EMULSIONS

Jing Liu, Tawei Mou and Jackie Pousset

Department of Physics and Astronomy,
California State University, Long Beach, CA 90840

### INTRODUCTION

The evolution of field-induced structure in a colloidal system through dipole interactions has recently been studied in cells with little or no confinement. For example, one study [1] was carried out in an electrorheological (ER) fluids. A dipole interaction aligns the particles into chains, and different chains further coalesce to form columns. The coarsening is believed to be due to thermal fluctuations along chains. The theory [1] has predicted that for an infinitely long chains the coarsening due to thermal fluctuation should follow a power law of the form: $d \sim t^{5/9}$. The experimental result for a cell thickness of 0.7 mm shows that the chain coarsening varies in time in accordance with the power law $d \sim t^{0.4}$, where $d$ is a characteristic length of the structure. This experiment has considered the effect of cell thickness or width along the field direction to be essentially infinite due to image dipoles formed at the boundaries. Therefore, in that case the system is unconfined.

In a cell with strong confinement, the chain coarsening and growing process has not yet been studied either theoretically or experimentally. Presumably the thermal fluctuation induced wiggling motion of a short chain is much smaller than that of long chains. This is because of the strong repulsive forces among the ends of short chains which is screened among long chains, and also higher thermal energy required to excite shorter wavelength modes than longer ones. Therefore, the chain-chain coarsening kinetics are expected to be much slower, if it occurs at all, in the confined geometries.

### EXPERIMENTAL

In our experiments, monodisperse ferrofluid emulsions are used [2]. This is an oil-in-water emulsion in which kerosene droplets contain about ten thousands of small iron oxide grains. Each grain has a permanent magnetic dipole moment. The oil droplets have an average diameter of 0.50 $\mu$m with a standard deviation of less than 0.05 $\mu$m. One major feature of these droplets arises from their super paramagnetic behavior. In the absence of external magnetic field, these droplets have no permanent dipole moments, as the small magnetic grains within each droplet are randomly oriented due to their thermal motion.

*Progress in Electrorheology*, Edited by K.O. Havelka
and F.E. Filisko, Plenum Press, New York, 1995

However, once an external magnetic field is applied, the orientation of magnetic grains is slightly rotated towards the field direction, which results in a dipole moment in each droplet. The magnitude of the magnetic dipole moment increases with the strength of the applied field until saturation is reached. Consequently these droplets can interact through dipole forces which can be controlled by the magnitude of the external field [3].

The behavior of the ferrofluid emulsion is studied using optical microscopy and static light scattering technique. The sample is loaded in a flat thin cell having a thickness in the range of 10 to 300 $\mu m$. A homogeneous external field is set perpendicular to the cell surface and is produced by a coil to prevent the magnetic boundary from forming image dipoles. The applied magnetic field is varied from 0 to about 400 G with a rise time constant of 70 ms. A pair of car batteries is used to supply the current to the coil. The static light scattering measurement is performed using He-Ne laser of light wavelength 632.8 nm directed perpendicular to the surface of the cell. Scattered light is displayed on a screen behind the cell, collected by a CCD camera, and recorded by a VCR. The signal from the super VHS video tape is then sent one frame at a time to a Macintosh computer for processing.

## RESULTS AND DISCUSSION

Figure 1 shows the evolution of the scattered light intensity measured as a function of the scattering wave vector, $q$, over a period of two seconds after the magnetic field is applied. In Fig. 1 the scattered light intensity is radially averaged over constant $q$. The scattering wave vector $q$ is defined to be $4\pi \mathrm{Sin}(\theta/2)/\lambda_0$, where $\theta$ is the angle between the incident and scattered light inside the sample and $\lambda_0$ is the light wavelength in the sample.

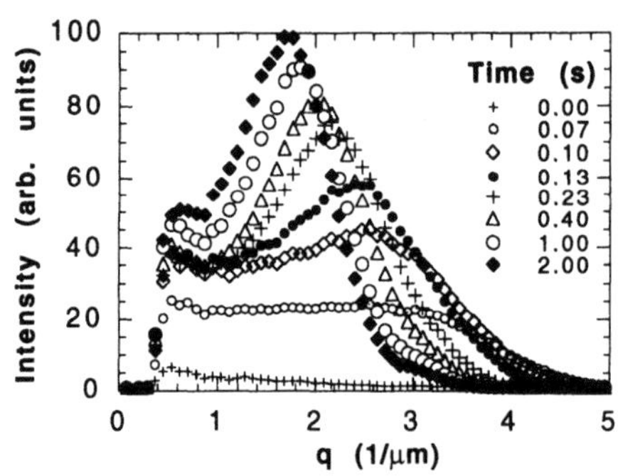


**Fig. 1.** Scattered light intensity is radially averaged over constant $q$ and plotted as a function of $q$, for various time after applying a magnetic field. The magnetic field $H$ is applied in a step function from 0 to 380G ($\lambda = 166$) with a rise time less than 70 ms; the volume fraction $\phi$ used is 0.03 and the cell thickness $L$ is 25 $\mu$m.

As shown in the Fig. 1, at $t = 0$ the scattering pattern is essentially flat, corresponding to a liquid phase with droplets dispersed randomly. At $t = 0.1$ s, a ring appears in the scattering pattern. The peak intensity of the ring increases with time, whereas the peak position, $q_{max}$, of the ring decreases with time. This indicates the occurrence of the phase transition from a random to a partially ordered "solid" structure. The new structure has a characteristic length scale, $d=2\pi/q_{max}$, which grows with time. After two seconds, no further change in the ring pattern is observed; the structure "freezes".

Figure 2 shows the growth of the solid structure as a function of time measured from the peak position, $q_{max}$, in scattering pattern of Fig. 1. These data are very different from results obtained by others in an unconfined system [1]. First, $q_{max}$ decreases sharply and quickly reaches the asymptotic value $q_s$ at $t \geq t_s$; here $t_s = 1$ s instead of the 10 min or more exhibited by the coarsening [1] of very long chains. This means that the phase transition from a randomly dispersed liquid phase to a partially ordered solid phase occurs within 1 s. Second, the formation of chains and the coarsening of chains to columns all occur within $t_s$. Therefore, from our data, the two processes cannot be distinguished. Finally, if we try to fit our results with a power law similar to those obtained in the system of reference 1, we find that the data in Fig. 2 can be fitted by a power law $q_{max} \sim t^{-0.16}$ for $0 \leq t \leq 2$ s, that is, in the region before complete saturation. The exponent which is smaller than that predicted and measured for ER fluid with very long chains, does reflect a slow phase transition process.

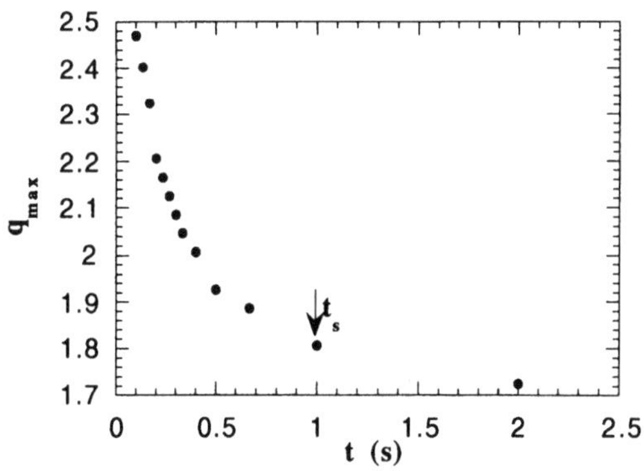


**Fig. 2.** The peak position, $q_{max}$, obtained from Fig. 1 as a function of time. The insert is the same plot in a log-log scale and solid line draw for guiding eyes only.

This is actually expected for a system with strong effect of repulsion between chains. The repulsive force occurs between the same poles from ends of chains or columns made of magnetic dipoles. In ER fluids, the chain lengths are effectively infinitely long due to the image dipoles formed at conducting boundaries. Thus, the effect of repulsion between ends of chains on the rest of chains are essentially zero. Whereas, in MR fluids without magnetic boundaries, the chain length is finite and limited by the cell thickness. The quick saturation of phase transition is probably due to the fact of only thin columns are formed in a thin cell, in contrast to the large columns formed in ER fluids where the chain coarsening supposes to lead to macroscopic phase separation.

The same measurement is performed for other volume fractions of the same cell thickness. We find that they all follow a similar power-law relation, $q_{max} \sim t^{-\alpha}$. Figure 3 shows the rate of coarsening, $\alpha$, as a function of volume fraction. $\alpha$ is essentially independent of the volume fraction for $0.015 \leq \phi \leq 0.13$. The saturation time $t_s$ is found to be independent of $\phi$ as well. The final value of $q_s$ varies but slightly with volume fractions.

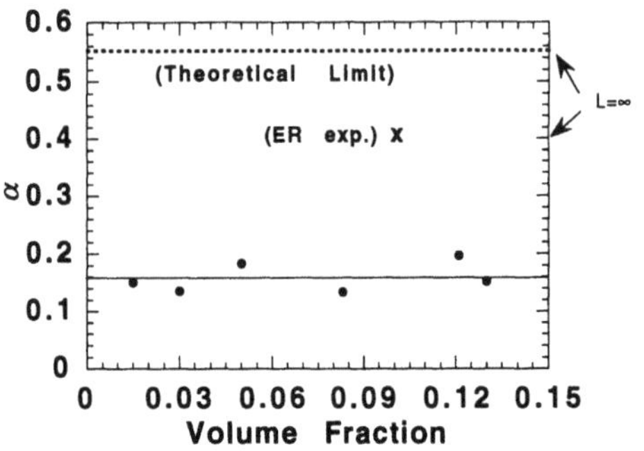


**Fig. 3.** Dependence of the exponent in the power law, $d \sim t^{\alpha}$, on droplet volume fraction. Here, the solid line is for guiding the eye only. The danshed line is the theoretical prediction for infinite long chains. x is the experimental result obtained for ER fluids.

We then varying the cell thickness to study the dependence of rate of coarsening and the saturation time on the cell confinement. Figure 4 shows the evolution of peak position as a function of time when the volume fraction is fixed at 3%. When the cell thickness varies from 10 to 300 $\mu$m, both the rate of coarsening before saturation and the saturation time increase with $L$. As seen from the plot on the log-log scale, a power-law relation, $d \sim t^{\alpha}$, can be approximated for all the data before saturation.

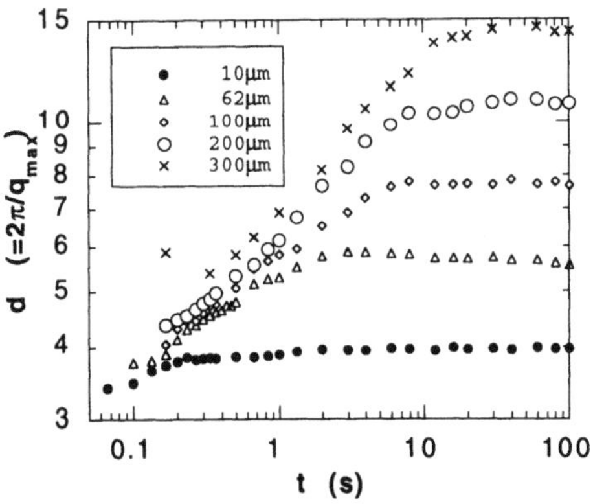


**Fig. 4.** Characteristic length scale, $d$, measured as a function of time at a constant volume fraction of 3% and various cell thicknesses.

Figure 5 shows the dependence of rate of coarsening on the cell thickness which is obtained from the exponent of the power law in Fig. 4. $\alpha$ increases with $L$ from 0.1 to 0.27 for $10 \leq L \leq 300$ $\mu$m. From fitting the data, a power law of $\alpha = 0.055\ L^{0.27}$ is obtained.

The experimental data show that $\alpha$ approaches asymptotically to the limit of infinite chain as one may expect when the confinement is reduced. It is interesting to note that in order to obtained $\alpha$ of 0.4, the measured value in ER fluids, we find that $L$ has to be 1500 $\mu$m which is about twice of the cell thickness used in ER fluid measurement. However, the behavior in chain coarsening in MR fluids may be quite different from that in ER fluids. Currently, we try to extend our measurements to larger cell thickness in order to compare with the ER fluid experiment and the existing theory. The saturation time is found to depend almost linearly on the cell thickness. The structure formation is complete or reaches saturation in a time of 0.2 to 12 seconds for $10 \leq L \leq 300$ $\mu$m after the application of the magnetic field. We do not have a quantitative explanation for these results yet.

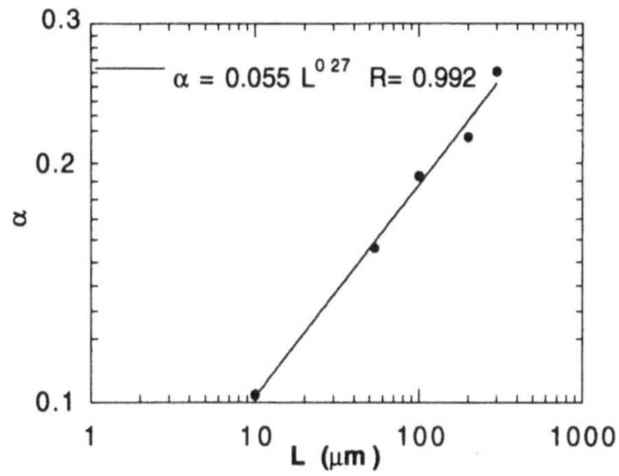


**Fig. 5.** The rate of coarsening vs. cell thickness obtained from Fig. 4.

## CONCLUSION

Our initial experiments demonstrate that the confinement of ferrofluid emulsions in the applied field direction strongly affects the rate of structure evolution. The stronger the confinement is, the slower the coarsening rate is and the quicker the saturation can be reached. The rate of coarsening depends on the confinement as a power law, $\alpha = 0.055\ L^{0.27}$. The saturation time is almost linearly proportional to $L$ in cells of thickness from 10 to 300 $\mu$m. The strong repulsive interactions between short chains or thin columns slow down the coarsening and result in thin columns. As confinement decreases, the rate and $t_s$ show a tendency to approach the result of infinite long chains predicted and obtained in ER fluid.


## ACKNOWLEDGMENTS

We are grateful to the technical supports provided by Prof. E.Y. Wong and Prof. L.J. Eliason, and to Dr. J. Bibette for supplying the sample. We wish to thank Prof. T. Halsey and Dr. Ron Rosensweig for helpful discussions and support. This research is partially supported by the NSF grant (DMR No. 9321201) and an award from Research Corporation. Acknowledgment is made to the Donors of The Petroleum Research Fund, administered by the American Chemical Society, for partial support of this research.

## REFERENCES


1. J. E. Martin, J. Odinek, T.C. Halsey, *Phys. Rev. Lett.* **69**, 1524 (1992). T.C. Halsey, *Science*, **258**,
   761, (1992).
2. J. Bibette, *J. of Magnetism and Magnetic Materials,* **122**, 37 (1993).
3. Jing Liu, Tawei Mou, and George A. Flores in *4th Proceedings of the International Conference on*
   *Electrorheological Fluids, Feldkirch, Austria, July, 1993*, edited by R. Tao (to be published by World Scientific, Singapore, 1994)

# THE DEPENDENCE OF SHEAR STRESS AND CURRENT DENSITY ON TEMPERATURE AND FIELD FOR MODEL ELECTRORHEOLOGICAL FLUIDS

Joseph W. Pialet and Denise R. Clark

The Lubrizol Corporation
29400 Lakeland Blvd.
Wickliffe, Ohio 44092

## INTRODUCTION

Electrorheological (ER) fluids have the potential of greatly improving the electronic control of mechanical devices. A lack of theoretical understanding of the ER effect has hindered commercialization of these materials. There has been considerable discussion in the literature on the nature of the ER response[1,2,3,4]. Performance data have been fit to simple equations by many researchers to gain insight into mechanisms and possible relationships between the variables. This can be particularly difficult since several mechanisms are probably contributing to the observed performance and the dominant mechanism probably varies with field strength, field frequency, temperature and shear rate. Most published work tends to deal with a single relationship for a single fluid. Because each author uses a unique combination of fluid, test device and operating conditions, it is difficult to combine data to get a complete picture. Examining several relationships for the same system under the same conditions may allow identification of additional underlying relationships.

In this paper, two model ER fluids were tested and the experimental data were fit to power law and Arrhenius equations. The relationships examined were:

1. Shear stress versus temperature at several electric fields.
2. Current density versus temperature at several electric fields.
3. Shear stress versus electric field at several temperatures.
4. Current density versus electric field at several temperatures.
5. Shear stress versus current density at several temperatures.

## EXPERIMENTAL

Data were collected on an electrically modified, horizontal Couette rheometer (gap 1.25 mm). The electric field was varied from 0 to 6 kV/mm (dc) and the temperature was varied from -20° to 120° Celsius. Data were collected on a Macintosh IIx using Labview 2.0 software at both constant field while varying shear rate (20 to $1000 s^{-1}$) and constant shear rate while varying field. The shear stress increase (shear stress under field minus shear stress at zero field) was measured in kilopascals and the current density was measured in milliamps per square meter. Data at approximately $500\ s^{-1}$ were used for the temperature and field dependence work. The blends were allowed to redisperse under zero field between each measurement under field. A Gamma High Voltage Research 15 kV/10 mA power supply was used to provide the field and the current was measured with a Keithley 196 System DMM. Stresses were measured with a GSE Inc. torque transducer and the temperature was controlled with a Brinkman RCS Lauda circulator. Blends were prepared by ball milling a mixture of the components for 24 hours. Two model systems, cellulose, representing water activated hydrophilic fluids and polyaniline, representing anhydrous semiconductor fluids, were used in these experiments.

A) Chromatography grade cellulose (Whatman ĊF-11), 15 wt%, plus a hydroxypropyl functionalized polydimethylsiloxane (Genesee Polymers EXP-69), 2 wt%, dispersed in a 10 cSt silicone oil (Dow Corning 200 Fluid). The water content of the cellulose was adjusted to 2.1 wt% (Karl Fischer) before blending.

B) Base washed, vacuum dried polyaniline[5], 20 wt%, plus a hydroxypropyl functionalized polydimethylsiloxane (Genesee Polymers EXP-69), 3 wt%, dispersed in a 10 cSt silicone oil (Dow Corning 200 Fluid).

## RESULTS AND DISCUSSION

### Shear stress versus temperature

Researchers have generally reported that, over a moderate temperature range, shear stress for most ER fluids increases as temperature (T) increases.[3] This would imply an increase in the degree of polarization of the particle. It is also observed in many cases that the shear stress will reach a maximum at some temperature and then decrease.[3] This may be a result of an inability to maintain particle polarization as the conductivity of the continuous phase increases. A simple equation that was used to fit the data over a temperature range is:

$$\text{shear stress} = ae^{b/T} \qquad a,b = \text{constants} \qquad (1)$$

In Figures 1A and 1B, ln shear stress increase is plotted versus 1/temperature (K) for three electric fields. If one tries to fit all of the data at each field to straight lines, one gets very different slopes and poor statistical fits. The data, as plotted, have an inflection around 293 K (20° C) for both fluids, so only the data at 20°C and above were fit to Equation 1 (Tables 1A and 1B). The inflections may be the result of multiple stress mechanisms with different activation energies. The data for the cellulose fluid give a good fit to the equation. There is an increase in slope (b) with field. The activation energy ($E_a$) was estimated by setting the slope (b) equal to $-E_a/R$, where R is the gas constant. The slopes obtained for the cellulose fluid correspond to activation energies of 3.2 to 4.4 kcal/mole. The fits for the polyaniline fluid are not as good because, although the inflection at 20°C seems to be present, the stress plateaus at higher temperatures. This second inflection results in a poor fit to the equation. As in the cellulose fluid, the slope for the polyaniline fluid increases with field.

**Table 1A**. Constants from ln shear stress increase versus 1/temperature at several fields for the cellulose fluid.

| FIELD kV/mm | b | ln a | adjusted $R^2$ |
|---|---|---|---|
| 3.0 | -1252 | 3.18 | 0.89 |
| 4.5 | -1685 | 5.24 | 0.96 |
| 6.0 | -2359 | 7.83 | 0.99 |

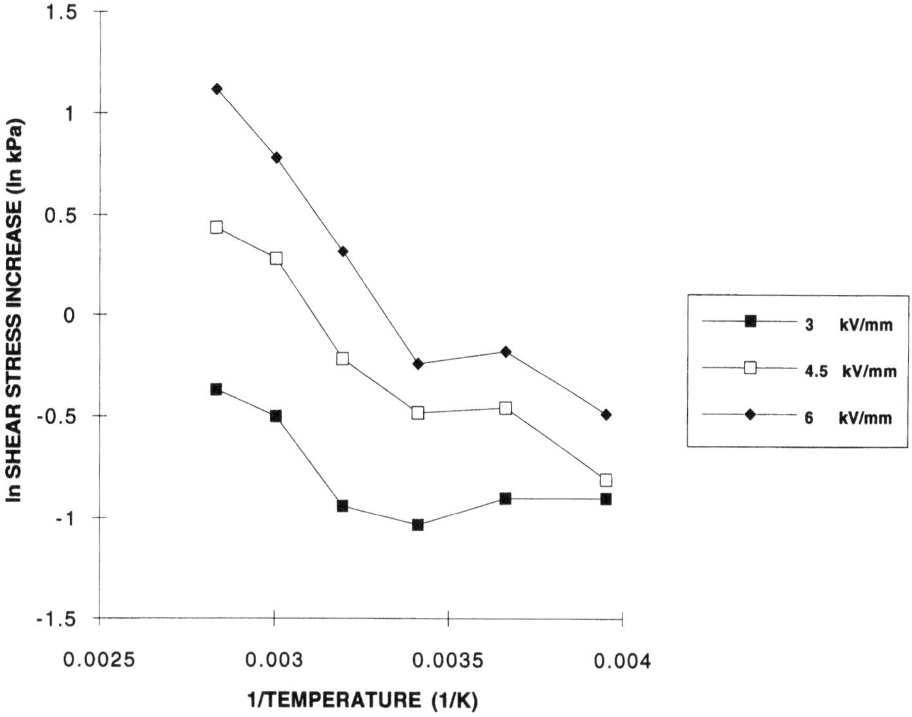


**Figure 1A**. Ln shear stress increase versus 1/temperature (253 to 353 K, -20 to 80°C) at several fields for the cellulose fluid.

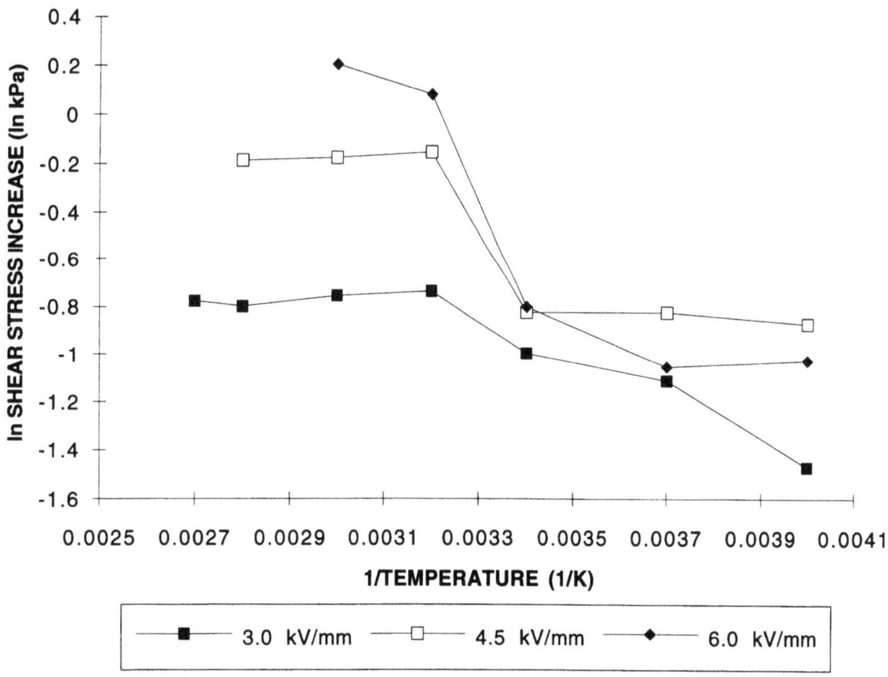


**Figure 1B**. Ln shear stress increase versus 1/temperature (253 to 373 K, -20 to 100°C) at several fields for the polyaniline fluid.

**Table 1B**. Constants from ln shear stress increase versus 1/temperature at several fields for the polyaniline fluid.

| Field kV/mm | b | ln a | adjusted $R^2$ |
|---|---|---|---|
| 3.0 | -215 | -0.16 | 0.13 |
| 4.5 | -940 | 2.58 | 0.33 |
| 6.0 | -2488 | 7.80 | 0.73 |

## Current density versus temperature

A major limitation to the commercial use of ER fluids in many applications is the rapid increase in the current density as the operating temperature increases. In contrast to the shear stress, the current density does not reach a maximum as temperature increases. Possible explanations for this increase in current density with temperature, which may apply to the cellulose system, polyaniline system or both, include the increased solubility of the polar activator (water) in the continuous phase, increased solubility of trace ionic materials, increased number of electrons in the conductance bands of semiconductor particles and increased ionic conductivity expected for the increase in ionic mobility due to a decrease in base fluid viscosity (Walden's rule). The current density versus temperature was fit to an equation similar to that used for the shear stress.

$$\text{current density} = ce^{d/T} \qquad c,d = \text{constants} \tag{2}$$

In Figures 2A and 2B, ln current density is plotted versus 1/temperature (K) for three electric fields. The cellulose fluid was tested over the range of -20 to 80°C and the polyaniline fluid was tested over the range of -20 to 100°C. In both fluids there are excellent fits to Equation 2 over the entire temperature range with an adjusted $R^2$ of 0.99 or greater in all cases (Tables 2A and 2B). In both fluids, the slope is independent of electric field. The rate of current density increase with temperature is much higher for the cellulose fluid. Setting the slope (d) equal to $-E_a/R$ yields an average activation energy of 16.1 kcal/mole for the cellulose fluid and 9.4 kcal/mole for the polyaniline fluid.

**Table 2A**. Constants from ln current density versus 1/temperature at several fields for the cellulose fluid.

| FIELD kV/mm | d | ln c | adjusted $R^2$ |
|---|---|---|---|
| 3.0 | -7744 | 25.73 | 0.99 |
| 4.5 | -8401 | 28.72 | 0.99 |
| 6.0 | -8226 | 28.92 | 0.99 |

**Table 2B**. Constants from ln current density versus 1/temperature at several fields for the polyaniline fluid.

| FIELD kV/mm | d | ln c | adjusted $R^2$ |
|---|---|---|---|
| 3.0 | -4861 | 18.42 | 0.99 |
| 4.5 | -4694 | 18.71 | 0.99 |
| 6.0 | -4605 | 19.03 | 0.99 |

## Shear stress versus electric field

In ER active systems, shear stress is usually reported to increase as the electric field (E) is increased. This increase is usually modeled as

$$\text{shear stress} = gE^m \qquad g, m = \text{constants} \tag{3}$$

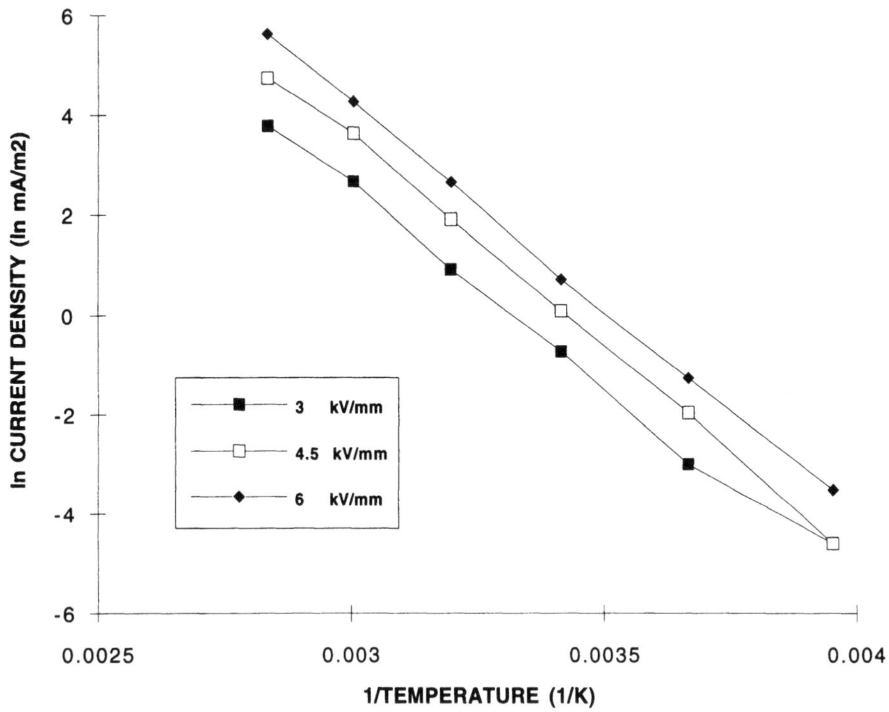


**Figure 2A**. Ln current density versus 1/temperature (253 to 353 K, -20 to 80°C) at several fields for the cellulose fluid.

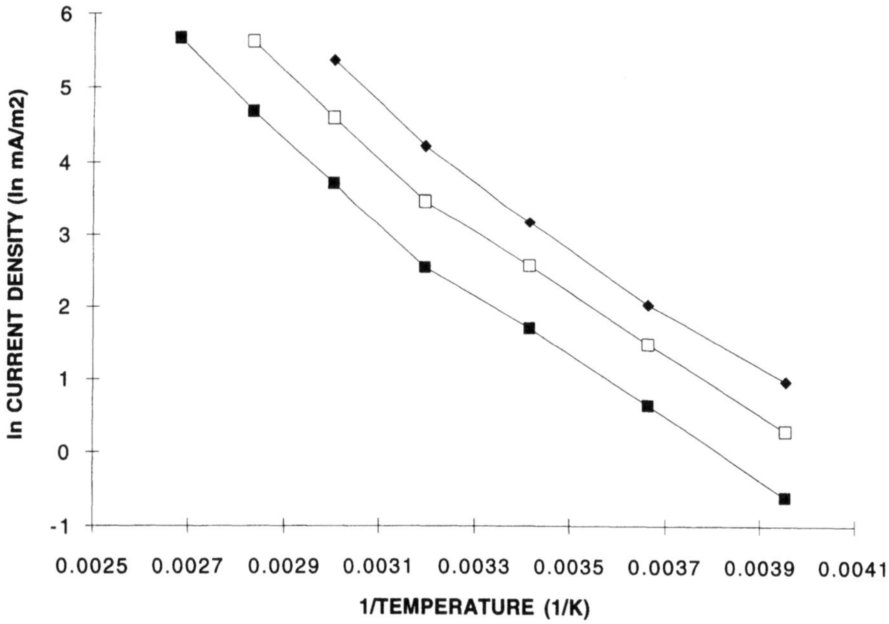


**Figure 2B**. Ln current density versus 1/temperature (253 to 373 K, -20 to 100°C) at several fields for the polyaniline fluid.

In the standard polarization model, application of an electric field results in an induced dipole on the particle which increases with the applied field.[1] The interaction of the induced dipoles with each other and the external field results in an increase in shear stress. From this polarization model, m would be expected to equal 2 for some range of E, with an expected decrease at high fields due to dipole saturation. From the conductivity model, m would be expected to be equal to approximately 1 for large differences in particle and base fluid conductivities.[6] A recent theory, taking into account the dielectric breakdown limit between particles, predicts an m of 1.5.[7] Experimental values for m in the literature range from 1 to 3 with values near 2 being the most common.[3,8,9,10]

In Figures 3A and 3B, ln shear stress increase is plotted versus ln field at several temperatures. At very low fields, there is almost no increase in shear stress with field. There is a large deviation from linearity for combinations of low temperature and low field. This is most obvious for the -20°C data. The shape of the -20°C cellulose curve is what would be expected if some minimum field, $E_0$, was required before a shear stress increase could be detected.

At most temperatures, shear stress increases could easily be measured above 1 kV/mm, so the data above 1 kV/mm were fit to Equation 3 (Tables 3A and 3B). Excluding the -20°C results for the polyaniline fluid, all of the slopes (m) are significantly less than 2. Zukoski has recently published a value of m for polyaniline which is consistent with the values reported here.[11] The -20°C cellulose curve was the only curve which had an inflection point significantly above 1 kV/mm. If the -20°C data above approximately 2.5 kV/mm are fit to the equation, the slope (m) increases to 2.5 which is much closer to the slope seen with the -20°C polyaniline. Only the 0°C polyaniline data had an adjusted $R^2$ less than 0.98.

**Table 3A**. Constants from ln shear stress increase versus ln field at several temperatures for the cellulose fluid.

| Temperature °C | m | ln g | adjusted $R^2$ |
|---|---|---|---|
| -20 | 1.33 | -3.23 | 0.98 |
| 20 | 1.47 | -2.75 | 0.99 |
| 40 | 1.54 | -2.36 | 0.99 |
| 60 | 1.43 | -1.81 | 0.98 |
| 80 | 1.36 | -1.48 | 0.99 |

**Table 3B**. Constants from ln shear stress increase versus ln field at several temperatures for the polyaniline fluid.

| Temperature °C | m | ln g | adjusted $R^2$ |
|---|---|---|---|
| -20 | 3.1 | -4.97 | 0.98 |
| 0 | 1.23 | -2.73 | 0.80 |
| 20 | 1.37 | -2.37 | 0.99 |
| 40 | 1.32 | -2.13 | 0.99 |
| 60 | 1.43 | -2.34 | 0.99 |
| 80 | 1.45 | -2.41 | 0.99 |
| 100 | 1.40 | -2.35 | 0.99 |

## Current density versus electric field

Current density increases as the electric field is increased. In contrast to the availability of several theoretical models for ER stress, there has been little published on the theoretical explanation for the currents seen in ER fluids or their relationship to ER stress. The equation used here for fitting the data is of the same form as that used with the shear stress.

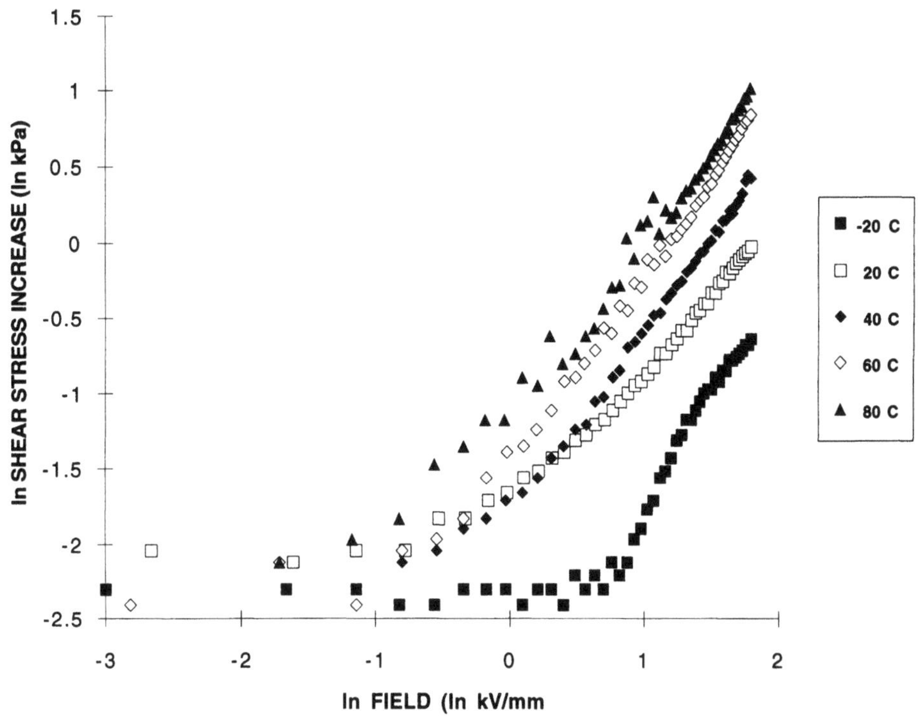


**Figure 3A**. Ln shear stress increase versus electric field at several temperatures for the cellulose fluid.

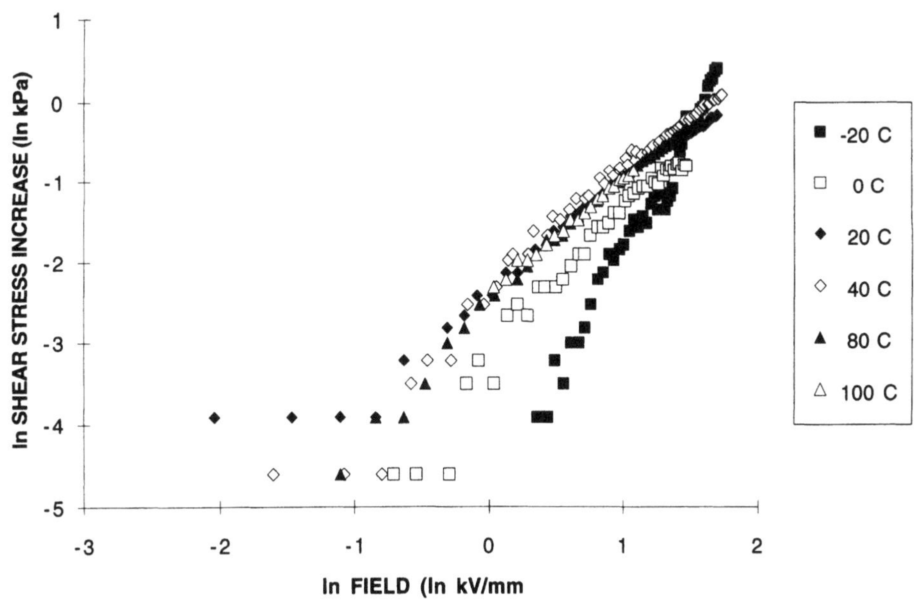


**Figure 3B**. Ln shear stress increase versus electric field at several temperatures for the polyaniline fluid.

$$\text{current density} = hE^n \quad h, n = \text{constants} \tag{4}$$

Values for n based on experimental data in the literature range from 0 to 5.[8,12,13,14] In Figures 4A and 4B, ln current density is plotted versus ln field at several temperatures. At combinations of low temperatures and low fields, there is a deviation from linearity. This deviation is more evident in the polyaniline fluid. All of the data are approximately linear above 1 kV/mm. Fitting the data with fields greater than 1 kV/mm to Equation 4 give slopes (n) with adjusted $R^2$ of 0.99 or greater in all cases (Tables 4A and 4B). Because of the high field needed to obtain a significant current density at -20°C for the cellulose fluid, a slope at that temperature was not determined. The field dependence (n) of current density for the polyaniline fluid was approximately 2. The field dependence of the cellulose fluid was slightly lower. In both fluids, the field dependence of the current density was larger than the field dependence of the shear stress increase.

**Table 4A**. Constants from ln current density versus ln field at several temperatures for the cellulose fluid.

| Temperature °C | n | ln h | adjusted $R^2$ |
|---|---|---|---|
| 20 | 1.73 | -2.65 | 0.99 |
| 40 | 1.89 | -0.91 | 0.99 |
| 60 | 1.75 | -0.98 | 0.99 |
| 80 | 1.65 | -2.41 | 0.99 |

**Table 4B**. Constants from ln current density versus ln field at several temperatures for the polyaniline fluid.

| Temperature °C | n | ln h | adjusted $R^2$ |
|---|---|---|---|
| -20 | 2.07 | -2.73 | 0.99 |
| 0 | 2.14 | -1.67 | 0.99 |
| 20 | 1.96 | -0.26 | 0.99 |
| 40 | 2.06 | 0.65 | 0.99 |
| 60 | 1.93 | 1.97 | 0.99 |
| 80 | 1.85 | 2.93 | 0.99 |
| 100 | 1.75 | 3.79 | 0.99 |
| 120 | 1.71 | 4.45 | 0.99 |

**Shear stress versus current density**

If both shear stress and current density are well described by similar models with similar exponents for E, then one would expect the relationship of shear stress to current density to be linear as field is increased. If the dependence of shear stress and current density on field have significantly different exponentials, then significant curvature would be expected. In Figures 5A and 5B, shear stress increase over the entire range of electric fields is plotted versus current density at several temperatures. In several cases, the data points at combinations of low shear stress increase and low current density are hidden by other data. Several of the lines show some curvature but, over the temperatures and fields used, there is reasonably good fit to a straight line. In most cases, the adjusted $R^2$ is greater than 0.95 for both fluids (Tables 5A and 5B). Since data over the entire range of electric fields (0 - 6 kV/mm) are used for this fit, some of the deviation may be from an $E_0$ effect at low fields and dipole saturation at high fields.

The slope of shear stress versus current density decreases as the temperature increases. This decrease in slope with increasing temperature can be fit to the Arrhenius equation

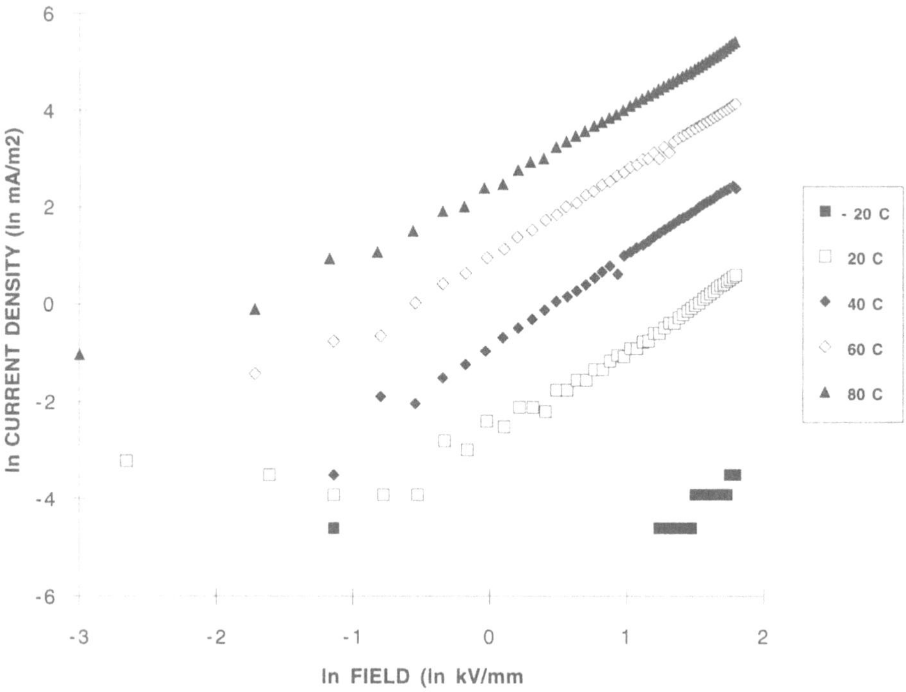


**Figure 4A**. Ln current density versus electric field at several temperatures for the cellulose fluid.

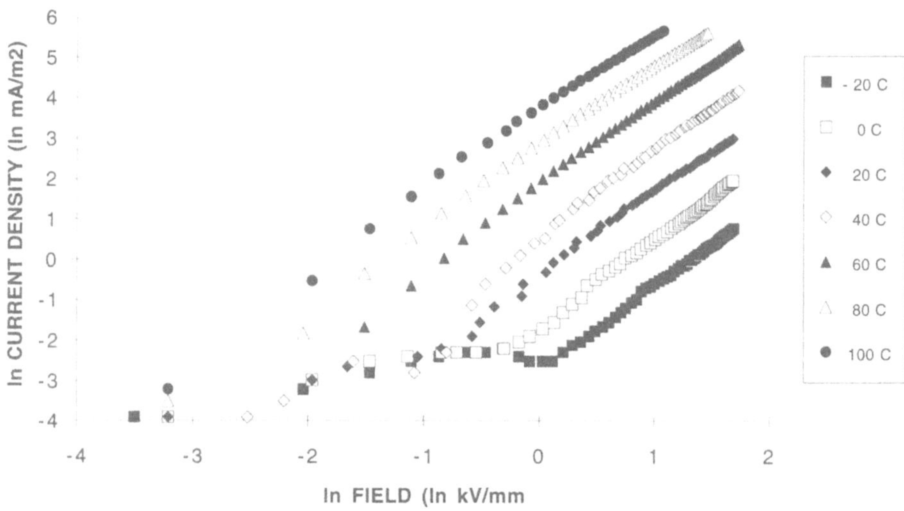


**Figure 4B**. Ln current density versus electric field at several temperatures for the polyaniline fluid.

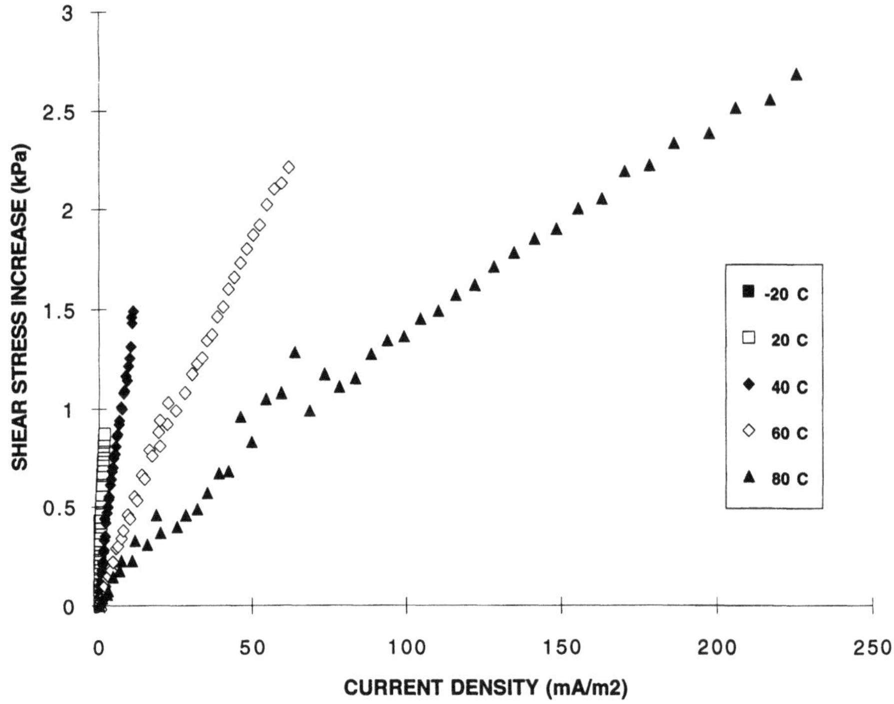


**Figure 5A**. Shear stress increase versus current density at several temperatures for the cellulose fluid.

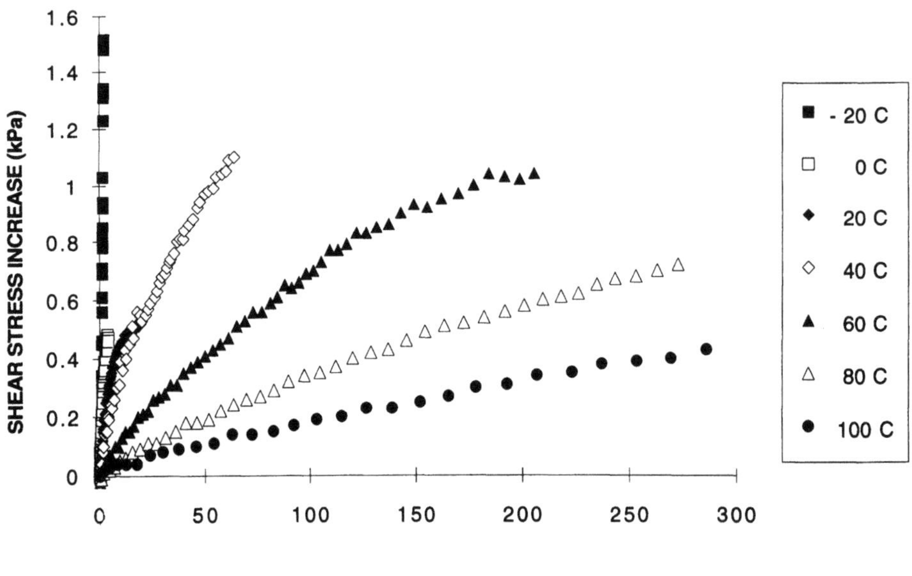


**Figure 5B**. Shear stress increase versus current density at several temperatures for the polyaniline fluid.

$$\text{slope} = ke^{-Ea/RT} \qquad k, E_a = \text{constants} \qquad (5)$$

The cellulose fluid fits Equation 5 with an adjusted $R^2$ of 0.999 while the polyaniline fluid has an adjusted $R^2$ of 0.986. The combination of processes responsible for the change in slope with temperature has an overall activation energy of -12.6 kcal/mole for the cellulose fluid and -9.4 kcal/mole for the polyaniline fluid.

**Table 5A**. Shear stress increase versus current density for the cellulose fluid.

| Temperature °C | slope | intercept | adjusted $R^2$ |
|---|---|---|---|
| -20 | 14.7 | 0.03 | .88 |
| 20 | 0.49 | 0.08 | .96 |
| 40 | 0.12 | 0.09 | .98 |
| 60 | 0.04 | 0.07 | .99 |
| 80 | 0.01 | 0.17 | .98 |

**Table 5B**. Shear stress increase versus current density for the polyaniline fluid.

| Temperature °C | slope | intercept | adjusted $R^2$ |
|---|---|---|---|
| -20 | 0.6645 | -0.158 | 0.93 |
| 0 | 0.1217 | 0.043 | 0.95 |
| 20 | 0.0255 | 0.130 | 0.84 |
| 40 | 0.0168 | 0.131 | 0.96 |
| 60 | 0.0053 | 0.107 | 0.97 |
| 80 | 0.0026 | 0.051 | 0.99 |
| 100 | 0.0014 | 0.030 | 0.99 |

## CONCLUSIONS

The performance of both the water activated cellulose and the anhydrous organic semiconductor polyaniline based electrorheological fluids generally fit the equations well at high fields and temperatures. The poorer fits for shear stress at low fields and temperatures may be the result of device limitations or an indication of a minimum field ($E_0$) below which stress increases are not detected.[3,9,15,16] The values for the stress dependence on field (m) obtained here were similar for both systems and were less than the m = 2 predicted by the polarization model.[1,3] This may be the result of a combination of an $E_0$, dielectric breakdown between particles, continuous phase conductivity, electrode polarization under dc which reduces the effective field, dipole saturation and displacement of water from the particle as the field increases. The values for the current density dependence on field (n) are slightly higher than the stress dependence (m) and in some cases approach 2. The values of m and n are close enough that shear stress versus current density at constant temperature is reasonably approximated by a straight line.

In contrast, the temperature dependence of current density is much larger than the temperature dependence of the shear stress and there is a significant difference between the values obtained for the two model systems. Because of the different dependencies, very different combinations of current densities and shear stresses are obtained at each temperature. This shows that the shear stress increase is not solely dependent on the current density. Because the temperature dependence of current density is larger than the temperature dependence of the shear stress, the slope of shear stress increase versus current density decreases as temperature increases. This decrease in slope with temperature can be described by activation energies which are significantly different for the two systems. The field dependence of ER performance does not seem to be greatly dependent on the solid used, while the temperature dependence of performance does seem to vary with the solid. The decreasing slope of shear stress increase versus current density with increasing temperature seen in Figures 5A and 5B means that the cost, in current density, of obtaining a given increase in shear stress will increase with temperature. If the slope did not change with temperature, one could obtain the same combination of shear stress increase and current density over a wide temperature range by simply varying the field with the temperature. Any decrease in the change of slope with temperature should result in a wider usable temperature range. This suggests that modification of the temperature dependence of stress and current density, separately or in combination, would be the preferred route to improvement of ER fluid performance.

## ACKNOWLEDGMENTS

We wish to thank Robert Pollack of Lubrizol and Alan Ross of RheoActive for help in sample preparation and data collection.

## REFERENCES

1. Gast, A.P.; Zukoski, C.F. *Adv. Coll. Int. Sci.* **1989** , 30, 153.
2. Block, H.; Kelly, J.P. *J. Phys D: Appl. Phys.* **1988**, 21, 1661.
3. Jordan, T.C.; Shaw, M.T. *IEEE Trans. Elect. Insul.* **1989,** 24, 849.
4. Block, H.; Kelly, J.P.; Qin, A.; Watson, T. *Langmuir* **1990**, 6, 6.
5. Pialet, J.W.; Bryant, C.P.; Lal, K. WO 9307244.
6. Foulc, J.N.; Atten, P.; Felici, N.J. In *Proc. 1993 IEEE Int. Conf. on Conductivity and Breakdown in Dielectric Liquids*; p 504.
7. Davis, L.C.; Ginder, J.M. *Polymer Preprints* **1994**, 64(2):321.
8. Conrad, H.; Chen , Y.; Sprecher, A.F. In *Proc. 2nd Int. Conf. on Electrorheological Fluids*; Carlson, J.D.; Sprecher, A.F.; Conrad, H., Eds., Technomic, Lancaster, 1990, p 252.
9. Winslow, W.M. *J. Appl. Phys.* **1949,** 20:1137.
10. Uejima, H. *Japan J. Appl. Phys.* **1972,** 11:319.
11. Katsikopoulos, P.V.; Zukoski, C.F. *Polymer Preprints* **1994,** 64:(2):385.
12. Bullough, W.A.; Peel, D.J.; Firoozian, R. In *Proc. 1st Int. Conf. on Electrorheological Fluids*; Conrad, H.; Sprecher, A.F.; Carlson, J.D., Eds., NCSU College of Engineering Publications, 1989, p 99.
13. Conrad, H.; Shamala, A.R.; Sprecher, A.F. In *Proc. 1st Int. Conf. on Electrorheological Fluids;* Conrad, H.; Sprecher, A.F.; Carlson, J.D., Eds., NCSU College of Engineering Publications , 1989, p 47.
14. Frood, D.G. In *Proc. 2nd Int. Conf. on Electrorheological Fluids*; Carlson, J.D.; Sprecher, A.F.; Conrad, H., Eds., Technomic, Lancaster, 1990, p 265.
15. Stangroom, J. *J. Stat. Phys.* **1991**, 64:1059.
16. Evans, L.; Harness, I.; Kermode, P.; Stangroom, J. In *Proc. Int. Conf. on Electrorheological Fluids*, Tao, R., Ed., World Scientific, Singapore, 1992, p 154.

# ER EFFECT ON THE NORMAL STRESS AND INDUCED NETWORK STRUCTURE OF LIQUID CRYSTAL POLYMER

Keiji Minagawa, Hiroshi Kimura, and Kiyohito Koyama*

Department of Materials Science and Engineering,
Yamagata University, Yonezawa 992, Japan

## INTRODUCTION

There have been many reports on electrorheological (ER) [1-4] suspensions because of their remarkable ER effect. Besides these two-phase systems, some one-phase system or homogeneous ER fluids [5-14], such as polymeric liquids, have been developed. The homogeneous ER fluids are free from the problems seen in suspensions due to dispersed particles, e.g., sedimentation, aggregation, friction, which reduce the ER effect. We have been studying ER effects in polymeric liquids to establish a model for understanding the mechanism of the ER effects in such homogeneous systems [6, 12-14]. Among the homogeneous ER fluids, polymers with flexible side chain connected with polar mesogenic group have a large ER effect [8, 9]. At present, the mechanism of the ER effect in these liquid crystal polymers (LCP) has been interpreted as the inhibition of the slipping between mesogenic domains by the flexible chains connecting them under an electric field. Detailed rheological examinations are necessary for further understanding of the mechanism.

In general, rheological properties of ER fluids are investigated by measuring shear viscosity and dynamic moduli which are obtained by measuring the shear stress. As a new method for obtaining additional rheological data, here we demonstrate normal stress measurement [12] to estimate the elastic properties of ER fluids under an electric field.

## EXPERIMENTAL

### Sample

A comb-shaped LCP [9] was supplied by Asahi Chemical Co. Ltd. The LCP is a random copolymer composed of flexible main chain of silicone and flexible side chains

*To whom correspondence should be addressed.

connecting mesogens of p-cyanophenyl benzoate residue (Fig. 1). The mesogen content is 75 weight %. The mesogens have a larger dipole moment along the long axis. Since the viscosity of this LCP is too high, we used a 67 weight % LCP solution in 20cSt dimethyl siloxane, and carried out the ER measurements at 80 ℃. It is known that the present LCP solution exhibits a significant ER effect at 50 ℃ [9]. Polarizing microscopy [12] indicated no apparent ordered structures at these temperatures, while a macroscopic structure, probably smectic phase, was observed at 25 ℃. From DSC measurements, the transition temperature was found to be around 36 ℃. Thus, the LCP solution in no electric field is macroscopically isotropic at the temperature for our rheological measurements.

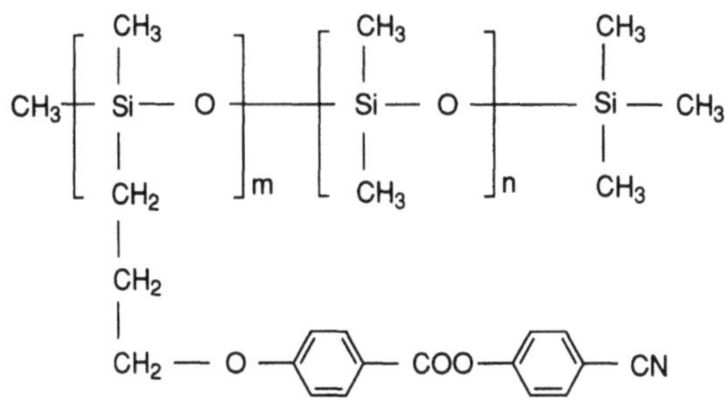

**Figure 1.** Chemical structure of the LCP; Average molecular weight is ca. 6,000 (n=13, m=17).

## Apparatus

A rotating parallel-disc type viscometer (MR-300 V2, Rheology Co., Ltd.) modified for ER measurements was used. Fig. 2 shows a schematic illustration of the measuring system. The two parallel plates play a role of electrodes to apply a homogeneous electric field to the sample fluid. Torque and normal force were detected simultaneously by the viscometer. The strength of applied electric fields ranged from 0 kV/mm to 2 kV/mm. The diameter of the two parallel discs was 32mm and the gap between the electrodes was 1 mm. In this measurement system, a positive normal stress value indicates the stress to extend the gap between the discs. For the evaluation of the strain and shear rate, we used the values at the edge of the parallel discs because these values vary with the distance from center of rotation.

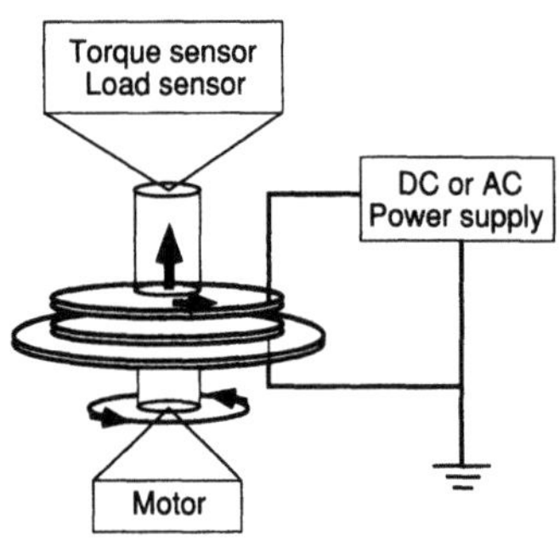


**Figure 2.** Schematic illustration of measuring system.

## RESULTS AND DISCUSSION

Although a parallel-disc rheometer can not apply a homogeneous shear to a fluid, the use of parallel-discs is the most convenient way to evaluate the normal stress effect of ER materials and also enables us to apply a homogeneous electric field to the sample. We evaluated the shear viscosity, dynamic moduli, and normal stress using parallel discs because a homogeneous electric field is preferentially necessary rather than homogeneous shear for the present purpose. Since the present fluid exhibited no apparent difference between DC and 50Hz-AC fields under the all conditions examined, here we show only the results obtained under DC fields.

### Shear Viscosity and Dynamic Moduli

Fig. 3 shows shear rate dependence of shear viscosity. The range of shear rate was from $10^{-1}$ to $10^{3}$ $sec^{-1}$. Without field, shear viscosity decreased at high shear rates, while under the electric field it did not decrease and showed constant value at all shear rates examined. Thus it is found that the LCP exhibits shear thinning property without field, and it has nearly Newtonian property under an electric field.

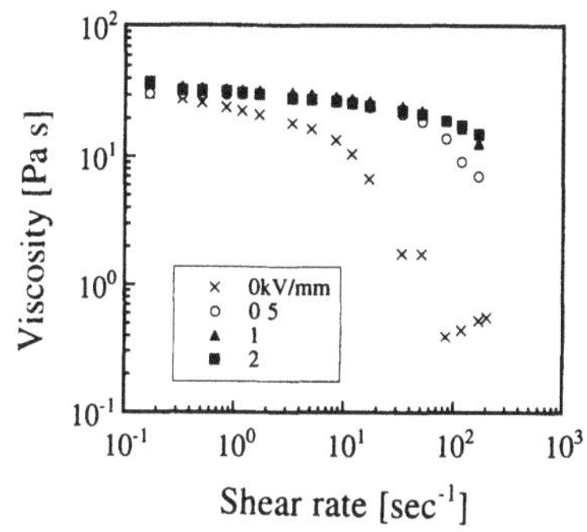


**Figure 3.** Shear rate dependence of shear viscosity.

The ER effect is then evaluated as the relative viscosity $\eta_E/\eta_0$, where $\eta_E$ and $\eta_0$ are the viscosities measured in the presence and absence of the field, respectively. The relative viscosity for 2kV/mm electric field is plotted against shear rate in Fig. 4. It is found that the ER effect tends to increase with increasing shear rate, although it slightly decreases at higher shear rate above $10^2$ $sec^{-1}$. The dependence of ER effect at high shear rate is due to the increase of $\eta_0$ at the high shear rate region, which is obviously seen in Fig. 3. This result might be an experimental error, partly caused by the complicated flow behavior of the fluid in the parallel-disc geometry. The dependence of relative viscosity on the applied field strength is shown in Fig. 5, in which typical results for three different shear rate conditions are plotted. The relative viscosity is dependent on the field strength at the high shear rate and is increased with the increase in the field strength, while it tends to reach a constant value at the lower shear rate conditions. Further increase of the field strength caused large current in the fluid and the measurement was hardly performed.

Figs. 6 and 7 show strain dependence of storage modulus G' and of loss modulus G", respectively. The range of measured strain was from $10^{-3}$ to $10^{1}$. The applied angular frequency was 31.4 rad/sec. Both values of G' and G" drastically decreased without field at large strain. Under 2kV/mm electric field, the decrease at large strain was restricted comparing to without field. Figs. 8 and 9 exhibit angular frequency ω dependence of G' and G", respectively. The applied strain was 2.7. Both G' and G" values increased in

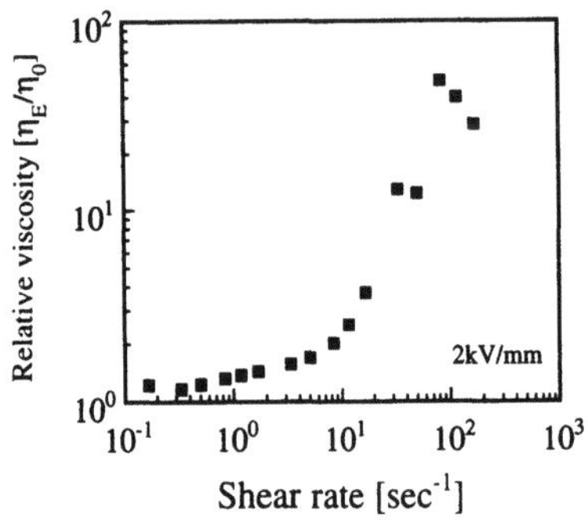


**Figure 4.** Shear rate dependence of relative viscosity.

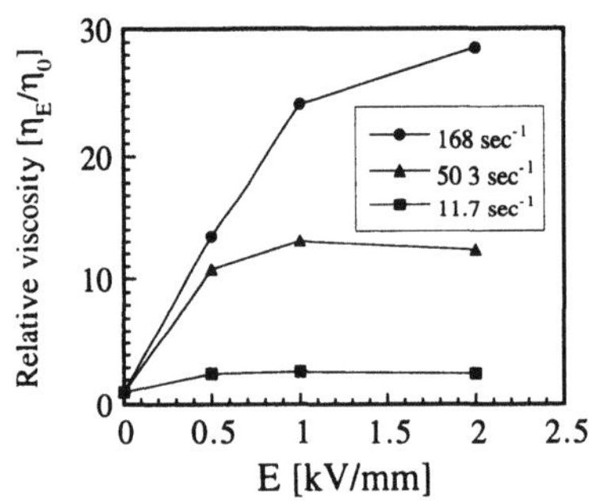


**Figure 5.** Relative viscosity vs. field strength.

proportion to ω at low frequency. At higher frequency, G' and G" decreased without field while they increased further under the electric field. These rheological data are consistent with the results reported by Inoue et al. [8, 9], and suggest that some structure which has resistance to shear appears under electric field. In particular, the behavior of G' implies that a three-dimensional structure such as network which exhibits elastic property is induced by the electric field.

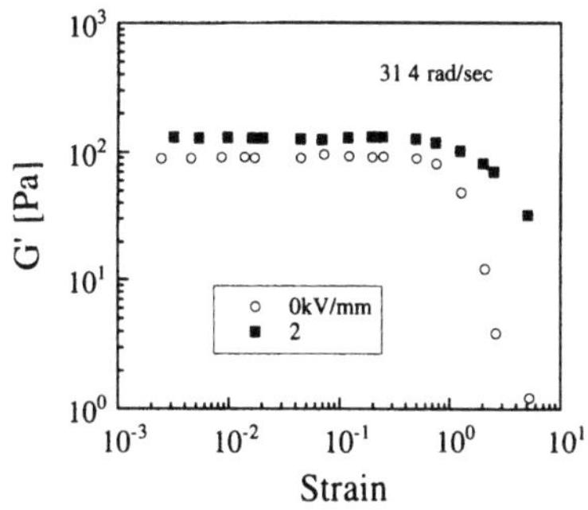


**Figure 6.** Strain dependence of storage modulus G'.

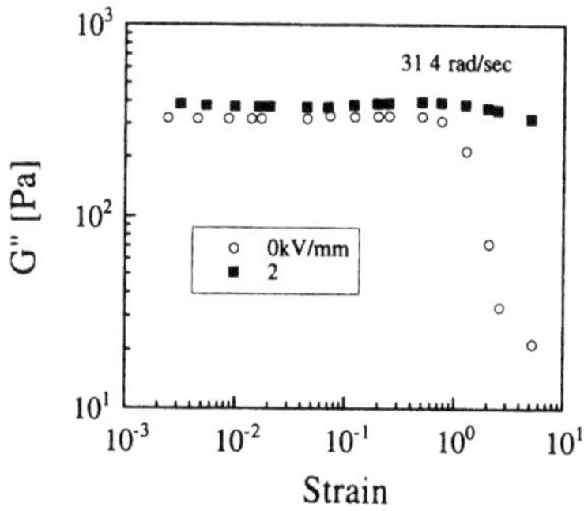


**Figure 7.** Strain dependence of loss modulus G".

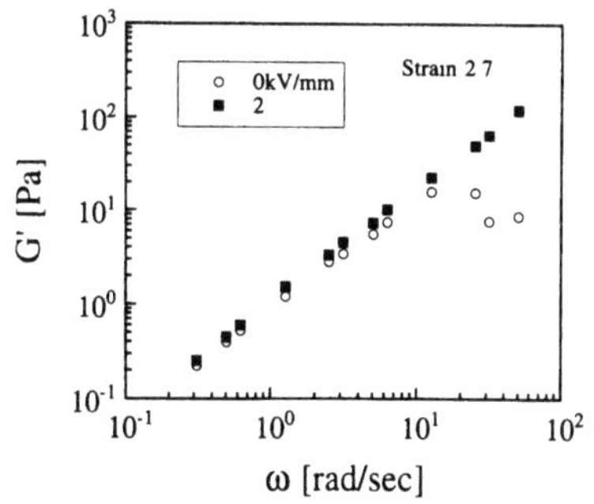


**Figure 8.** Angular frequency dependence of G'.

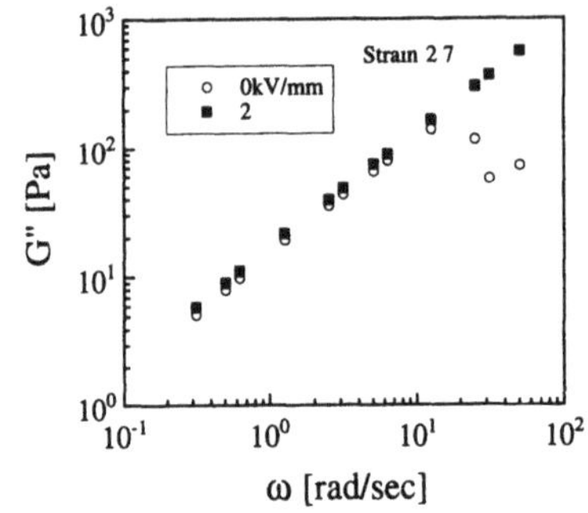


**Figure 9.** Angular frequency dependence of G".

## Normal Stress

To confirm the existence of the structure, we measured normal stress under steady shear flow. Since the normal force in parallel-disc system is parallel to the direction of electric field applied by the disc-electrodes, the electrostatic force between the discs would significantly affect the stress data. Thus, the effect of electric field on the normal force detected by the present apparatus was examined prior to the rheological measurements. Fig. 10 shows the apparent normal stress obtained for no sample, silicone oil, and LCP solution, measured without shear. These data are plotted against the square of field strength. For each condition the normal stress under electric field is negative and the absolute value of stress change is in proportion to the square of field strength. It is clear that the apparent negative values of normal stress are directly related to the electrostatic force. Therefore, the normal stress measurements under an electric field should be performed taking into account the electrostatic effect on the stress data.

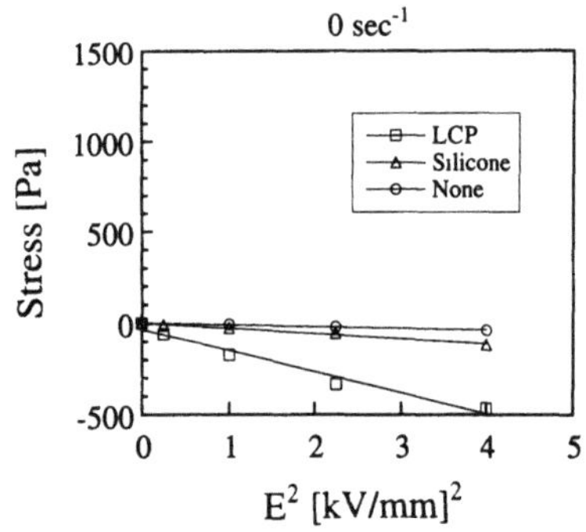


**Figure 10.** Attractive force between electrodes under electric field.

Normal stress in the steady shear was measured in the presence and absence of electric field. The stress was determined with the following simplified equation without correcting for the inhomogeneous shear.

$$\Delta n_1 = \frac{2F}{\pi R^2} \tag{1}$$

Here $\Delta n_1$ is the primary normal stress difference, $F$ the thrust, and $R$ the radius of the parallel discs. The $\Delta n_1$ value calculated from eq.(1) is, thus, not the precise primary normal stress difference but it is a simple estimation for the force normal to the parallel discs. The range of shear rate was from $10^{-1}$ to $10^3$ $sec^{-1}$. Fig. 11 shows shear rate dependence of the apparent normal stress. Without field, the values of normal stress were almost zero at all shear rates. On the other hand, a drastic change is seen for the normal stress under the field. The stress value is negative at low shear rate, while at the shear rates higher than $10^1$ $sec^{-1}$ large normal stress appears. As mentioned above, the apparent negative stress at low shear rate region is due to the electrostatic attractive force between the electrodes. It is difficult to precisely remove the effect of the electrostatic stress from this result because the dielectric property of the ER fluid changes depending on the shear rate.

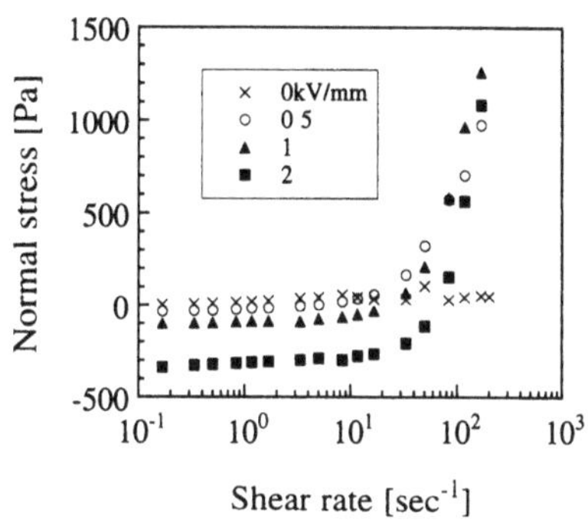


**Figure 11.** Shear rate dependence of apparent normal stress.

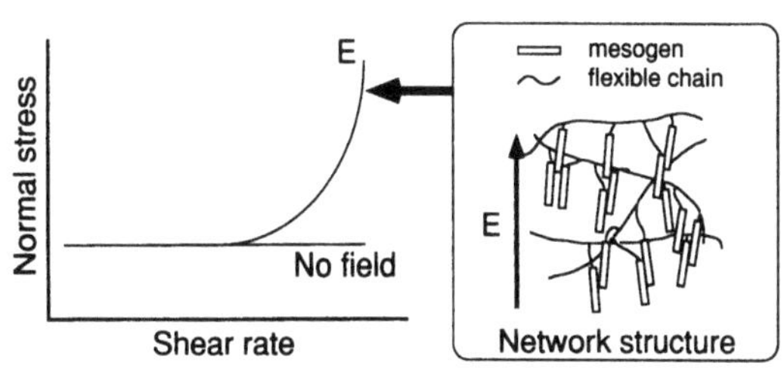


**Figure 12.** Schematic representation of network structure of LCP and stress behavior.

A simple approximation to correct the electrostatic effect is to subtract the negative stress shown in Fig. 10 from the all data [12]. With this correction the general tendency of the normal stress is schematically represented in Fig. 12, where a plausible structure of LCP under the field is also illustrated. Here the difference of electric field strength is neglected because there was little electric field dependence in the corrected normal stress within our experimental conditions. The normal stress effect of the LCP solution is clearly observed in the behaviors at the high shear rate region. The large normal stress at high shear rate region directly relates to elastic properties of the material. One of typical examples is a well-known phenomenon called the poynting effect of elastic materials. The electrically induced normal stress of the LCP solution can be understood by considering the existence of rubber-like elasticity arising from a network structure in the LCP (Fig. 12). By applying an electric field, the mesogens attract each other and form electrically induced domains. This structure well explains the large ER effects of the LCP.

The microdomain structures in ER fluids based on liquid crystals and polymeric liquids are interesting to consider the essentials of the ER effects in these systems. Recently some "negative ER effects" have been found for liquid crystals of low molecular-weight [7] and for solutions of polymers having both hard and soft segments within the polymer chain [13, 14]. The ER effects in these fluids are presumably related to the change in the domain structure. For example, we have suggested an induced phase separation in the polymer solution as a preliminary interpretation for the negative ER effect. In any case, the ER properties should be dependent on the equilibrium structures of the fluid in shear flow. Theoretical analyses of the distribution of mesogen direction [10, 11] and microscopic observations of domain structures together with the present normal stress study will be helpful for the elucidation of the mechanism.

## CONCLUSION

By measuring normal stress as an estimation of elastic property of ER fluids, the existence of network structure under electric field has been confirmed. The normal stress measurement gives useful information on the structure on homogeneous liquid type ER fluids.

### Acknowledgments

We are grateful to Dr. Akio Inoue (Asahi Chemical Co. Ltd.) for providing the LCP samples. We also wish to thank one of the reviewers for valuable comments. This work was partly supported by the Grant-in-Aid from the Ministry of Education, Science and Culture, Japan.

## REFERENCES

1. W.M. Winslow, Induced fibration of suspensions, *J. Appl. Phys.*, **20**, 1137 (1949).
2. H. Block and J. P. Kelly, Electro-rheology, *J. Phys. D: Appl. Phys.*, **21**, 1661 (1988).
3. T.C. Jordan and M.T. Shaw, Electrorheology, *IEEE Trans. Elect. Insul.*, **24**, 849 (1989).
4. T.C. Halsey, Electrorheological fluids, *Science,* **258**, 761 (1992).
5. T. Honda, T. Sasada and K. Kurosawa, Electro-viscous effect in the MBBA liquid crystal, *Jpn. J. Appl. Phys.*, **17** (9), 1525 (1978).
6. K. Tanaka, A. Fujii and K. Koyama, Electro-rheology of poly(vinylidene fluoride-trifluoroethylene) solution, *Polym. J.,* **24**, 995 (1992).

7. M. Fukumasa, K. Yoshida, S. Ohkubo and A. Yoshizawa, Decrease in apparent viscosity in electric field found in chiral smectic liquid crystals, *Ferroelectrics,* **147**, 395 (1993).
8. A. Inoue and S. Maniwa, EP 478,034A (1992).
9. S. Maniwa and A. Inoue, Electrorheological effect of liquid crystal polymers (2): Dependence on molecular structure of polymer, *Polym. Prepr. Jpn.,* **42** (4), 1591 (1993).
10. I.-K. Yang and A. D. Shine, Electrorheology of a nematic poly(n-hexyl isocyanate) solution, *J. Rheol.,* **36** (6), 1079 (1992).
11. K.-L. Tse and A. D. Shine, Electrorheology of poly(n-hexyl-L-glutamate), *Polym. Prepr. (ACS),* **35** (2), 383 (1994).
12. H. Kimura, K. Minagawa and K. Koyama, Induced network structure in liquid crystalline polymer evidenced from electrorheological normal stress measurements, *Polym. J.,* **26** (12), 1402 (1994).
13. T. Uemura, K. Minagawa and K. Koyama, Novel electro-responsive property of polyether-polycarbamate solution, *Chem. Lett.,* **1994**, 1995 (1994).
14. T. Uemura, K. Minagawa and K. Koyama, Negative ER effect in polymeric liquids, *Polym. Prepr. (ACS),* **35** (2), 360 (1994).

# STRAIN DEPENDENCE OF DYNAMIC VISCO-ELASTIC PROPERTIES OF ER FLUIDS WITH IRON(III) HYDROXIDE COATED SILICA PARTICLES

Katsufumi Tanaka[1], Kiyohito Koyama[2], and Touru Watanabe[2]

[1]Department of Polymer Science and Engineering,
Kyoto Institute of Technology, Matsugasaki, Kyoto 606, Japan
[2]Department of Materials Science and Engineering
Yamagata University, Yonezawa 992, Japan

## INTRODUCTION

The electro–rheological (ER) effect is thought to be closely related to the aggregation and yield of polarized particles. These polarized particles aggregate parallel to the electric field to form pairs of particles (short clusters). Then, the short clusters are elongated to bridge between electrodes to form long clusters. Flow resistance of the clustered suspension increases as much as the resistance to break the long clusters which corresponds to the yield stress. On the basis of induced dipole–dipole interaction models[1,2], it is expected that the yield strain and stress of the clusters are affected by an external electric field, the dielectric constant of particles and that of disperse phase of the suspension, and the intensity of hydrodynamic deformation, which are included in a general dimensionless parameter called Mason's number[3]. Experimentally, the yield strain and stress can be obtained by measuring the stress–strain relation directly at a constant strain rate[4]. In this method, the strain rate immediately after the onset of deformation should be precisely controlled and the high resolution of small strains should be needed. There can be another method to obtain the yield strain and stress. That is, the yield strain and stress can be estimated from the strain dependence of dynamic viscoelastic properties under sinusoidal oscillating shear flow[5,8]. The periodic measurement of stress response to the strain stimulus is expected to be the high resolution of small strains, and the oscillatory deformation with a sufficiently low frequency can be assumed to the quasi–static shear deformation.

In the present paper, the strain dependence of dynamic viscoelastic properties of ER suspensions is reported and the yield behavior estimated from the dynamic properties is discussed by the yield strain and stress which are related to Mason's number.

## EXPERIMENTAL

### Samples

Silica particles covered with iron(III) hydroxide, and the particles of their components, silica and iron(III) hydroxide, were kindly supplied from Fuji–Davison Chemical,

*Progress in Electrorheology*, Edited by K.O. Havelka
and F.E. Filisko, Plenum Press, New York, 1995

Ltd. They were dried at 130 °C under vacuum prior to use. The dried particles were dispersed in a solvent of 30 cSt silicone oil so that the volume fraction of these particles was 0.04 assuming the volume additivity of the homogeneous particles and solvent.

### Measurement of Dielectric Properties

The dielectric constant of the sample suspensions and that of silicone oil alone were measured at a frequency of 1 kHz using an LF impedance analyzer (Hewlett and Packard 4192A). Relative dielectric mismatch of the particles and silicone oil, which is needed for the calculation of Mason's number and defined here as $\beta = (\varepsilon'_p - \varepsilon'_c)/(\varepsilon'_p + 2\varepsilon'_c)$, was calculated from the following equation,

$$(\varepsilon' - \varepsilon'_c)/(\varepsilon' + 2\varepsilon'_c) = \phi\beta \tag{1}$$

where $\varepsilon'$ is the real part of dielectric constant of sample suspensions, $\varepsilon'_p$ is that of particles alone, and $\varepsilon'_c$ is that of continuous phase of silicone oil alone.

### Measurement of Dynamic Viscoelastic Properties

The strain amplitude dependence of dynamic viscoelastic properties under DC electric fields up to 3 kV/mm was measured by a coaxially rotational rheometer (Rheology Engineering Co., Ltd., MR–300). A sufficiently rigid torsion wire with a wire constant of $1.3350 \times 10^7$ dyn·cm/rad was used as a torque sensor. In our rheometer, the storage modulus and the loss tangent at each preset strain amplitude can be calculated using Markovitz's method[6].

Similar as reported previously[7], it was found that the wave form of the stress response to a sinusoidal strain at large amplitude under an electric field was so distorted and thus contained higher order harmonics[8]. In our rheometer, $G_1'$, $G_1''$ ($\tan\delta_1$)[9] as well as the fundamental component of the torque amplitude ($T_1$) can be obtained by extracting the fundamental component from the distorted torque wave using the digital filtering technique (FFT analysis). Then, the amplitude of strain which could be applied to the sample suspensions ($\gamma_a$) was calculated from the following equations,

$$\sigma_1 = T_1 / 2\pi R_b^2 L \tag{2}$$

$$\gamma_a = \sigma_1 / G_1'(1 + \tan^2\delta_1)^{1/2} \tag{3}$$

where $\sigma_1$ is the fundamental component of stress amplitude, L is immerse length (14 mm), and $R_b$ is radius of inner cylinder of the rheometer (17 mm).

The frequency of preset sinusoidal strain (f) was 1 Hz and the amplitude of the preset strain was varied from 0.005 to 2.5. The sample suspensions were shear deformed sufficiently after application of a DC electric field so that long clusters could elongate and span the electrodes.

## RESULTS

Figure 1 shows the strain amplitude ($\gamma_a$) dependence of storage modulus ($G_1'$) and loss tangent ($\tan\delta_1$) of the iron(III) hydroxide ($Fe(OH)_3$) suspension under an electric field of 2 kV/mm. In Fig.1, $G_1'$ and $\tan\delta_1$ are almost constant below a critical strain ($\gamma_c$). Above $\gamma_c$, $G_1'$ decreases while $\tan\delta_1$ increases steeply. Therefore, the yield of the long clusters elongated in the suspension could occur at $\gamma_c$. Similar tendencies of $G_1'$ and $\tan\delta_1$ as $Fe(OH)_3$ suspension could be seen for the silica suspension and $Fe(OH)_3$ coated silica suspension at the same experimental conditions.

Figures 2 and 3 show, respectively, the strain amplitude dependence of storage modulus and that of loss tangent for $Fe(OH)_3$ coated silica suspension measured at various electric fields from 0 to 3 kV/mm. In Fig.2, $G_1'$ at a given strain amplitude increases with an increase in the electric field strength. Two kinds of plateau at smaller strains of $10^{-3}$

order and at larger strains of $10^{-1}$ order can be seen in the curves which were measured at electric fields over 2 kV/mm. Also, $\tan\delta_1$ measured at electric fields over 2 kV/mm in Fig.3 shows two kinds of the critical strain in the corresponding strain region of the plateau in Fig.2. The loss tangent measured at the smaller strains and at electric fields over 2 kV/mm is much smaller than the others. The smaller critical strain measured at 3 kV/mm is slightly larger than that measured at 2 kV/mm. On the other hand, the larger critical strain can be seen in the curves measured at each electric field strength and it is not much affected by the electric field strength.

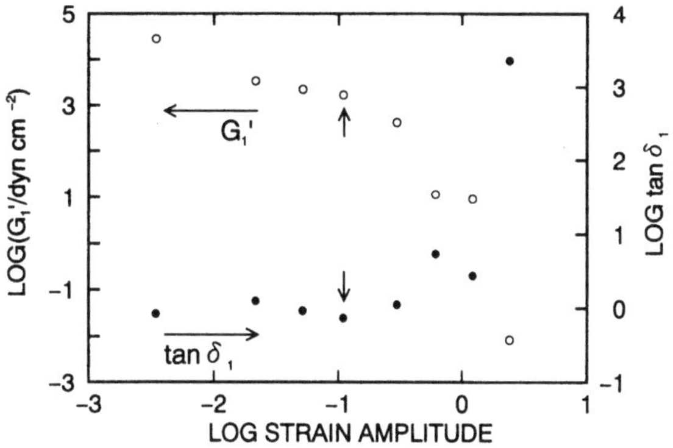


**Figure 1.** Storage modulus ($G_1$') and loss tangent ($\tan\delta_1$) plotted against strain amplitude for $Fe(OH)_3$ suspension at 2 kV/mm.

Figure 4 shows the relation between stress amplitude and strain amplitude of the suspensions which was calculated from Eqs.(2) and (3) with the dynamic storage modulus and loss tangent at each strain amplitude. The stress amplitude of $Fe(OH)_3$ and silica suspensions shows a maximum peak at a strain which corresponds to a yield strain of each suspension. The stress amplitude of $Fe(OH)_3$ coated silica suspension does not show the maximum peak. At larger strains, the stress amplitude of the investigated suspensions is saturated, and the saturated stress correlates with the fundamental component of yield stress of the suspensions. The larger critical strain and the saturated stress amplitude of $Fe(OH)_3$ coated silica suspension are much larger than those of the other suspensions under the same experimental conditions. The similar stress–strain relation of $Fe(OH)_3$ coated silica suspension under various electric fields could be also calculated. The saturated stress amplitude increased with an increase in the electric field strength although no maximum peak of the stress amplitude could be seen in the relation. These results suggest that the interaction among $Fe(OH)_3$ coated silica particles is much larger than that of the $Fe(OH)_3$ and silica particles, and that the interaction of $Fe(OH)_3$ coated silica suspension is closely related to the induced dipole of the particles. Also, the ionic or interfacial polarization between the silica core and $Fe(OH)_3$ layer may contribute to the interaction among $Fe(OH)_3$ coated silica particles.

## DISCUSSION

See and Doi investigated the interaction between a pair of particles displaced horizontally (x–direction) under combined action of the shear flow and the pulse electric field applied perpendicular to the x–direction[2]. They derived the x–component of the induced dipole–dipole interaction between two particles ($F_x^{elec.}$) as a function of x–component of their center to center separation ($R_x$), and also derived the shearing force on a particle

($F_x^{flow}$) using a simplifying approximation. We can easily rewrite $F_x^{elec.}$ as a function of strain $\gamma$.

$$F_x^{elec.} = (3a^2/64)\varepsilon_c\varepsilon_0(\beta E)^2\gamma(4 - \gamma^2)/(1 + \gamma^2)^{7/2} \tag{4}$$

$$\gamma = R_x/a \tag{5}$$

$$F_x^{flow} = 3\pi\eta_c a^2(d\gamma/dt) \tag{6}$$

where a is the diameter of the particles, $\varepsilon_0$ is the permittivity of free space, $\varepsilon_c$ and $\eta_c$ are, respectively, the dielectric constant and the viscosity of continuous phase, and E is the electric field strength which is assumed here to be independent of time.

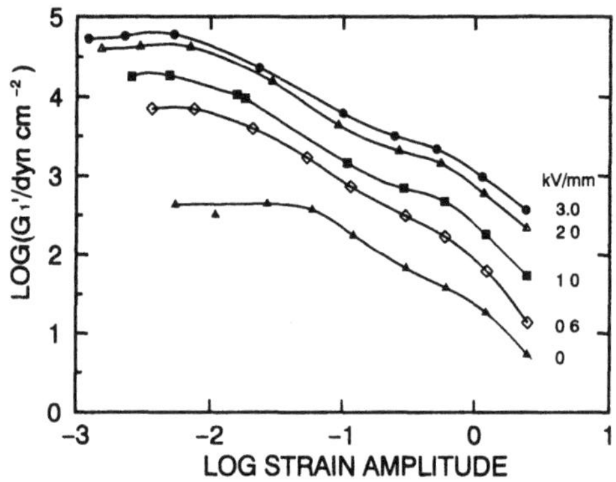


**Figure 2.** Storage modulus plotted against strain amplitude for $Fe(OH)_3$ coated silica suspension at different electric field strength.

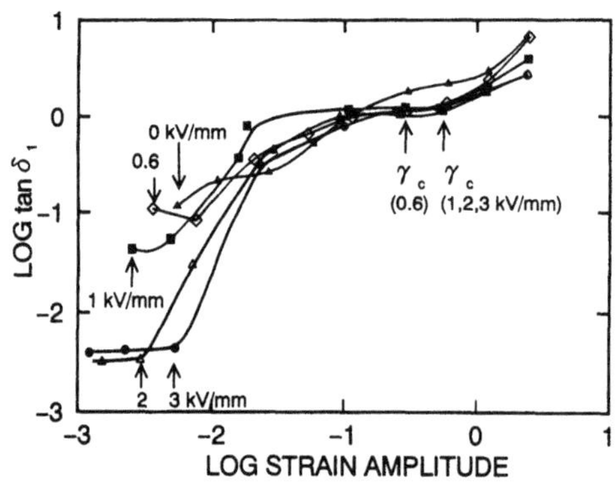


**Figure 3.** Loss tangent plotted against strain amplitude for $Fe(OH)_3$ coated silica suspension at different electric field strength. Symbols are as in Fig.2.

Figure 5 shows the strain dependence of $F_x^{elec.}$. $F_x^{flow}$ is also shown in the figure assuming that $F_x^{flow}$ is constant and given. In Fig.5, three typical cases, (a), (b), and (c), of the maximum of $F_x^{elec.}$ are plotted. Three characteristic strains can be seen which are the inception points of $F_x^{elec.}$ and $F_x^{flow}$. It is considered that the yield of the short cluster occurs when $F_x^{flow}$ exceeds $F_x^{elec.}$. Therefore, the yield will occur at $\gamma_{c2}$ (in the case of (a)) or $\gamma_{c1}$ (case (b)). In the case of (c), the yield will occur immediately after beginning of shear flow showing indefinite yield strain. In the case of (a), the yield strain increases with an increase in the electric field strength. The yield strain can be related to Mason's number[3] (Mn) because $F_x^{elec.}$ and $F_x^{flow}$ balance at a critical point $\gamma_c$.

$$S(\gamma_c) \equiv \gamma_c(4 - \gamma_c^2)/(1 + \gamma_c^2)^{7/2} \propto \mathrm{Mn} \tag{7}$$

$$\mathrm{Mn} \equiv \eta_c(d\gamma/dt)/2\varepsilon_c\varepsilon_0(\beta E)^2 \tag{8}$$

From the Eqs.(7) and (8), we can know the relation between the critical strain and Mason's number which is predicted from the model of See and Doi.

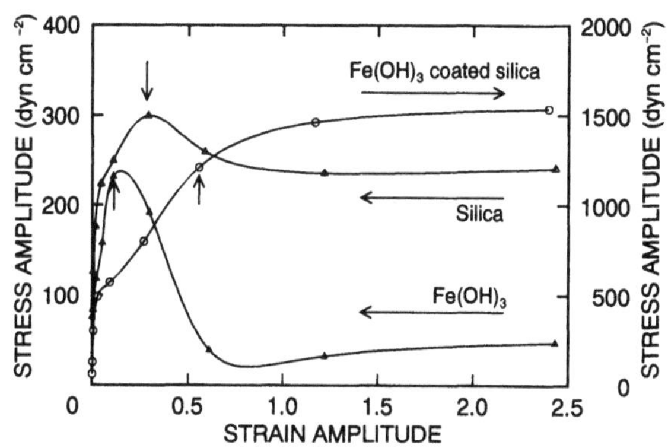


**Figure 4.** Relation between stress amplitude and strain amplitude at 2 kV/mm calculated from Eqs. (2) and (3) for investigated suspensions. The left and right side of the longitudinal axes represent the stress amplitude of silica and $Fe(OH)_3$ suspensions and that of $Fe(OH)_3$ coated silica suspension, respectively.

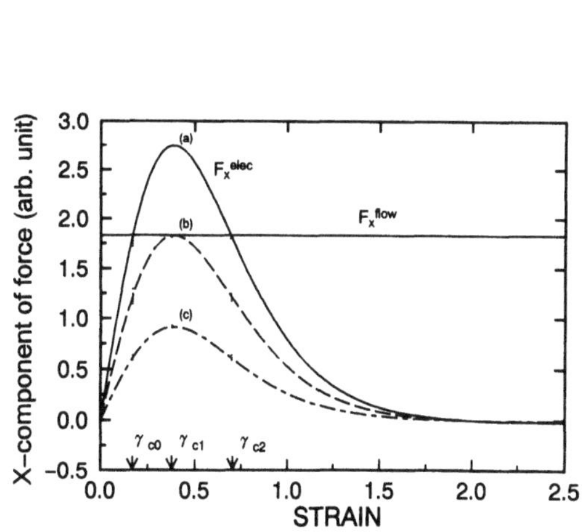


**Figure 5.** Electrostatic force ($F_x^{elec.}$) calculated from a dipole-dipole interaction model[2] plotted against strain. Shearing force ($F_x^{flow}$) is also plotted. Three cases of maximum of $F_x^{elec.}$($F_{xmax}^{elec.}$), (a) $F_{xmax}^{elec.} > F_x^{flow}$, (b) $F_{xmax}^{elec.} = F_x^{flow}$, and (c) $F_{xmax}^{elec.} < F_x^{flow}$, are superimposed in the figure.

At the same time, the yield stress is expected to be constant from this model, which cannot explain the experimental results. This discrepancy may arise from the simple assumption of $F_x^{flow}$ which is constant throughout the quasi-static shear deformation. $\gamma_{c2}$ may correlate with onset of the dynamic yield phenomena. Also, it must be stated more clearly that the shear deformation is assumed to be steady in the model of See and Doi while the shear deformation in the present experiments was oscillatory flow. As mentioned above, the oscillatory flow can be assumed to the quasi-static shear deformation when the frequency is sufficiently low. The frequency of oscillatory flow was 1 Hz in the present experiments and it was relatively low. In the present paper, the oscillatory flow is assumed to the quasi-static deformation over a period although it is not necessarily clear. (There is a report which shows the extension of the Cox-Merz rule to concentrated suspensions and other materials with a yield stress using an "effective shear rate", the angular frequency of oscillatory flow times the strain amplitude, instead of the angular frequency[10]. If the extended Cox-Merz rule can be applied to ER suspensions, the steady shear rate is thought to be equivalent to the effective shear rate.)

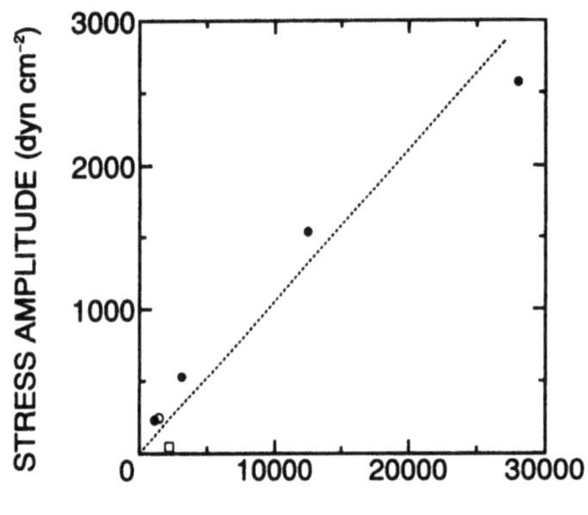


**Figure 6.** Saturated stress amplitude of the investigated suspensions plotted against inverse of Mason's number[3]. ●: $Fe(OH)_3$ coated silica suspension; o: silica suspension; and ▯: $Fe(OH)_3$ suspension.

Another concept of the yield process under the quasi-static deformation like the present experiment is that the yield occurs at a strain to reach the maximum of $F_x^{elec.}$ ($\gamma_{c1}$). In this case, the yield strain is independent of the electric field strength. On the other hand, the yield stress is proportional to $(\beta E)^2$, which is also proportional to $Mn^{-1}$. In Fig.6, the saturated stress amplitude of the investigated suspensions is plotted against $Mn^{-1}$. In the calculation of Mn, $(d\gamma/dt)$ was assumed to be the "effective shear rate", $2\pi f\gamma_a$. It can be seen in Fig.6 that the saturated stress amplitude of $Fe(OH)_3$ coated silica and silica suspensions tends to be proportional to $Mn^{-1}$, as is expected from the yield at $\gamma_{c1}$. The saturated stress amplitude of $Fe(OH)_3$ suspension scatters relatively from the line. In the relation between the strain function of the larger critical strain, $S(\gamma_c)$, and Mn, $S(\gamma_c)$ was not much affected by Mn. Therefore, it is considered that the larger critical strain corresponds to the strain at the maximum of $F_x^{elec.}$ ($\gamma_{c1}$). However, the smaller critical strain as shown in Fig.3 cannot be explained by the yield at the maximum of $F_x^{elec.}$. The smaller critical strain may correspond to the elastic limit of the clustered particles because loss tangent of $Fe(OH)_3$ coated silica suspension at electric fields of 2 and 3 kV/mm increases steeply above the critical strain. Indefinite smaller critical strain of $Fe(OH)_3$ coated silica suspension and relatively large value in the loss tangent at smaller strains and electric fields under 1 kV/mm suggest that the short clusters could not be elongated enough at the electric fields. The sufficiently rigid and long cluster of $Fe(OH)_3$ coated silica suspension could be elongated at electric fields over 2 kV/mm in the present experiment.

Finally, the difference of the larger critical strain between sample suspensions in an electric field cannot be explained by Eq.(4) because $F_x^{elec.}$ is not shifted to the horizontal direction by changing β. Klingenberg et al. reported[1] the dependence of $F_x^{elec.}$ on strain as well as on the ratio of the dielectric constant of the particles ($\varepsilon_p$) to that of continuous phase ($\varepsilon_c$) (α) which correlates with β. They calculated $F_x^{elec.}$ by eliminating the point-dipole approximation and reported that the critical strain at the maximum of $F_x^{elec.}$ could depend on α (or β).

## CONCLUSIONS

The yield strain and stress of the suspensions of $Fe(OH)_3$ coated silica particles and their components were estimated from the strain dependence of dynamic viscoelastic properties. Two kinds of the yield strain of $Fe(OH)_3$ coated silica suspension, the smaller one and the larger one, were found at electric fields over 2 kV/mm. The larger yield strain of $Fe(OH)_3$ coated silica suspension was not much affected by the electric field strength. Also, the fundamental component of yield stress of $Fe(OH)_3$ coated silica suspension tended to be proportional to the inverse of Mason's number, which was changed by electric fields. The larger yield strain and the yield stress of $Fe(OH)_3$ coated silica suspension under an electric field of 2 kV/mm were larger than those of silica and $Fe(OH)_3$ suspensions under the same electric field. The yield strain and stress which were related to Mason's number suggested that the larger critical strain at the yield of $Fe(OH)_3$ coated silica suspension corresponded to the strain at the maximum of induced dipole–dipole interaction.


## ACKNOWLEDGEMENTS

One of the authors, K.T., thanks to the Ogasawara Foundation for the Promotion of Science and Engineering and Yoshida Foundation for Science and Technology for the joint support of his presentation at the ER Materials and Fluids Symposium held in the 208th National ACS Meeting at Washington, D.C. This work was partially supported by a Grant–in–aid for Scientific Research from Ministry Education, Science and Culture of Japan (No. 04750728).


## REFERENCES


1. D. J. Klingenberg and C. F. Zukoski, Studies on the steady–shear behavior of electrorheological suspensions, *Langmuir*, 6:15 (1990).
2. H. See and M. Doi, Shear resistance of electrorheological fluids under time–varying electric fields, *J. Rheol.*, 36:1143 (1992).
3. L. Marshall, C. F. Zukoski and J. W. Goodwin, Effects of electric fields on the rheology of non–aqueous concentrated suspensions, *J. Chem. Soc., Farady Trans. 1*, 85:2785 (1989).
4. H. Conard, A. F. Sprecher, Y. Choi and Y. Chen, The temperature dependence of the electrical properties and strength of electrorheological fluids, *J. Rheol.*, 35:1393 (1991).
5. T. C. Jordan, M. T. Shaw and T. C. B. McLeish, Viscoelastic response of electrorheological fluids. II. Field strength and strain dependence, *J. Rheol.*, 36:441 (1992).
6. H. Markovitz, A property of Bessel functions and its application to the theory of two rheometers, *J. Appl. Phys.*, 23:1070 (1952).
7. K. Tanaka, T. Yoshida and K. Koyama, Transient response of ER fluids, *in*: "Electrorheological Fluids - Proc. 3rd Int. Conf. ER Fluids", R.Tao, ed., World Scientific, Singapore (1992).
8. D. R. Gamota and F. E. Filisko, High frequency dynamic mechanical study of an aluminosilicate electrorheological material, *J. Rheol.*, 35:1411 (1991).
9. S. Onogi, T. Masuda, T. Matsumoto, Non–linear behavior of viscoelastic materials. I. Dispersed systems of polystyrene solution and carbon black, *Trans. Soc. Rheol.*, 14:275 (1970).
10. D. Doraiswamy, A. N. Mujumdar, I. Tsao, A. N. Beris, S. C. Danforth and A. B. Metzner, The Cox–Merz rule extended: A rheological model for concentrated suspensions and other materials with a yield stress, *J. Rheol.*, 35:647 (1991).

# TRANSIENT STRESSES IN ELECTRORHEOLOGICAL FLUIDS: ROLE OF PARTICLE POLARIZATION DYNAMICS

J. M. Ginder,[a] L. C. Davis,[a] and S. L. Ceccio[b]

[a]Research Laboratory, Ford Motor Company
20,000 Rotunda Drive, MD 3028 SRL
Dearborn, MI 48121-2053 USA

[b]Department of Mechanical Engineering and Applied Mechanics
The University of Michigan
Ann Arbor, MI 48109-2125 USA

## INTRODUCTION

While the magnitude of the field-induced shear stress is a critical parameter in designing semiactive electrorheological (ER) devices, the stress response time is also relevant. Furthermore, sharp transients in the stress response may lead to perceptions of harshness that is undesirable, for example, in automotive applications. In light of the importance of these transient effects, a variety of time-dependent rheological phenomena were explored experimentally in several model ER fluid systems.[1,2] The character of the transient stress response was found to depend strongly on the composition of the ER system under test. The detailed time dependence of the stress and its variation with material properties can be understood semiquantitatively by using a simple model that accounts for the polarization dynamics of the suspended particles. The character of the transient response depends largely on whether particle polarization involves primarily a particle-fluid conductivity mismatch or a dielectric constant mismatch. The transient stress response is therefore a powerful tool to uncover the mechanisms by which ER activity arises. The shear-rate dependence of the stress response was also investigated; perhaps counterintuitively, the characteristic times for stress growth and decay were largest at low shear rates and varied inversely with shear rate at high shear rates. This relationship is qualitatively consistent with the contributions of particle polarization and the growth and strain of the field-induced structure to the observed transient stresses.[2]

*Progress in Electrorheology*, Edited by K.O. Havelka
and F.E. Filisko, Plenum Press, New York, 1995

## EXPERIMENTAL TECHNIQUES

The ER fluid systems utilized in this study included a dispersion of crystalline aluminosilicate (zeolite) particles at a volume fraction $\phi = 0.3$ in poly(dimethyl siloxane) or silicone oil. The bulk dielectric constant $K_p$ of many crystalline aluminosilicates[3] is roughly 2 while the dielectric constant $K_f$ of silicone oil is 2.5. A dispersion of amorphous aluminosilicate particles[4] at $\phi = 0.3$ in paraffin oil ($K_f = 2.2$) was tested. Finally, a suspension of barium titanate ($BaTiO_3$) particles ($K_p \approx 2000$) at $\phi = 0.2$ in dodecane ($K_f = 2.0$) was studied.[5-7] The particulates used in each model fluid were nominally anhydrous, as they were dried in vacuo at elevated temperatures in order to minimize adsorbed water. In light of the low dielectric constants of the aluminosilicates, ER activity in these systems is ascribed to a surface or bulk conductivity mechanism involving the transport of ions.[3] By contrast, the dielectric constant of bulk barium titanate should produce ER activity in alternating electric fields by a dielectric polarization mechanism. Indeed, the predictions of an electrostatic model for high-dielectric-constant ER fluids agree with a variety of rheological measurements on barium-titanate-based ER fluids.[6]

Rheological experiments were performed using a custom-made Couette or concentric cylinder (cup-and-bob) rheometer. Fluid samples were electrified using a high-voltage amplifier having 25 kHz bandwidth driven by a signal generator. The temporal resolution of the stress measurement was limited[8] by the torsional resonant frequency of the apparatus, ~ 400 Hz. All measurements were performed at room temperature, nominally 22 ± 1 °C. Time series of the motor rotation speed, the torque signal, the applied voltage, and the current were recorded and averaged using a digital storage oscilloscope. For more details on sample preparation and experimental techniques, see References 2 and 6.

## EXPERIMENTAL RESULTS

### Unipolar Excitation: Turn-On Effects

Transient shear stresses were generated by imposing steady shear on ER fluid samples while subjecting them to stepwise changes in the applied electric field. This experimental procedure is a convenient approximation to real-life conditions under which fast response is desired. The response of the amorphous aluminosilicate suspension at a constant shear rate $\dot{\gamma} = 5\ s^{-1}$ to excitation with a unipolar square wave of peak amplitude 1 kV/mm and frequency 5 Hz was measured; the applied field, current density, and shear stress are shown in Fig. 1. A substantial response time after the field was turned on was required for the stress to obtain its maximum value, which was reached under steady-state conditions. The character of the stress growth and decay in this system is similar to that predicted in a recent molecular dynamics simulation of the stresses in ER fluids.[9] While qualitatively similar, the response of the crystalline aluminosilicate system, Fig. 2, is much faster than that of the amorphous aluminosilicate system; very fast initial stress growth is followed by a somewhat slower approach to the steady state. When the field is removed and the damping effect of the fluid is much reduced, transient oscillations of the rheometer bob are excited; they appear as oscillations of the stress in Fig. 2.

The transient stress response of the barium titanate fluid, Fig. 3, is remarkably different than those exhibited by the aluminosilicate fluids. A unipolar square wave of peak amplitude 2 kV/mm and frequency 10 Hz was utilized; the time dependence of the applied field, current density, and shear stress with $\dot{\gamma} = 5\ s^{-1}$ were recorded. It is

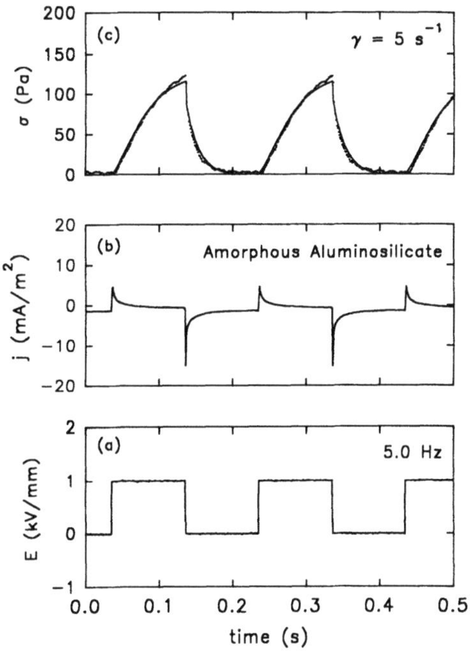


Figure 1. The time dependence of (a), the applied field E; (b), the current density j; and (c), the shear stress $\sigma$ (dots) in the amorphous aluminosilicate ER fluid. The sample was sheared in a Couette cell at a constant rate $\dot{\gamma} = 5\ s^{-1}$ and was excited by a 1 kV/mm peak amplitude, 5 Hz unipolar square wave. The solid line in (c) is a fit to the stress by the dynamic polarization model, Eqs. 1, 2, and 4.

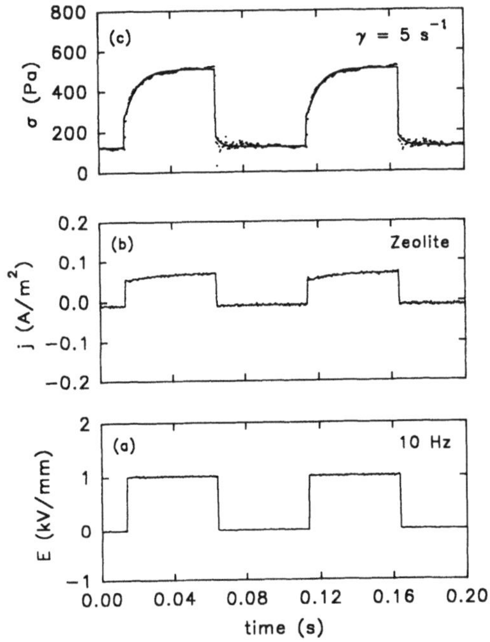


Figure 2. The time dependence of (a), the applied field E; (b), the current density j; and (c), the shear stress $\sigma$ (dots) in the crystalline aluminosilicate, or zeolite, ER fluid. The sample was sheared in a Couette cell at a constant rate $\dot{\gamma} = 5\ s^{-1}$ and was excited by a 1 kV/mm peak amplitude, 10 Hz unipolar square wave. The solid line in (c) is a fit to the stress by the dynamic polarization model, Eqs. 1, 2, and 4. The relatively high conductivity of this fluid is reflected in the high current density in (b).

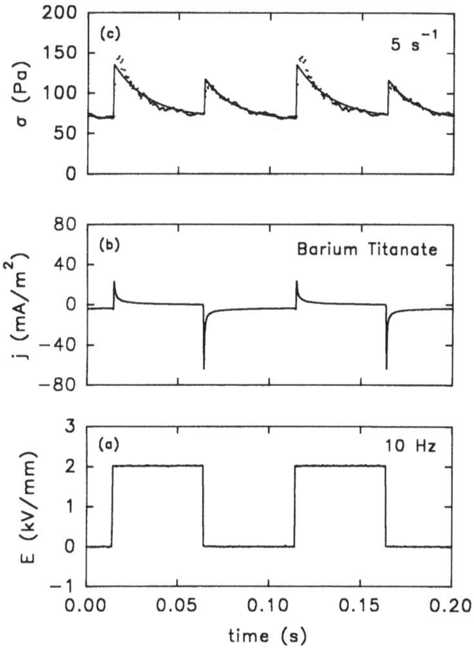


Figure 3. The time dependence of (a), the applied field E; (b), the current density j; and (c), the shear stress $\sigma$ (dots) in the barium titanate ER fluid. The sample was sheared in a Couette cell at a constant rate $\dot{\gamma} = 5\ s^{-1}$ and was excited by a 2 kV/mm peak amplitude, 10 Hz unipolar square wave. The solid line in (c) is a fit to the stress by the dynamic polarization model, Eqs. 1, 2, and 4.

seen that both the application and removal of the field induces essentially instantaneous (or beyond the limits of resolution of the apparatus) peaks in the shear stress, which then decay to some steady-state level. The shear stress peak induced by the turn-on of the field is larger than that produced by the turn-off.

### Bipolar Excitation: Effect of Field Reversal

The stress response of these fluids to bipolar square wave excitation, i.e., to reversals of the applied field, were also studied. While field reversals are not encountered in most actual applications, the reaction of the fluids to such transients can nonetheless provide insight into the origins of the stress response. In Fig. 4, the current density and shear stress generated in the amorphous aluminosilicate fluid by bipolar square wave excitation with a peak amplitude of 1 kV/mm at a frequency of 1 Hz is shown. The shear rate is again $\dot{\gamma} = 5\ s^{-1}$. As might be expected intuitively, a reversal of the direction of the field produces a temporary reduction of the stress, which recovers its steady-state value after a nonzero recovery time.

By applying a bipolar square wave of 1 kV/mm peak amplitude and 10 Hz frequency to the barium titanate fluid as it is sheared at 5 $s^{-1}$, the current density and stress transients shown in Fig. 5 result. In distinct contrast to the amorphous aluminosilicate fluid, each reversal of the field direction produces a transient *peak* in the shear stress. Qualitatively similar results are obtained using ER fluids based on particles of bismuth titanate,[7,10] another high-dielectric-constant ceramic.

## DISCUSSION

### The Transient Stress Response: A Model

It can be seen that each ER fluid system possesses a distinctive transient stress signature; of course, the detailed response depends on the shear rate, electric field, and frequency.[1,2] By considering the mechanisms by which polarization is induced in ER particulates, and thus how ER activity is promoted, one may semiquantitatively understand the transient stress response. To see this, an ER fluid is modeled as identical, isolated spherical particles of radius a, dielectric constant $K_p$, and conductivity $S_p$ suspended in a fluid with dielectric constant $K_f$ and conductivity $S_f$. These assumptions form the basis of the Maxwell-Wagner-Sillars treatment of the dielectric response of inhomogeneous materials.[11,12] We wish to extend the generally-accepted polarization model for ER fluids, which associates the particle polarization with the ER response,[13] to transient electrorheology. The time dependence of the dipole moment $\mathbf{p} = p(t)\hat{\mathbf{z}}$ on each particle as it is subjected to step changes in the electric field is thus of interest; here $\hat{\mathbf{z}}$ is the direction of the applied field. By solving for the electric field, current, and the resulting particle polarization, one finds that the time-dependent dipole moment induced by a step increase in the electric field from 0 to $E_0$ at a time t = 0 is given by

$$\begin{aligned} p(t) &= 0\ , & t<0\ , \\ p(t) &= 4\pi\epsilon_0 K_f a^3 E_0[\beta_c+(\beta_d-\beta_c)e^{-t/\tau}]\ , & t\geq 0\ , \end{aligned} \tag{1}$$

where $\epsilon_0 = 8.85 \times 10^{-11}$ F/m is the permittivity of free space. The relative polarizability or dielectric mismatch $\beta_d$ is given by $\beta_d = (K_p - K_f)/(K_p + 2K_f)$, while the analogous conductivity mismatch is defined as $\beta_c = (S_p - S_f)/(S_p + 2S_f)$. The characteristic time for charge transport $\tau$ is given by $\tau = \epsilon_0(K_p + 2K_f)/(S_p + 2S_f)$. If the steady state is achieved in an electric field $E_0$ that is removed at t = 0, the time-dependent dipole moment is given by

$$p(t) = 4\pi\epsilon_0 K_f a^3 E_0 \beta_c\ , \qquad t<0,$$
$$p(t) = 4\pi\epsilon_0 K_f a^3 E_0 (\beta_c - \beta_d) e^{-t/\tau}\ , \qquad t\geq 0\ . \qquad (2)$$

Finally, if the field amplitude is changed from $-E_0$ to $+E_0$ at t = 0, the evolution of the dipole moment is

$$p(t) = -4\pi\epsilon_0 K_f a^3 E_0 \beta_c\ , \qquad t<0,$$
$$p(t) = 4\pi\epsilon_0 K_f a^3 E_0 [\beta_c - 2(\beta_c - \beta_d) e^{-t/\tau}]\ , \qquad t\geq 0\ . \qquad (3)$$

Note that the time dependences of the dipole moment expressed in Eqs. 1 - 3 can be written as $p(t) = p_0 f_i(t)$ for Eq. i = 1 to 3, where $p_0 \equiv 4\pi\epsilon_0 K_f E_0 \beta_c$ is the steady-state dipole moment. The $f_i(t)$ are dimensionless functions involving the transport time, $\tau$, and the relative size of the dielectric and conductivity mismatches, $\beta_d/\beta_c$. For example, $f_1(t) \equiv 1 + (\beta_d/\beta_c - 1)e^{-t/T}$ describes the time dependence of the dipole moment when the field is increased from zero, as in Eq. 1.

The electrostatic force between pairs of particles having identical dipole moments p separated by a distance R scales as $p^2/R^3$.[13] Since these interparticle forces are responsible for the field-induced increase in shear stress, one may infer that the time-dependent stress is related to the dipole moment p(t) by

$$\sigma(t) - \sigma_0 \propto p^2(t)\ , \qquad (4)$$

where the viscous contributions to the stress[14], as well as any systematic offsets in the measured stresses, are lumped into $\sigma_0$. The polarization dynamics described by Eqs. 1 - 4 can then be cast into the general form $\sigma(t) = \sigma_0 + \sigma_1 f_i^2(t)$ where $\sigma_1$ is a constant into which the unknown effects of geometrical factors such as the interparticle spacing are lumped. The most important prediction of this model -- and a central point of this paper -- is that the shape of the transient stress curves depend sensitively on the relative contribution of the dielectric and conductivity mismatches to the particle polarization mechanism. If polarization occurs predominantly by a conductivity mismatch ($\beta_d/\beta_c << 1$), a sudden increase in the applied electric field is predicted to lead to growth in the shear stress over a characteristic time $\tau$ and the eventual maximization of the stress in the steady state. If the polarization is instead produced by a dielectric mismatch ($\beta_d/\beta_c >> 1$), an increase in the field produces an instantaneous peak in the stress that decays over a time $\tau$ due to screening, i.e., because of the residual conductivity mismatch. *These distinct qualitative behaviors enable a determination of the mechanisms of electrorheological activity in ER fluids.*

The predictions of the dynamic polarization model were fitted to the measured stress transients using a Levenberg-Marquardt nonlinear-least-squares curve-fitting algorithm; the solid lines in Figs. 1-5 demonstrate that qualitatively good fits were obtained. The resulting fitted parameters $\beta_d/\beta_c$, $\tau$, $\sigma_0$, and $\sigma_1$ are shown in Table I. Good agreement is obtained between the values of $\beta_d/\beta_c$ and $\tau$ obtained in independent fits of the unipolar and bipolar transients in the amorphous aluminosilicate and barium

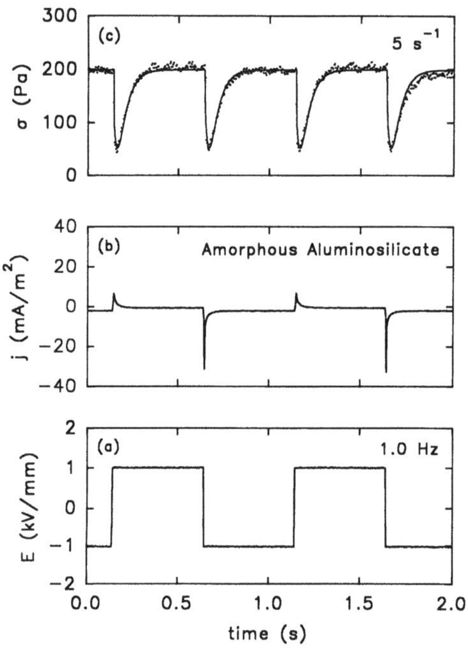


Figure 4. The time dependence of (a), the applied field E; (b), the current density j; and (c), the shear stress $\sigma$ (dots) in the amorphous aluminosilicate ER fluid. The sample was sheared in a Couette cell at a constant rate $\dot{\gamma} = 5\ s^{-1}$ and was excited by a bipolar electric field of 1 kV/mm peak amplitude at a frequency of 1 Hz. The solid line in (c) is a fit to the stress by the polarization model, Eqs. 3 and 4.

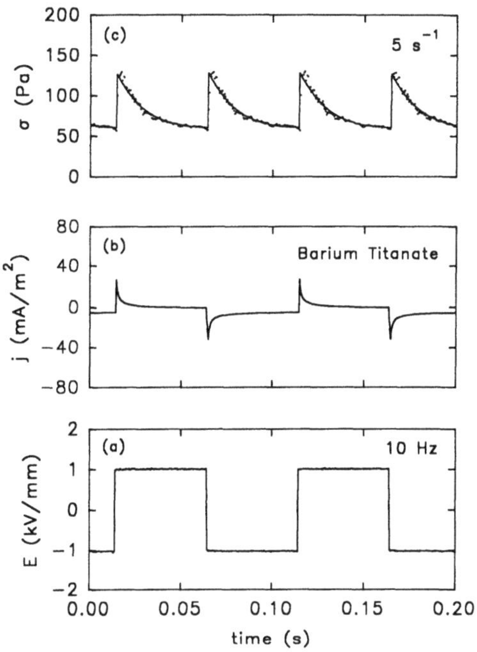


Figure 5. The time dependence of (a), the applied field E; (b), the current density j; and (c), the shear stress $\sigma$ (dots) in the barium titanate ER fluid. The sample was sheared in a Couette cell at a constant rate $\dot{\gamma} = 5\ s^{-1}$ and was excited by a bipolar electric field of 1 kV/mm peak amplitude at a frequency of 10 Hz. The solid line in (c) is a fit to the stress by the polarization model, Eqs. 3 and 4.

Table I. Parameters obtained by fitting the measured stress transients to the polarization model.

| ER system | excitation | $\beta_d/\beta_c$ | $\tau$ (ms) | $\sigma_0$ (Pa) | $\sigma_1$ (Pa) |
|---|---|---|---|---|---|
| amorphous aluminosilicate | unipolar (Fig. 1) | 0.15 | 34 | 2 | 129 |
| | bipolar (Fig. 4) | 0.16 | 40 | 52 | 147 |
| zeolite | unipolar (Fig. 2) | 0.61 | 6.2 | 124 | 382 |
| barium titanate | unipolar (Fig. 3) | 7.2 | 33 | 66 | 1.32 |
| | bipolar (Fig. 5) | 7.3 | 27 | 56 | 0.40 |

titanate fluids. Because both the viscous and electrostatic stress components give rise to time-independent stresses, the values of the parameters $\sigma_0$ and $\sigma_1$ (and to some extent $\beta_d/\beta_c$) are somewhat uncertain; the quality of the fits is in any case not affected.

For the amorphous aluminosilicate fluid, the fits imply that the conductivity mismatch $\beta_c$ is much larger than the dielectric mismatch $\beta_d$. As stated above, a conductivity-mismatch-based polarization mechanism is associated with the finite time required for the growth of the shear stress in this system, Figs. 1 and 4. That $\beta_c >> \beta_d$ is expected in ER systems in which the transport of real charge carriers -- ions, electrons, or holes -- to the interface between particle and fluid promotes the ER effect.[15-17] Within the present model, such a situation arises from a substantial mismatch between the particle and fluid conductivities. It may also result from surface conductivity effects due to adsorbed polar impurities. When the field direction is reversed (Fig. 4), a finite time is required for the charge carriers to redistribute and thus for the ER response to recover. The parameters obtained by fitting the field reversal data again suggest that $\beta_c >> \beta_d$ as expected for a system dominated by the conductivity mismatch. The transport time obtained in the unipolar fit, $\tau \approx 34$ ms, can be used to infer an upper limit on the conductivity of the suspended particles; a plausible value, $S_p \sim 1.8 \times 10^{-9}\ \Omega^{-1}m^{-1}$, is obtained.

In the barium titanate system, by contrast, the fitted dielectric mismatch is larger than the conductivity mismatch: $\beta_d >> \beta_c$. A large dielectric mismatch is consistent with the high dielectric constant of bulk barium titanate. Because dielectric polarization occurs instantaneously in this model, it is the substantial dielectric mismatch that gives rise to the initial particle polarization and to the shear stress maximum observed as the field is applied, Figs. 3 and 5. Because a particle-fluid conductivity mismatch also exists, charge is carried to the particle surface on a time scale $\tau$, eventually screening the dielectric polarization and thus causing the shear stress to decay to its steady-state value. When the field is rapidly returned to zero, the dielectric polarization drops to zero nearly instantaneously. The residual polarization due to the screening charge (caused by $\beta_c$) gives rise to a smaller transient stress peak, Fig. 3, that also decays due to charge transport on the time scale $\tau$.

The unipolar stress transients in the crystalline aluminosilicate system, Fig. 2, suggest that in general *both* dielectric and conductivity mismatches contribute to ER activity, since the fit to the polarization model implies that $\beta_d \sim 0.6\ \beta_c$. The dielectric mismatch gives rise to the initial fast component of the stress growth, while the slower growth toward the steady state stress is assignable to the conductivity mismatch.

It has been shown that the very different transient responses of the aluminosilicate and barium titanate systems can be understood qualitatively by applying this dynamic polarization model. Correlated measurements of the dielectric spectra of these ER systems and their constituents are necessary for a more complete quantitative test of the model. The model does not include several factors that may influence the transient response. For example, a nontrivial frequency dependence of the dielectric constants and conductivities would lead to more complex transient effects. Rotations of the particles in shear flow may assist charge transport and thus influence the transport time $\tau$.[18] Nonlinear conductivity effects associated with ionization and breakdown of the suspending fluids[19] may give rise to field-dependent transient phenomena. The time-dependent growth of particle chains or columns induced by the application of the electric field has not been treated. The observed transient stresses might be associated with the buildup of stress as these structures are sheared, although these effects will be shown below to contribute mainly at shear rates much higher than that for which the model was applied (5 $s^{-1}$). The time dependence of the dipole moment is calculated as if the particles were isolated in an external field; local field and multipole effects due to neighboring particles are clearly relevant in dense ER suspensions. An integral equation solution of the time-dependent forces in a linear chain of particles[20] is still qualitatively described by the form of Eqs. 1-4; the general conclusions of the dynamic polarization model are therefore unaltered by these effects.

## The ER Response: Frequency Dependence

These interpretations of the time dependence of the stress transients are in agreement with those made by analyzing the stresses in the frequency domain. Specifically, the ER activity of the amorphous aluminosilicate fluid is substantial in static fields and decreases as the frequency of the applied ac field is increased; the barium titanate fluid, on the other hand, exhibits virtually no ER activity with static fields and shows an increasing shear stress with increasing frequency.[1,2] To reiterate, these contrasting behaviors reflect two different mechanisms by which ER activity is achieved:[15-17] ER activity in the amorphous aluminosilicate fluid (and many other model and commercial fluids) is promoted by charge transport. The ER response drops off as the frequency of the applied field exceeds a rate $\sim 1/\tau$ characteristic of this transport process. In the barium titanate system, the conductivity mismatch is relatively small, so this fluid is not ER active in static fields. As the frequency is increased above $\sim 1/\tau$, the polarization is dominated by the large particle/fluid dielectric constant mismatch and the shear stress increases, as is observed.

## The Stress Response Time: Shear Rate Dependence

The dependence of the ER response on shear rate is clearly important for device design. The rise time required for the shear stress to reach 1 - 1/e or 63.2 % of its steady-state value was measured as a function of shear rate in the amorphous aluminosilicate system, Fig. 6. Perhaps counterintuitively, the stress response is faster at higher shear rates in this and all of the ER systems studied.[1,2] The rise time is large and nearly independent of shear rate at low rates, while it varies inversely with shear rate at high rates.

The various physical processes that determine the rheological response time can be identified. First of all, it is likely that some degree of field-induced structure formation or particle rearrangement is required in order to generate substantial ER shear stresses. The time required for the growth of the applied field, the polarization

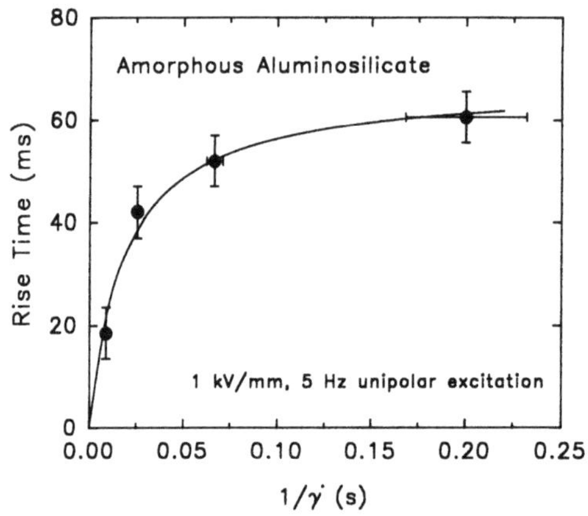


Figure 6. The strain rate dependence of the stress rise time in the amorphous aluminosilicate fluid. The solid curve is a fit to Eq. 7.

of the particles, and the electrostatic forces necessary for structure formation can belabeled $\tau_{el}$. The initial aggregation time $\tau_{agg}$ required for the formation of particle chains in quiescent, i.e., unsheared ER fluids, has been examined in molecular-dynamics simulations[21] and in diffuse-optical experiments.[22,23] The overall structural response time can be estimated as

$$\tau_{structure} \approx \tau_{el} + \tau_{agg} \ . \tag{5}$$

Other potentially relevant time scales include the eventual coarsening of the chains into columns,[24] other structural processes, or dielectric relaxations. The application of shear might also be expected to modify the particle aggregation process.[25]

Even if the particles are polarized and they aggregate into a network structure, no shear stress will result unless the network is strained. Substantial stresses are achieved as the network structure is strained at levels approaching a critical strain (the yield strain) $\gamma_y$. Assuming that the structure deforms affinely, the characteristic time scale required to approach this level of strain is[14]

$$\tau_{strain} \approx \frac{\gamma_y}{\dot{\gamma}} \ . \tag{6}$$

The overall rheological response time is controlled by the time scales for particle polarization, particle aggregation, and the growth of the applied strain. Assuming that the structure formation and strain processes occur simultaneously and thus in parallel, the stress response rate would be[1,2]

$$\frac{1}{\tau_{rise}} \approx \frac{1}{\tau_{structure}} + \frac{\dot{\gamma}}{\gamma_y} \ . \tag{7}$$

Equation 7 predicts a relationship between the rheological response time and the shear rate that is at least qualitatively consistent with experiment, as shown in Fig. 6. At low strain rates, the rheological response time is nearly independent of shear and is dominated by particle polarization and structure formation processes. At high strain rates, the response time is controlled by shear. The measured dependence of the stress rise time on shear rate for the amorphous aluminosilicate ER fluid was fitted to Eq. 7, as shown in the solid curve in Fig. 6; the fitted parameters are $\tau_{structure} \approx 70$ ms and $\gamma_y \approx 3.5$. Open questions surrounding the nature of the shear-rate dependence of the response time include the role of shear-rate-dependent aggregation processes, particle or cluster rotations, and cluster deformation in the transient rheology of ER fluids.

## CONCLUSIONS

The transient rheological response of ER fluids in time-dependent electric fields was investigated. The sudden changes in electric field associated with square wave excitation induce stress transients that unambiguously reflect the mechanisms by which ER activity is achieved. For example, the stress induced by the application of an electric field in an amorphous aluminosilicate ER fluid grows over a characteristic rise time until it reaches a maximum in the steady state. By contrast, the stress induced by the application of a field in the barium titanate system appears almost instantaneously, but decays over some characteristic time toward the steady state. These disparate transient behaviors can be explained by appealing to a model for the dynamics of the particle polarization in ER fluids. This model reveals that the transient behavior of the aluminosilicate systems is consistent with the dominance of a conductivity mismatch between particles and fluid, while that of the barium titanate system is in agreement with the expected large dielectric mismatch. These differences also appear in the frequency domain: those ER fluids that incorporate a mismatch in particle and fluid conductivities are active with static and low-frequency excitation, while those that rely solely on a dielectric constant mismatch are ER active only with ac excitation. The transient response is thus a powerful tool to uncover the physical origins of electrorheological activity.

The characteristic time scale for the stress transients was found to vary inversely with the applied shear rate. Consideration of the role of the time scales for particle polarization, structure formation, and the strain and yielding of the structure leads to a predicted shear-rate dependence qualitatively consistent with observation. These results imply that high shear rates must be incorporated into ER devices that require a fast response.

## ACKNOWLEDGEMENTS

The authors thank Frank Filisko for providing an ER fluid sample, Larry Elie for technical assistance, and Dan Klingenberg and Alan Wineman for useful discussions.

## REFERENCES

1. J. M. Ginder and S. E. Ceccio, Transient stresses in electrorheological fluids, *Polym. Prepr.* 35:319 (1994).
2. J. M. Ginder and S. E. Ceccio, The effect of electrical transients on the shear stresses in electrorheological fluids, *J. Rheol.*, in press.

3. F. E. Filisko, Rheological properties and models of dry ER materials, *in*: "Proceedings of the Third International Conference on Electrorheological Fluids," R. Tao, ed., World Scientific, Singapore, p. 116 (1992).
4. F. E. Filisko, unpublished.
5. T. Garino, D. Adolf, and B. Hance, The effect of solvent and particle dielectric constants on the electrorheological properties of water-free ER fluids, *in*: "Proceedings of the Third International Conference on Electrorheological Fluids," R. Tao, ed., World Scientific, Singapore, p. 167 (1992).
6. J. M. Ginder and L. C. Davis, Viscoelasticity of electrorheological fluids: role of electrostatic interactions, *in:* "Proceedings of the Fourth International Conference on Electrorheological Fluids," R. Tao, ed., World Scientific, Singapore, p. 267 (1994).
7. C. A. Randall, D. E. McCauley, C. P. Bowen, T. R. Shrout, and G. L. Messing, High dielectric constant particulate materials for electrorheological fluids, *in:* "Proceedings of the Fourth International Conference on Electrorheological Fluids," R. Tao, ed., World Scientific, Singapore, p. 60 (1994).
8. Z. Lou, R. D. Ervin, and F. E. Filisko, The influence of viscometer dynamics on the characterization of an electrheological fluid under sinusoidal electric excitation, *J. Rheol.* 37:55 (1993).
9. H. See and M. Doi, Shear resistance of electrorheological fluids under time-varying electric fields, *J. Rheol.* 36:1143 (1992).
10. C. A. Randall and J. M. Ginder, unpublished.
11. J. C. Maxwell, "A Treatise on Electricity and Magnetism," vol. 1, Dover, New York, p. 452 (1954).
12. See, e.g., L. K. H. van Beek, Dielectric behavior of heterogeneous systems, *in:* "Progress in Dielectrics," vol. 7, CRC Press, Cleveland, p. 69 (1967).
13. See, e.g., A. P. Gast and C. F. Zukoski, Electrorheological Fluids as Colloidal Suspensions, *Adv. Coll. Int. Sci.* 30:153 (1990).
14. R. T. Bonnecaze and J. M. Brady, Yield stresses in electrorheological fluids, *J. Rheol.* 36:73 (1992).
15. L. C. Davis, Finite-element analysis of particle-particle forces in electrorheological fluids, *Appl. Phys. Lett.* 60:319 (1992).
16. L. C. Davis, Polarization forces and conductivity effects in electrorheological fluids, *J. Appl. Phys.* 72:1334 (1992).
17. R. A. Anderson, Effects of finite conductivity in electrorheological fluids, *in:* "Proceedings of the Third International Conference on Electrorheological Fluids," R. Tao, ed., World Scientific, Singapore, p. 81 (1992).
18. H. Block, K. M. W. Goodwin, E. M. Gregson, and S. M. Walker, Stimulated resonance between electrical and shear fields by a colloid system, *Nature* 275:632 (1978).
19. N. Felici, J.-N. Foulc, and P. Atten, A conduction model of electrorheological effect, *in:* "Proceedings of the Fourth International Conference on Electrorheological Fluids," R. Tao, ed., World Scientific, Singapore, p. 139 (1994).
20. L. C. Davis, unpublished.
21. K. C. Hass, Computer simulations of nonequilibrium structure formation in electrorheological fluids, *Phys. Rev. E* 47:3362 (1993).
22. J. M. Ginder and L. D. Elie, Optical probes of structure formation in electrheological fluids, *in:* "Proceedings of the Third International Conference on Electrorheological Fluids," R. Tao, ed., World Scientific, Singapore, p. 23 (1992).
23. J. M. Ginder, Diffuse optical probes of particle motion and structure formation in electrorheological fluids, *Phys. Rev. E* 47:3418 (1993).
24. T. C. Halsey and W. Toor, Structure of electrorheological fluids, *Phys. Rev. Lett.* 65:2820 (1990).
25. D. J. Klingenberg, private communication.

# REQUIREMENTS AND TARGETS FOR ER FLUIDS IN ELECTRICALLY FLEXIBLE HIGH SPEED POWER TRANSMISSIONS

W.A. Bullough, J. Makin, and A.R. Johnson

Department of Mechanical and Process Engineering
The University, Mappin St.
PO Box 600, Sheffield, S1 4DU
United Kingdom

## ABSTRACT

The paper explains the comprehensive characterisation data required on ER Fluids if they are to be applied optimally to flexibly operated, fast machines for light loads. Yield or electro-stress, electron-hydraulic time delay and mixture viscosity (zero volts) and their relationship are dealt with by reference to the inertial design of a representative machine. Associated temperature/heat generation/cooling and electrical design/control/supply problems are cited as important parallel modelling necessities towards achieving the best machine performance on a given fluid. The need for equivalent fluid performance versus materials/operational parameters/constituents models is emphasised, performance predictions are made and non-yield stress limiting factors to the flexible machine concept are described. The advantages of a 'technotronic' approach to the machine and fluid developer are shown by example.

## THESIS

The concern implicit in the stated aim of this meeting, that fluid development is the most important factor in the well being of the FRF (field responsive fluids) area and should get corresponding attention, is well justified. Much time has been spent on often seemingly hit or miss trials of devices using only partly specified fluids. On the other hand it could be claimed that with a priori perception of application requirements more pertinent fluid characterisation and development would have taken place. The purpose of the present act is to clarify current targets for one particular class of FRF application - high speed, flexibly operated/electronically designated, variable configuration machines.

The use of field responsive fluids appears likely to divide into two streams. These will be categorised: (a) 'apparently' very light duty, rapidly responding (fast yield stress response after trigger control signal i.e. a low electron-hydraulic time constant) devices for

*Progress in Electrorheology*, Edited by K.O. Havelka
and F.E. Filisko, Plenum Press, New York, 1995 

which electro-rheological fluids (ERF) seem well suited and (b) heavier duties where external or inertial loads are high and the supply excitation time constant is consequently less important. Magneto-rheological suspensions (MRS) currently appear to be pre-eminent in this second class. How much so will depend on their speed of response to forced field coil excitation, toleration of the size of the associated magnetic circuit and serviceability evaluations. Nevertheless, from this basis the development aims and operational targets for electro-rheological fluids can be clarified.

It is important to realise that the properties of FRF are only a part of the engineering problem. Very often the dead weight of the transmission parts which lie between a load and a controlling ERF, which rapidly connects the load to the driver, have a much greater inertia than the load. An optimal low inertia transmission geometry and lightweight materials can reduce the demands on the fluid considerably[1,2]. However, this approach cannot be followed altogether independently and the dynamics of the whole system, including the time domain properties of the ERF[3], a heat transfer estimation[4] plus electrical supply, control and consumption[5,6] situations need to be simulated in parallel. The data required for these exercises, analyses, etc. determines the scope of fluid development and characterisation.

Firstly, the electro shear stress $\tau_e$ versus shear rate $\dot{\gamma}$ and field strength E characteristics need to be determined with respect to fluid temperature $\theta$. Radial effects of electric and velocity fields need to be designed out of the apparatus used[7] for this; temperature gradients are not usually severe. At the same time the rise time or electron-hydraulic time delay ($t_m$*) of $\tau_e$ (with respect to excitation) can be established. Since this is of very short duration there is usually no great necessity to further establish the form $\tau_e$ (t). This is as well since it is difficult to get a noiseless torque reading (Couette apparatus) or totally eliminate flow system (Poiseuille test rig) effects. If a high resolution, high speed transient, multi-channel recorder (typically 12 bit x $10^6$ sample/sec) is used, electrical time domain characteristics can be taken simultaneously. It is very important to make the resulting data as free from spurious effects as is possible; these can be much slower than ER effects and can distort the characteristics thus derived.

$\tau_e$ ($\dot{\gamma}$, $\theta$, E), defined as total excited stress minus the viscous stress obtained at zero excitation (but otherwise under the same conditions) are very important characteristics for the designer who will often be compelled to confront both a high field strength and shear rate and design for a high fluid temperature, where a near uniform shear stress level (or falling or rising) versus shear rate can occur[3]. Within this sector some near wall effect is present; the fluid is not a true continuum so results are required to be taken as close as possible to realistic engineering conditions. Since flexible machine power transmissions will switch on and off regularly, $\tau_e$ at $t^*_m$ will satisfactorily describe the electro stress component for the 'on' interval.

An equivalent circuit of parallel and series resistances and capacitances is needed to establish switching times and rates. In order to achieve the parametric data from the $\tau_e$, $t_m$* tests and to isolate hysteresis effects, on-off step excitation will prove useful. From the model thus produced supply and control strategies can be deduced. Though conductance has importance in terms of heating, capacitance largely determines the supply circuit.

Throughout these multi-variable exercises, repeatability in controlled tests is the prime aim. Results from these must be so far as possible edited by comparison with well defined fluid properties. Serviceability data can only be achieved at realistic application conditions and this often sets operation at near discharge/breakdown stress, and $\dot{\gamma} \rightarrow$ 10,000 $s^{-1}$ (at least). Air and/or water susceptibility of the fluid can be a problem with some materials and centrifugal separation behaviour at 100g or more should be sought.

In high speed machines the zero volts viscosity can be more important in terms of heating than how it effects the dynamic performance of the device. Viscous drag should not be allowed to grow too far at the expense of acceleration level attainment. Only a mechatronic/thermal assessment of the overall/specific device can determine the full effects of $\mu$ and $\tau_e$. Whatever thermodynamic transport properties of the fluid and its constituents are available should be given on every specification sheet, as a function of temperature and pressure.

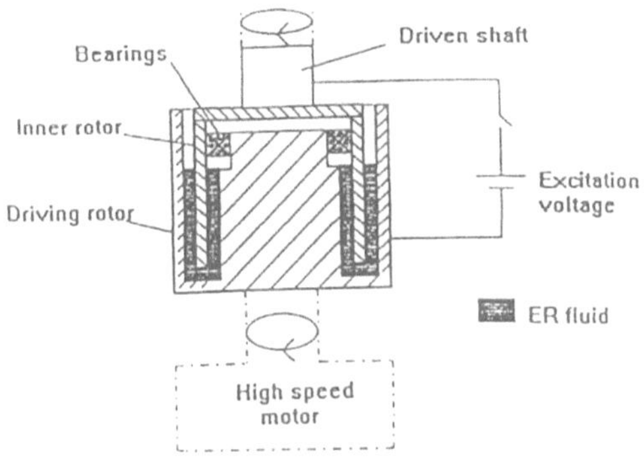


**Figure 1** Schematic Double Acting Clutch
Outpulley Pulley is located on Driven Shaft
Two laterally displaced, vertical axis clutch/pulley arrangements are connected by a belt

## TEST MECHANISM

By way of an example, the order of performance improvements which can be achieved by the total mechatronic design approach to a device can be highlighted by a study of a reversing apparatus which is driven by two contra-rotating ER clutches, each similar in construction to that shown in Fig. 1. The original prototype built to test the concept was more influenced by a (then) necessary quasi-steady philosophy - in ERF devices non-linear inertial considerations dominate the problem.

The two input rotors are connected via ERF to two output clutch/pulley connections which are in turn connected via a belt. As the controlling excitation is switched from one clutch to the other the belt is driven backwards and forwards. If high accelerations in the linear motion are to be achieved and the device prove satisfactory, each part must be examined in an inertial sense and improved as far as is possible, in every aspect of performance.

Originally, the prototype device has three pulleys, one an idler which serves to tension the belt and provides an encoder for measurement purposes, with the other two spindles additionally carrying a short clutch, made so for ease of manufacture, reduction of overhang load on the bearings, and inverted so as to keep fluid from them etc. The motion (from constant steady state velocity of the belt to the right and then to the left) was based on a then best fluid for the purpose of circa 2 kPa nominal $\tau_e$ and 45 mPas $\mu$ at 30°C with $\dot{\gamma}$ max being approximately from -6000 to +6000 $s^{-1}$ (at one point 12,000 $s^{-1}$) and gave turn round times > 200 msec on test. A fluid of $\tau_e$ approximately 15 kPa would have been required to reduce this to 10 ms turn round time. An examination of the dynamic equations gives

$$d\omega = \frac{\left(\tau_e - 2\mu\omega A r^2 / h\right)}{2I + mR^2} dt \qquad (1)$$

and turn round time $\overleftarrow{u}$ to $\overrightarrow{u} = \int_{-\omega_1}^{+\omega_1} \frac{\left(2I + mR^2\right) d\omega}{\tau_e + 4\pi\mu\omega r^3 \ell / h}$ (2)

Equation 2 shows the importance of the output inertial load (fluid mass has been neglected). $\omega_1$ is a constant input clutch speed, $\omega$ the instantaneous output clutch speed, I is the rotating inertia of one output spindle assembly, m the mass of the belt at pulley radius R, A the area of one electrode surface ($2\pi r\ell$), and h the inter electrode gap.

The analysis was proved by experiment with the low $\tau_e$ fluid, and a computer game then played viz: (i) the idler pulley was removed and the other pulleys inertia lowered by two thirds; this reduces turn round time in proportion to output side inertia. (ii) an exercise on the best pulley or clutch diameter (turn round time basis) leads to an increase in pulley diameter, a reduction in clutch speed and a lengthening of the ER clutch - see Table 1 - a further 30% improvement in ERF requirement is established.

**TABLE 1** The effect of changing the cup radius and pulley radius on the turn round time. Cup length 45.5 mm, $\mu$ = 170 mPas, $\tau_e$ = 10 kPa U = 5 m/s, h = 0.5 mm

| R Pulley Radius (mm) | Turn Round Time (ms) r Cup radius (mm) 25 | 30 | 35 | 40 |
|---|---|---|---|---|
| 30 | 6.47 | 6.69 | 7.3 | 8.15 |
| 35 | 6.23 | 6.2 | 6.58 | 7.21 |
| 40 | 6.27 | 5.98 | 6.26 | 6.6 |
| 45 | 6.54 | 5.98 | 5.95 | 6.22 |
| 50 | 7.01 | 6.24 | 5.91 | 6.02 |
| 55 | 7.64 | 6.47 | 6.01 | 5.96 |

Also gleaned from the interplay is the primary importance of the fluid zero volts viscosity, in heating. In Table 2, above a certain value of $\mu$ relative to $\tau_e$ the machine will not start up. When the size of $\tau_e$ is large relative to $\mu$ then (within reason) the latter has relatively little dynamic effect. Whilst the size of the inter electrode space can be increased to reduce the effect of $\mu$ on heating the excitation level must be increased to maintain $\tau_e$. This elevates the device resistance and reduces its capacitance (no bad thing) the electrical model indicating that (near classically) the RC of the dielectric circuit remains constant i.e. without deterioration of the switching duration.

**TABLE 2** - Interactions Between Various Parameters of Reversing System
- effect on TRT msec and (maximum linear acceleration of cup x g)
Conditions U = 5 m/s, h = 0.5 mm with constant output inertia

(a) Effect of $\tau_e$ and $\mu$ aluminium clutch/rotor 2 mm thick

| ↓ $\tau_e$ kPa | 50 | 100 | 200←$\mu$mPas |
|---|---|---|---|
| 2 | 47.8 (29.4) | 56.5 (37) | No run up |
| 5 | 18.4 (62.2) | 18.8 (69.8) | 20.7 (85) |
| 10 | 9.2 (116.8) | 9.2 (124.4) | 9.4 (139.6) |

(b) Effect of U on TRT for $\tau_e$ = 10 kPa $\mu$ = 50 mPas -
1 mm thick aluminium cup/clutch

| Um/s | TRT |
|---|---|
| 2 | 2.75 (147.8) |
| 5 | 6.88 (155.9) |
| 10 | 13.82 (166.2) |

There are other corollaries to this change (in h). The shear rate (fall) effect on $\tau_e$ is reduced by virtue of the lower $\dot{\gamma}$ (at low temperatures) or can be negated by a higher working temperature - $\tau_e$ ($\dot{\gamma}$) can have a level or rising characteristic there. The viscous part of the load will be affected; $t^*_m$ improves with rise in $\theta$. However conductance increases with $\theta$ and if high temperature working is to be feasible then fluid conductance must be low and less sensitive to $\theta$ than hitherto seen in commercial fluids of high $\tau_e$ capability. Of greatest importance is the tendency to thermal runaway that can easily be brought about by a seemingly slight (and unguided) change of parameters; the heat transfer rate and hence thermal equilibrium depends on all of these parameters

## LIMITS TO CONCEPT

Given an optimisation on the above basis only an improvement in fluid properties can further aid performance. However there are perhaps limitations on the overall concept of flexible operation of mechanisms which come from sources other than the direct $\tau_e$ limitation viz: If a high $\tau_e \sim 50$ kPa is postulated for the mechanism then, from a simple (mass x acceleration) calculation, acceleration of the solid parts rises to 1,000g and over (fluid inertia neglected). Much above this level problems of materials availability and secure construction of lightweight parts (slightness of form) loom. The lighter the solid power transmission parts and the load, the more important becomes the inertia of the fluid. For example, the pick up of a 1mm thick aluminium, double acting output clutch with 1mm electrode separation at either side at 1000g surface acceleration would take a yield stress of circa 15 kPa for the clutch/cup/fluid alone i.e. with the assumption of linear velocity profile and a density of 1000 kg/m$^2$. Hence counter to previous indications a smaller inter-electrode gap and a light (low specific gravity) low viscosity fluid are inferred. These paradoxically may limit the use of high percentage concentrations of solid and hence $\tau_e$.

Other potentially limiting factors (with respect to high speed flexible machines) or limiting aspects of fluid constituent parameters that remain unquantified are: (a) the natural relaxation time of particle hysteresis, and electro-phoresis - if these are significant it will be necessary to drive down the remnant charge/structure (in a more expensive excitation system), (b) the influence of polarisation properties and conductivity on $t^*_m$[8] and $\tau_e$; do better fluids in these respects need to have an associated charging problem for fast operation ($i \alpha Cdv/dt$) because the requirement for a large dipole moment leads to high capacitance and charging currents and, a high steady state conductance respectively? This could make the electrical problem the dominant aspect of a design and could make it a larger part of the heating problem than the viscous load (c). the effect of particle size needs resolving by experimentation and modelling - what are its effects (for the same concentration) on separation and on $\tau_e$, $t^*_m$ ? If small gap electro-rheological devices are to be built then particles small enough to form a continuum will presumably be required.

In a valve-flow mode controller the level of $\tau_e$ will be crucial in reducing the volumetric size of the valve/piston combination and the compliance effect on frequency/time response[9]. In this case any potential advantage of particle size on dipole moment level can seemingly be taken advantage of since the aperture size will tend to be large for reasons of controlling pressure drops due to acceleration alone[10].

## DISCUSSIONS

As an example of the problems confronting a design engineer re the interpretation of fluid characteristic data or, the fluid developer in choosing what parameters to concentrate on (albeit incomplete) the mechanism of figure 1 is further considered. This is done so as to show how nearly an intuitive process based on quasi-steady state experiences can lead one astray.

Table 2 shows that provided the mechanism will 'run up' or operate at all, zero volts viscosity seems to have but a secondary effect (within reason) on turn round time. Also the effect of raising $\tau_e$ diminishes (even without inclusion of the fluid load). Though an increase in inertial load increases the turn round time linearly (equation 2) higher belt speeds tend to indicate seemingly prohibitive yield stresses. Perhaps it is fortuitous that a lack of materials from which to manufacture the too highly accelerated solid parts is a prior limitation.

Consideration of heat dissipation overturns this philosophy. Parallel heat transfer simulations would quickly show that viscosity increase leads to thermal runaway, on account of the magnitude of the negative resistivity coefficient. This plus the $\mu(\theta)$ characteristic must be available to the designer.

Table 1, on the face of it, shows how heavily the optimisation depends on the relative values of the various parameters. Initially the larger cup/clutch gives a higher proportion of its torque to the external load, its own torque/inertia ratio remaining at unity, despite the larger/higher inertia pulley seeming to be consistent with a poor turn round time. If these setbacks to the intuitive process were not enough, the picture changes completely when only one clutch is involved (in a rotary drive) and the acceleration of the fluid film is considered[11]. In this case little is to be gained by having very high yield stresses whilst a high viscosity is an aid to reducing the time taken to run up to high speed, but is not beneficial to temperature control.

All of this indicates that very careful study of the application in static (start up) kinematic (output motion requirement) and dynamic (full acceleration accounting) must be carried out before the optimum fluid/mechanical/electrical combination can be determined. This is why more general characteristics than have been hitherto sent out by fluid manufacturers, must be delivered. This situation in turn appears to lead to the conclusion that if the benefits of field responsive fluids are to be brought into commercial operation quickly then collaboration between scientists and application engineers must be effective!

The elements of the design process considered here should not be considered exhaustive. Having said that it is obvious that much more work in time and frequency domain characterisation is called for. This may be no bad thing since the modus operandi of the fluids themselves are more likely to be determined in this area of operation than in the steady or quasi-steady flow modes.

## CONCLUSIONS

In conclusion, it has been seen that for inertial loading, careful design can limit or greatly reduce the demand on the characteristics required of an ER fluid for successful applications. This makes the case for more detailed and wider scope than hitherto electrical and mechanical characterisation of commercial fluids with $\tau_e$, $t^*_m$ ($\theta$, E, $\dot{\gamma}$) models and R, C ($\dot{\gamma}$, $\theta$, E,) equivalent circuits (for high frequency time domain step input) becoming necessary if advantages are to be taken of a full mechatronics approach to an applications problem; this will of necessity involve a multi-variable, non-linear, parallel simulations approach. So far, with improvements of fluids included, a 2,200% improvement in turnround time of the specific device seems achievable. This may be at the cost of an increased power consumption. The interplay between these two parameters is a crucial feature of the often intense non-linearity of the various aspects of the design.

The approach described is adequate for a fluid of known time and steady domain performance. Ideally, a similar fluid parameters model (s) R, C, $\tau_e$, $t^*_m$, $t^*_e$, $\mu$ ($\dot{\gamma}$, $\theta$, % c, d/h) would be available (d is particle size, % c is solid concentration and $t^*_e$ is the electrical charging equivalent to $t^*_m$). Then full 'technotronic' optimisation will be in vogue. No doubt the effect of particle size distribution and other efforts to reduce viscosity plus dielectric/polarisation/conductance properties would eventually be added to the data.

In the long run the velocity profile in acceleration and the effect of $\tau_e$ ($\dot{\gamma}$, $\theta$) and $t^*_m$ ($\dot{\gamma}$, $\theta$) need to be added into the existing dynamic model and the various models run interactively. Only then will the critical $\theta$ runaway be effectively predicted. Before this the precision of the mechanism needs to be established in terms of repeatability of position control - this involves fluid and solid part stiffness considerations and testing.

## ACKNOWLEDGEMENTS

The authors wish to thank the Science and Engineering Research Council and the Department of Trade and Industry for their financial sponsorship of associated work through the LINK scheme and for discussions held over the years with the authors many colleagues.

## REFERENCES

1. W. A. Bullough, J. Makin, A. R. Johnson and R. Firoozian, "A Variable Duty ER Linear Reversing System: Concepts, Analysis, Proof and Predictions". Proc. Actuator '94, Bremen (June 1994). Publ. Axon GmbH, Germany - p302-305.
2. J. R. Yates, D. Lau and W. A. Bullough, "Inertial Materials for Flexible, High Speed Machines: Perspective, Review, Future Requirements". Ibid. p275-278.

3. J. Makin, W. A. Bullough, A. R. Johnson, R. Firoozian and A. H. Sianaki, "The Effect of Solid Fraction Concentration on the Performance of an ER Fluid in the Shear Mode". Proc. 4th Int. Conf. ER Fluids, Feldkirch, Austria, p372-392. World Sci. Publ. Co. (July 1993).
4. R. Smyth, K. W. Tan and W. A. Bullough, "Heat Transfer Modelling of an ER Catch". Ibid p657-675.
5. R. Tozer, C. T. Orrell and W. A. Bullough, "On-Off Excitation Switch for ER Devices". Ibid p597-607.
6. A. H. Sianaki, W. A. Bullough, R. Tozer and M. Whittle, "Experimental Investigations into Electrical Modelling of ER in Shear Mode". Proc. Inst. Elect. Engs., London. Sci. Meast. & Tech. Vol. 141, No. 6, p531-537.
7. R. Atkin, Xiao Shi and W. A. Bullough, "Solution of the Constitutive Equations for the Flow of an ER Fluid in Radial Configurations". Jnl. Rheology Vol. 35, p1441-1461 (1991).
8. M. Whittle, D. J. Peel, R. Firoozian and W. A. Bullough, "Dependance of ER Response Time on Conductivity and Polarisation Time". Phys.Rev.E Pt 6A Vol. 49 p5249-5259 June 1994.
9. M. Whittle, W. A. Bullough, D. J. Peel and R. Firoozian, "Electro-Rheological Dynamics Derived from Pressure Response Experiments in the Flow Mode". Jnl. Non-Newtonian Fluid Mechanics (to be publ. Vol. 57 No. 1 Apr/May 1995).
10. W. A. Bullough, D. J. Peel, R. Stanway, D. L. Rodgers and J. Sproston, "A Long Stroke Damper for Vehicle Suspension Applications". ASME Winter Annual Meeting, Chicago (Dec. 1994). Proc. Devs. ER Flows and Measurements, p41-50.
11. M. Whittle, R. J. Atkin and W. A. Bullough "Fluid Dynamic Limitations on the Performance of an Electro-rheological Clutch". Jnl. Non Newtonian Fluid Mechanics (to be publ. Vol. 57 No. 1 Apl/May 1995.

# EFFECTS OF DC AND AC ELECTRIC FIELDS ON THE RESPONSE OF ELECTRORHEOLOGICAL FLUIDS COMPRISING CYLINDRICAL PBZT PARTICLES

Rex C. Kanu[1], and Montgomery T. Shaw[2]

[1]Institute of Materials Science, Polymer Science Program
[2]Department of Chemical Engineering and Institute of Materials Science
University of Connecticut
Storrs, Connecticut 06269-3136

## INTRODUCTION

Electrorheological fluids (ERF's) are suspensions consisting of micron-sized polarizable particles dispersed in non-conducting suspending media. ERF's undergo liquid-to-solid transitions on the application of an electric field. These reversible transitions are on the order of milliseconds and are accompanied by orders-of-magnitude increase in yield stress and dynamic modulus (stiffness). This phenomenon is known as electrorheology (ER), and it is important because it permits the direct use of electricity without intermediate transformations. Therefore, devices based on ERF's can be more economical and energy efficient than mechanical devices.[1] ERF-based applications include shock absorbers[2-4], clutches[5-7], valves[7-9], engine mounts[2,11-13], brakes[6,14], biomedical devices[15] and control of high-speed machines.[16] Shul'man et al.,[17] have listed several ERF-based devices. However, in spite of its vast potential, relatively few ERF-based devices are in use.

ER technology comprises the ER device, the ER fluid and the power/circuitry control. Of these, it is generally accepted that ERF's are the weak link in the ER technology. In particular, the reproducibility and performance of ERF's have been hindered by inadequate knowledge of the fluids' fundamental behavior. A thorough understanding of ER mechanisms is needed to resolve these problems.

Many investigators have made substantial contributions to elucidating the ER phenomenon since its discovery by Winslow.[18] These contributions include the role of conductivity[19-21], relative permittivity and frequency of electric field[22-24], temperature[25,26], moisture[24,27] and particle surface morphology[28] on ER behavior. Most of the ERF's studied have had particles with irregular or spherical shape. To the best of our knowledge, no

*Progress in Electrorheology*, Edited by K.O. Havelka
and F.E. Filisko, Plenum Press, New York, 1995 

systematic investigation of the role of particle shape (geometry) on the performance of ERF's has been undertaken, although it has been suggested by several authors that such studies are necessary. Mason et al.[29-32] have studied the rheology of dilute suspensions containing ellipsoidal particles in combined shear and electric fields. They observed small increases in the suspensions' intrinsic viscosity that they attributed to the orientation of the particles. The effect of particle interactions was absent in their studies because the suspensions were dilute. Cognizant of ER phenomenon, Chaffey and Mason[29] advised that "in suspensions of finite concentrations, particle interactions will greatly influence the viscosity behavior of the suspensions." Gast and Zukoski[33] suggested that "the ideal shape of a constituent particle for electrorheology has not yet been determined; however, the added degree of freedom afforded by choice of particle geometry may be an important design parameter." See et al.[34] predicted that "the dependence of the ERF's behavior on particle size and shape is clearly of great practical interest, and it is hoped that systematic experimental investigations will be carried out in the near future." A recent assessment of the advances made in ER technology by the U.S. Department of Energy[35] identified the role of the particle geometry in ER responses as an area deserving attention. The significance of particle shape in ER response stems from the fact that an ellipsoidal particle of same volume and bulk polarizability of a spherical particle will have greater induced dipole moment than that of the sphere if the ellipsoidal particle aligns with its major axis along the applied field. A second advantage is that fewer particles are needed to form chains between the electrodes, allowing for a lower viscosity in the "off" state. This also implies cost savings.

One of the goals of this study is to elucidate ER mechanisms by examining the impact of particles' shape on the ERF's behavior. Our hypothesis is that the enhancement of dipole-dipole interactions of the dispersed-phase will result in improved ER performance, viz., greater dynamic modulus and yield stress. Although our hypothesis is based on the dipole theory, we recognize that some overlapping of ellipsoidal particles can occur during chaining of the particles. The approach employed in this work is to vary systematically the geometric aspect ratio (length of particle/diameter of particle) of rodlike particles.

Another objective of this work is to examine the influence of electric field frequency on the of ERF's consisting of rodlike particles. While many studies[22,23,36-43] have examined the effects of electric field frequency on ER response, few studies[44] have considered the conbined influence of electric field frequency and particle geometry on ER response. Therefore, this study will examine the consistency of our hypothesis under dc and ac excitations. For convenience, an ac frequency of 60 Hz was chosen.

## EXPERIMENTAL

### Materials and Sample Preparations

The model ERF's consisted of cylindrical particles of poly(p-phenylene-2,6-benzobisthiazole) [PBZT] dispersed in mineral oil. PBZT is a highly conjugated liquid crystalline polymer with a ladder-like structure. Figure 1 shows the repeat unit of PBZT. It has favorable dielectric and electrical properties. The estimated average polarizability of the PBZT repeat unit is $34.2 \times 10^{-24}$ $cm^3$ [45] compared to $10 \times 10^{-24}$ $cm^3$ for benzene.[46] The only reported experimental value[47] of a band gap for trans-PBZT is 0.76 eV. The calculated electronic band gap of PBZT is 1.73 eV, a value similar to that determined experimentally in trans-polyacetylene (1.4 - 1.9 eV).[48] Trans-polyacetylene is considered to have semiconductive properties.[49] PBZT is chemically and thermally stable. Thermogravimetric analysis (TGA) performed on heat-treated fibers showed virtually no weight loss until they began to degrade at 750 $^0$C.[50] The reported densities of PBZT fibers range from 1.44 to 1.6 g/cm$^3$.[51,52] PBZT was supplied by the Department of the Air Force, Wright-Patterson Air Force Base, Ohio, USA. Light mineral oil was purchased from Fisher Scientific.

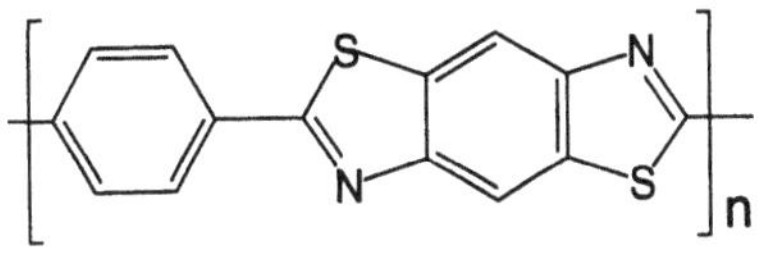


Figure 1. Repeat unit of poly(p-phenylene-2,6-benzobisthiazole) [PBZT].

PBZT fibers have moduli as high as 280 GPa and strengths of 2.7 GPa.[53] Because these fibers are high strength materials, they are difficult to cut. To obtain the desired particle size and aspect ratio we embedded the PBZT fibers in a polystyrene matrix and cut them with a modified sledge microtome. The microtome was modified by attaching the knife to a motor through a set of gears that caused the knife to vibrate parallel to its cutting-edge. Figure 2 is a schematic diagram of the modified microtome. The fibers were embedded by polymerizing styrene monomer (with 0.5 wt% benzoyl peroxide) in a glass tube containing strands of PBZT fibers. Polymerization of the monomer was accomplished in 48 hours at 50$^0$C. The microtomed fibers were recovered by dissolving the polystyrene with toluene. The recovered fibers were dried and characterized with an optical microscope. Suspensions of PBZT in mineral oil were placed under vacuum to remove trapped air and moisture. The mineral oil was dried over molecular sieves before the suspensions were made.

### Rheological Measurements

Rheological characterization was done with modified Rheometrics Mechanical Spectrometer [RMS 7200] and Rheometrics System 4 using the parallel plate fixtures shown in Figure 3. The RMS has a 10,000 g cm transducer while the System 4 is equipped with a 10, 100 and 2,000 g cm transducers.

The top plate of the parallel plates was made of titanium while the bottom plate was made of stainless steel. Because the top plate was attached to the sensitive transducer (which is stationary), it was necessary to design the top plate with a light weight material such as titanium. Both stainless steel and titanium have good electrical conductivity. The plates were rounded at the edges to reduce electric field fringing. Voltage [dc] was supplied with a Bertran High Voltage power supply Model 205A-05R capable of delivering 0-5 kV at 0-5 mA. The ac high voltage power supply was built by the electronic shop at the Institute of Materials Science, University of Connecticut. It's capable of delivering 0-5 kV at 60 Hz.

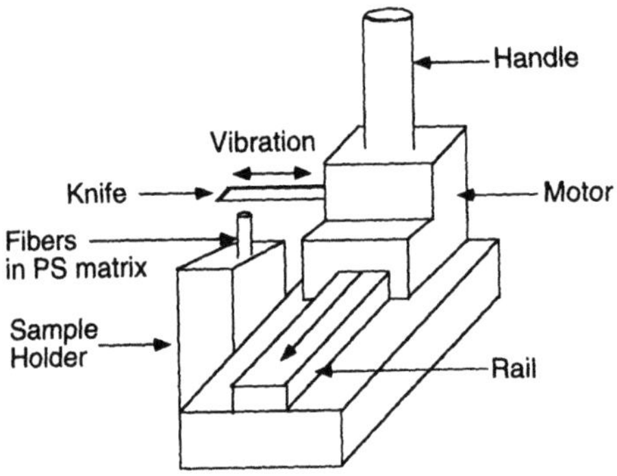


Figure 2. Modified sledge microtomed.

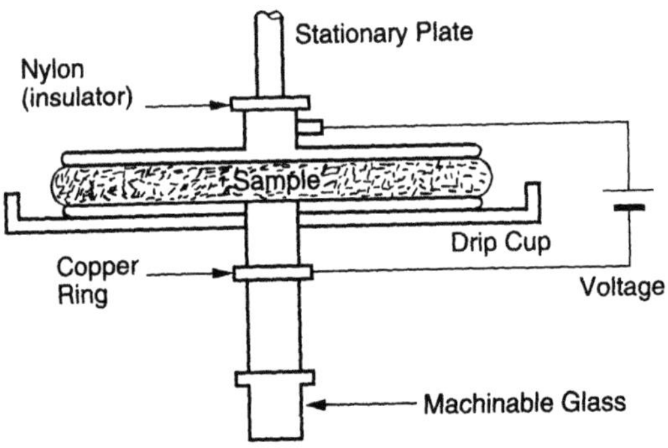


Figure 3. Parallel plate fixture for the fluids motor of Rheometrics System IV.

**Dielectric Measurements**

Dielectric characterization of the fluids was done with an IMASS Time Domain Dielectric Spectrometer [TDDS]. Measurements were made with the fibers randomly dispersed and aligned, respectively. The TDDS was modified for *in situ* measurements of the dielectric properties of the fluids after the application of an electric field. Figure 4 shows a schematic diagram of the dielectric measuring assembly. To cause the particles to chain along the electric-field direction, the high voltage from the dc power source (Bertran) was applied to the electrodes by engaging the leads to the electrodes. The ground deactivator (shown in diagram) depressed a microswitch rendering the ground of the top electrode inactive. This precaution was taken to isolate the TDDS assembly during the

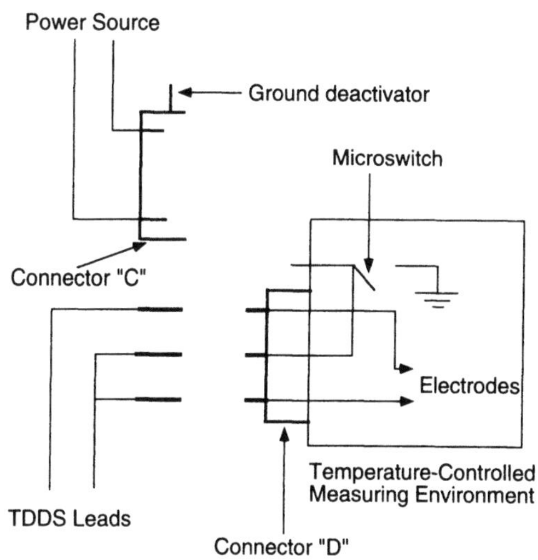


Figure 4. Time Domain Dielectric Spectrometer modified to measure *in-situ* the dielectric properties of aligned and randomly dispersed PBZT particles.

application of electric field. Voltage was applied to the samples between the electrodes for about 30 seconds, whereupon connector "C" was removed. The leads from the TDDS was then connected to the electrodes as shown in the diagram and measurements were taken. The electrodes were designed for measuring the dielectric properties of fluids. The bottom electrode consisted of a cup with a diameter of 84 mm. The top electrode had a diameter of 60 mm. The top electrode was attached to a micrometer gauge used in setting the gap between the electrodes. Both electrodes were made with stainless steel. The bottom electrode was held in place with plastic clamps so that any force of attraction developed between the electrodes on the application of electric field did not alter the gap between the electrodes, and consequently, the sample thickness. The grounding of the top electrode ensured that accurate and reproducible data were obtained for all measurements.

## RESULTS AND DISCUSSION

### Optical Microscope

An Olympus optical microscope with a custom-built microscope stage was used to observed the ERF's in combined shear and electric fields. A detailed description of the microscope stage is given elsewhere.[54] Figure 5 shows that at zero electric field, most of the rodlike particles were aligned parallel to the flow direction. Figure 6 and Figure 7 show the chaining of particles on the application of 1000 V/mm dc and ac, respectively. These

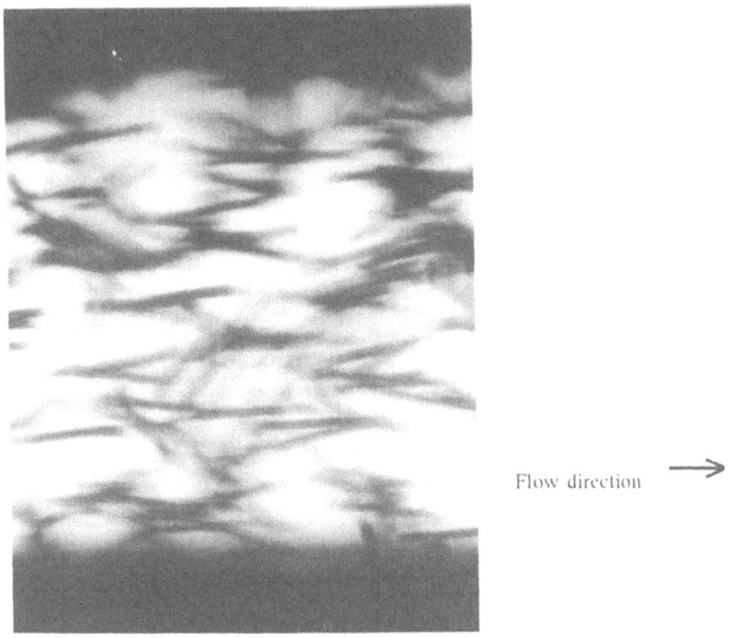


Figure 5. PBZT particles in mineral oil at 0.5 vol.% between two electrodes. E = 0 kV/mm.

micrographs show that the PBZT rodlike particles did respond to dc and ac fields. On the application of the electric fields, the particles formed chains that were parallel to the fields and perpendicular to the flow field. The micrographs also show that the particles chained in a head-to-tail fashion. However, there is evidence of some particle overlapping in spite of the low concentration. The particle overlapping seems to be more prevalent with dc than with ac. The overlapping of particles can lead to greater stress bearing capabilities because such an arrangement reduces particles slipping past one another[33] thus enhancing momentum transfer among particles.

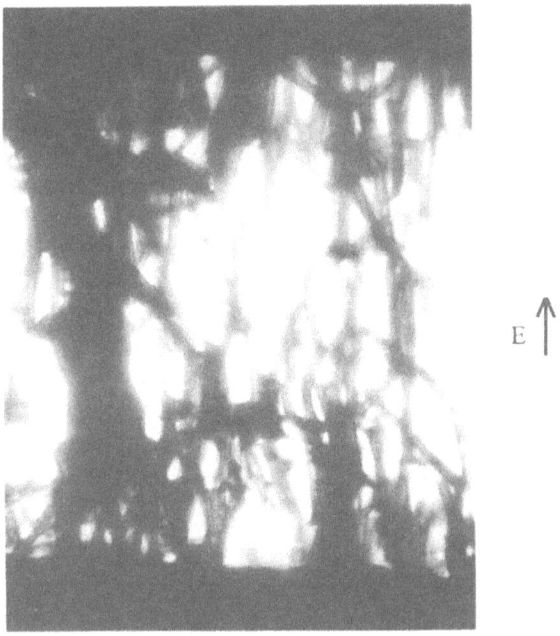


Figure 6. Chaining of PBZT particles in mineral oil at 0.5 vol.%.
E = 1.0 kV/mm dc.

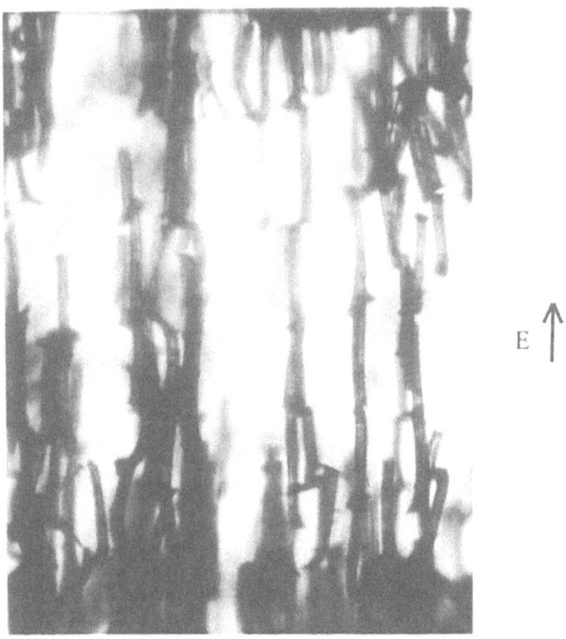


Figure 7. Chaining of PBZT particles in mineral oil at 0.5 vol.%.
E=1.0 kV/mm ac (peak).

### Steady Shear

Figure 8 shows that the apparent shear stress at low shear rates increased by about three orders of magnitude on the application of 2 kV/mm dc. Figure 8 also shows that the ERF with the larger aspect ratio particles exhibited the greater apparent stress. The apparent shear stress decreased with increasing shear rate. Figure 9 shows that with ac, the apparent shear stress at low shear rates increased by about two orders of magnitude on the application of 2 kV/mm. Figure 9 also shows some difference in apparent shear stress due to the influence of particle aspect ratio. However, this difference was not as large as that observed with dc. Compared with dc, apparent shear stress increased slightly with shear rate. Figure 10 shows that the influence of the particles aspect ratio was more significant with dc than with ac at low shear rates. With dc, as the shear rate increased, the impact of particle aspect ratio was reduced because of the degradation of the chains spanning the electrodes. However, with ac the difference remained relatively unchanged with shear rate.

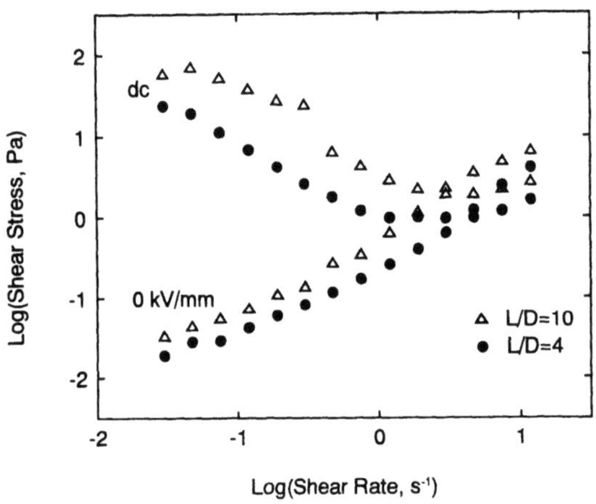


Figure 8. PBZT in mineral oil at 15 vol.%. 2.0 kV/mm dc.

The reason for the difference in the shape of the flow curve of our systems with dc and ac is not presently known. However, Kumakura et al.[44] observed that ERF's comprising needle-like iron particles responded differently to dc and ac excitations, respectively, while there was no difference with spherical iron particles. The needle-like particles had aspect ratio of 8.9 with a long axis length of 0.5 $\mu$m while the diameter of the spherical particles ranged from 1 to 10 $\mu$m. The particles were coated with a thin nonconducting layer. They speculated that the decrease in shear stress with time at 2.8 $s^{-1}$ under dc might be due to either the break-up of cluster structure or the aggregation of particles at one of the electrodes, while the slight increase in shear stress with time under ac was attributed to the thickening of the chain structures.

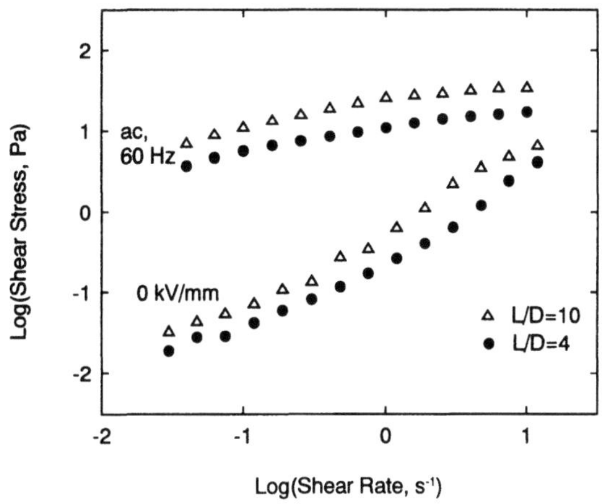


Figure 9. PBZT in mineral oil at 15 vol.%. 2.0 kV/mm ac (peak).

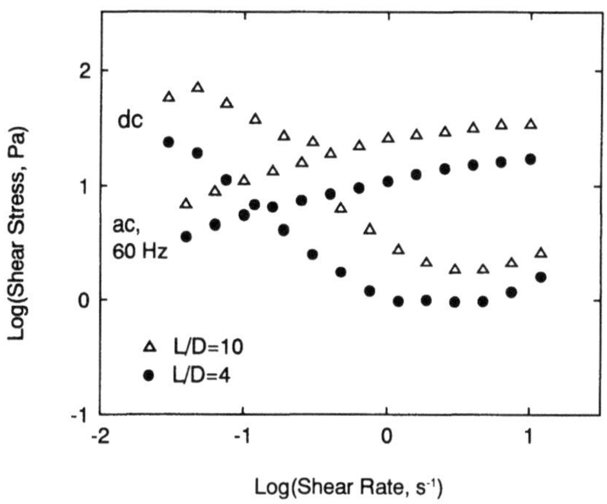


Figure 10. PBZT in mineral oil at 15 vol.%.
2.0 kV/mm, dc and ac (peak), respectively.

**Dynamic Shear**

Figure 11 shows that the dynamic moduli of the fluids decreased with strain, as has been observed elsewhere.[54] It also shows that for the strain range covered the fluid with the larger aspect ratio particles had the greater dynamic modulus. The difference in dynamic moduli between the aspect ratios studied increased with strain. With ac, the dynamic modulus was relatively independent of strain, and the influence of particle aspect ratio was less pronounced (Figure 12). The dynamic shear response of the fluids showed distinct patterns with dc and ac. With dc, Figures 13 and 14, the dynamic moduli were relatively independent of deformation frequency, a response characteristic of an elastic

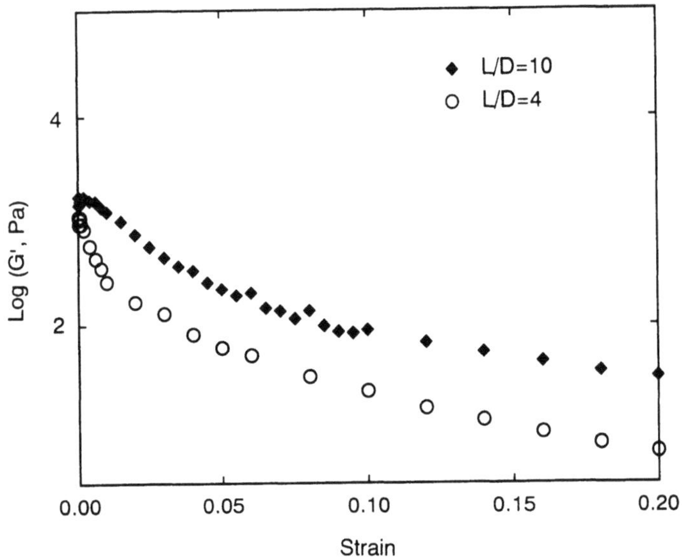


Figure 11. PBZT in mineral oil at 15 vol.%.
2.0 kV/mm dc, deformation frequency = 6.28 rad/s

solid. The loss moduli and the loss factor decreased with increasing deformation frequency. Figures 15 and 16 show that with ac, the dynamic moduli and the loss moduli increased with deformation frequency. The loss factor decreased with deformation frequency but went through a minimum between 1 and 10 Hz.

Since the dynamic modulus measures the stiffness or the extent of structure formation in a material, it is apparent that the degree and strength of structures formed with dc were greater than with ac. Figures 13 and 14 show that the dynamic moduli with dc were two orders of magnitude greater than with ac at low deformation frequency, and an order of magnitude greater at high deformation frequency.

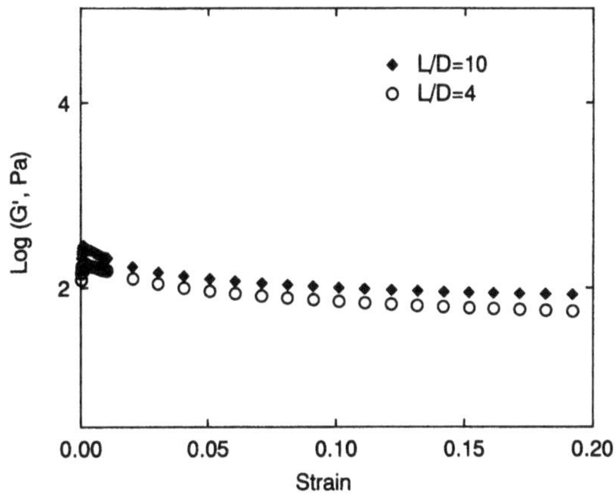


Figure 12. PBZT in mineral oil at 15 vol.%.
2.0 kV/mm ac (peak), deformation frequency = 6.28 rad/s.

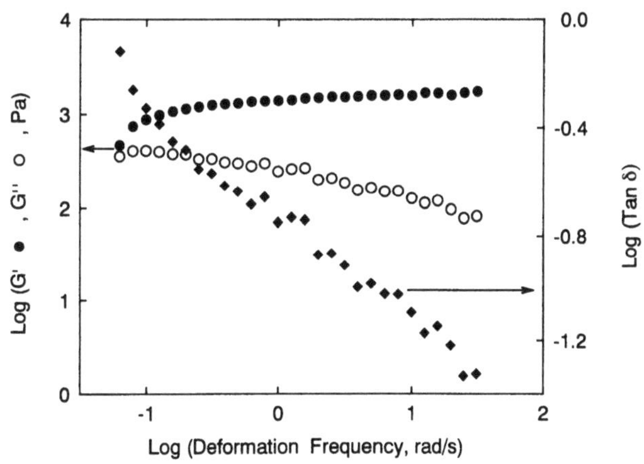


Figure 13. PBZT in mineral oil at 15 vol.%., L/D=4, 2.0 kV/mm dc, strain = 0.002.

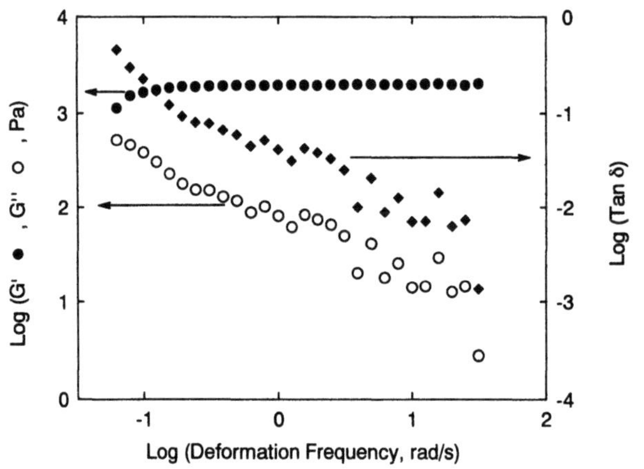


Figure 14. PBZT in mineral oil at 15 vol.%, L/D=10, 2.0 kV/mm dc, strain = 0.002.

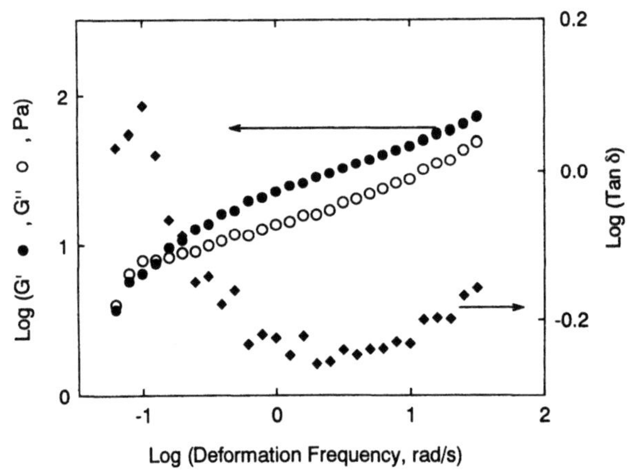


Figure 15. PBZT in mineral oil at 15 vol.%, L/D=4, 2.0 kV/mm ac, strain = 0.002.

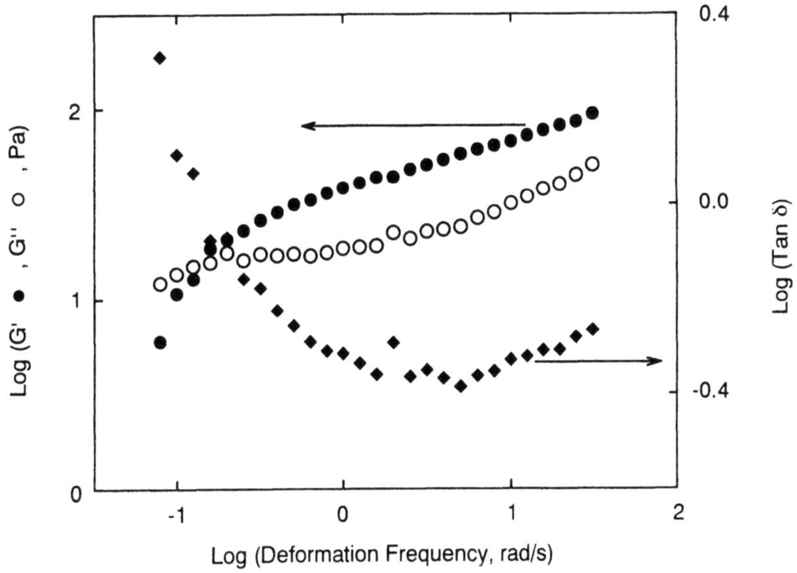


Figure 16. PBZT in mineral oil at 15 vol.%, L/D=10, 2.0 kV/mm ac, strain = 0.002.

## Dielectric Properties

Figure 17 shows the static dielectric constant of the suspending medium, light mineral oil before and after the application of 3 kV/mm dc electric field. Essentially no effect was observed, demonstrating that the technique was free of artifacts. Figures 18 and 19 show the dielectric spectra of the fluids. These spectra show a distinct interfacial polarization at 100 Hz. The dielectric measurements were made after the particles were aligned with 3 kV/mm field. It was assumed that the particles remained aligned while the dielectric measurements were being made. Figure 20 shows the effects of particle alignment and particles aspect ratio on the static dielectric constant. When the particles were aligned, the dielectric constant was substantially higher than when they were randomly dispersed. These observations are consistent with the theory of permittivities of heterogeneous systems[55]; the dielectric constant of a material should increase increase with particle aspect ratio and alignment for the frequency range covered. The dielectric constant of both the aligned and randomly dispersed particles decreased with frequency. However, the decrease with aligned particles was steeper, due to the stronger relaxation processes.

Based on point dipole approximation[54,56], it has been shown that for isolated particle chains, shear stress is given by

$$\tau = \frac{12\varepsilon_o\varepsilon_c\phi_p\beta^2E^2a^2}{r^4}[(2-3\gamma^2)\gamma+2\gamma] \qquad (1)$$

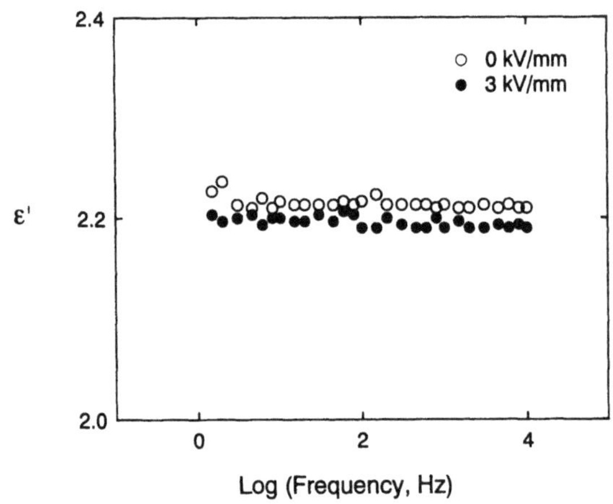


Figure 17. Dielectric constants of mineral at 0 kV/mm, and after the application of an electric field of 3.0 kV/mm. Measurements were made at 30$^0$C.

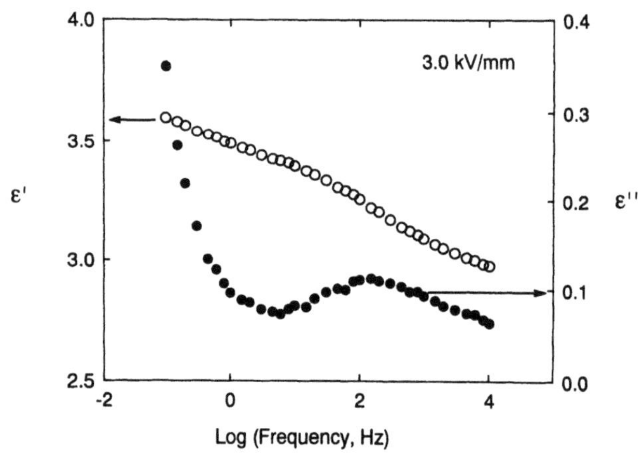


Figure 18. Dielectric spectra of PBZT in mineral oil at 15 vol.%, L/D=4 30$^0$C. Measurements were made after aligning the particles with 3.0 kV/mm dc.

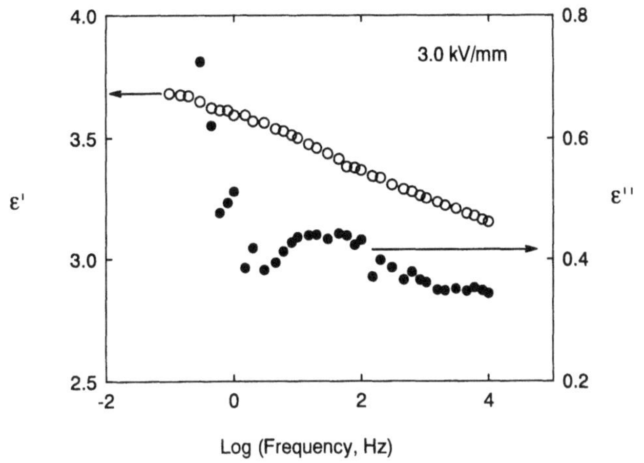


Figure 19. Dielectric spectra of PBZT in mineral oil at 15 vol.%, L/D=10 30°C. Measurements were made after aligning the particles with 3.0 kV/mm dc.

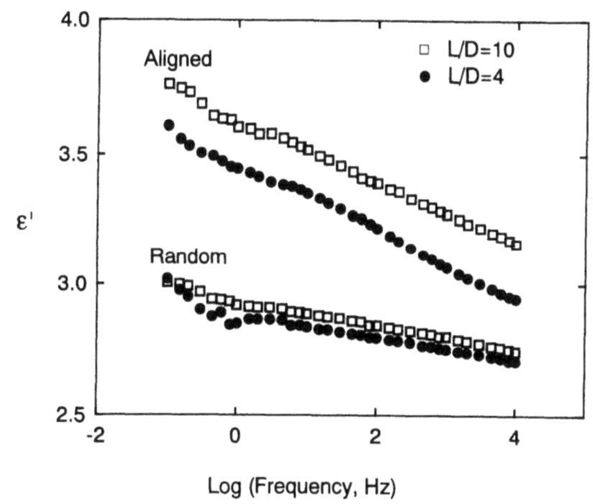


Figure 20. Dielectric constants of PBZT in mineral oil at 15 vol.%, 30°C. Particles were aligned with 3.0 kV/mm dc.

and shear modulus[54] is expressed as

$$G = \frac{3\varepsilon_o\varepsilon_c\phi_p\beta^2E^2}{4[1+2\gamma]}[4-3\gamma^2] \quad (2)$$

where

$$\beta = (\varepsilon_p - \varepsilon_c) / (\varepsilon_p + 2\varepsilon_c) \quad (3)$$

$\varepsilon_p$'s used in equation 3 were determined with a simple rule of mixtures

$$\varepsilon = (1 - \phi_p)\varepsilon_c + \phi_p\varepsilon_p \quad (4)$$

The static dielectric constants $\varepsilon_s$ for the PBZT-based ERF's were estimated from Cole-Cole plots, and the $\varepsilon$'s at 60 Hz were estimated from Figure 20. Table 1 shows $\varepsilon_p$ as a function of particle geometric aspect ratio and electric fields.

Table 1. Dielectric constants of aligned PBZT particles

| Electric Field | L/D | $\varepsilon_p$ |
|---|---|---|
| dc | 10 | 11.9 |
| dc | 4 | 10.8 |
| ac (60 Hz) | 10 | 9.7 |
| ac (60 Hz) | 4 | 8.7 |

From equation 2, we expect G to increase with $\beta^2$. Figure 21 shows that G′ increases monotonical with $\beta^2$ suggesting that the increases in dielectric constants as a function of particle geometric aspect ratio adequately explain the observed increases in the ER response.

Figure 21 also shows that the point-dipole approximation underestimates the dynamic moduli, even after multipole effects were factored in, following Jones et al.[57] The shortfalls of the polarization model have been discussed elsewhere.[58]

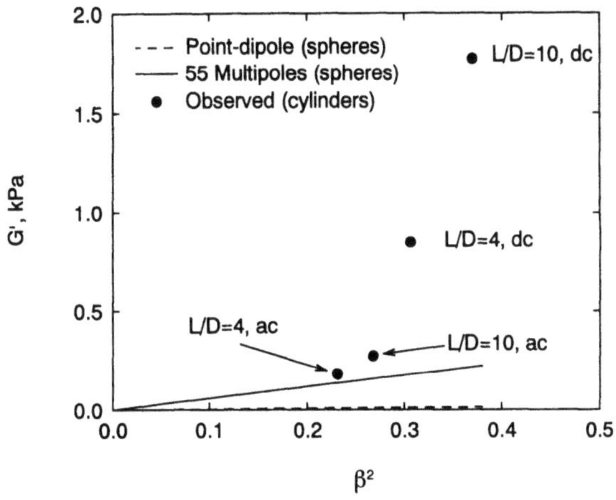


Figure 21. PBZT in mineral oil at 15 vol.%, 2.0 kV/mm strain=0.002, deformation frequency=6.28 rad/s

## CONCLUSION

ERF's comprising PBZT rodlike particles showed chaining under dc and ac fields. With dc, the fluid with the larger aspect ratio showed greater apparent shear stress for the shear rates covered. The increase in apparent shear stress at the application of 2 kV/mm electric field was three orders of magnitude. With ac, the influence of particle geometric aspect ratio was reduced, resulting in an increase in apparent shear stress by only two orders of magnitude over the same range of electric field strengths. Dynamic shear response showed that with dc, the fluid with the larger aspect ratio particles had a higher dynamic modulus for the strains and frequencies investigated. With ac, the influence of particle aspect ratio was reduced. With dc, the dynamic properties of the fluids were similar to those of an elastic solid while with ac there was a significant frequency dependency of the modulus. The dynamic modulus was one to two orders of magnitude greater with dc than with ac. These differences in apparent shear stress and dynamic moduli were attributed to the strength of the structures formed in dc and ac fields, respectively. The reduced impact of particle aspect ratio on ER response with ac was probably due to the time-averaged reversal of the direction of the induced dipole caused by the ac excitation that led to a decrease in polarization with electric field frequency. The findings of this work suggest that the particle aspect ratio is an important design parameter of ERF's, particularly in dc fields.

## DEFINITION OF SYMBOLS

a: particle radius.
E: electric field.
G: shear modulus.

$r$: distance between particle centers.
$\beta$: dielectric difference between particle and medium.
$\varepsilon_o$: permittivity in a vacuum=$8.854 \times 10^{-12}$ F/m.
$\varepsilon_c$: permittivity of dispersing medium.
$\varepsilon_p$: permittivity of particle.
$\phi_p$: volume fraction of dispersed phase in the suspension.
$\gamma$: strain.
$\tau$: shear stress.

## REFERENCE


1. R.G. Gorodkin, E.V.Korobko, G.M., Blokh, V.K., Gleb, G.I. Sidorova, and Ragotner,M.M., Application of the electrorheological effect in engineering practice, *Fluid Mech. - Sov. Res.*, 8:48 (1979).
2. N.K. Petek, An electronically controlled shock absorber using electrorheological fluid, *in*: SAE Technical Paper Series 920275, Int. Congress & Exposition, Detroit, Michigan(1992).
3. E. Westkamper, and J. Meschke, Study of the flow behavior of electrorheological fluids (ERF's) and their use in passive absorbers, *Rheology*, 243 (1993).
4. A. Schuelke, K. Engelsdorf, and K-H. Haegele, U.S. Patent Specification 5,293,968 (1994).
5. A.R. Johnson, W.A. Bullough, R. Firoozian, A. Hosseini-Sianaki, J. Makin, and S. Xiao, Testing on a high speed electro-rheological clutch, *in*: "Proceedings of the International Conference on Electrorheological Fluids: Mechanisms, Properties, Structure, Technology, and Applications," R. Tao ed., World Scientific Publishing Company, Singapore (1992).
6. J.D. Carlson, and T.G. Duclos, ER fluid clutches and brakes - fluid property and mechanical design considerations, *in*: "Proceedings of the Second International Conference on Electrorheological Fluids," J.D. Carlson et al., ed., Technomic Publishing Company, Inc., Pennysylvania (1989).
7. M.J. Egyed, U.S. Patent Specification 5,342,258 (1994).
8. Z. Lou, R.D. Erwin, and F.E. Filisko, Behavior of electrorheological valves and bridges, *in*: "Proceedings of the International Conference on Electrorheological Fluids: Mechanisms, Properties, Structure, Technology, and Applications," R. Tao ed., World Scientific Publishing Company, Singapore (1992).
9. A.D. Matsepuro, and V.I. Kordonsky, Some properties of a dispersed phase of Electrorheological Fluids, *in*: "Proceedings of the International Conference Electrorheological Fluids: Mechanisms, Properties, Structure, Technology, and Applications," R. Tao ed., World Scientific Publishing Company, Singapore (1992).
10. D.C. Reuter, U.S. Patent Specification 5,322,484 (1994).
11. E.W. Williams, S.G. Rigby, J.L. Sproston, and R. Stanway, Electrorheological fluids applied to an automotive engine mount, *J. Non-Newtonian Fluid Mech.*, 47:221 (1993).
12. N.K. Petek, R.J. Goudie, and F.P. Boyle, Actively controlled damping in electrorheological fluid-filled engine mounts, *SAE Technical Paper Series 881785, Passenger Car Meeting and Exposition*, Dearborn, Michigan (1988).
13. H. Aoki, U.S. Patent Specification 5,236,182 (1993).

14. D.P. Colvin and J.D. Carlson, Control of a fall-safe tether using an ER-fluid brake, *in*: "Proceedings of the Second International Conference on Electrorheological Fluids," J.D. Carlson et al., ed., Technomic Publishing Company, Inc., Pennysylvania (1989).
15. W.A. Bullough, Electro-rheological fluids: An introduction for biomedical applications, *J. Biomed. Eng.*, 13:234 (1991).
16. W.A. Bullough, Electrorheological fluids and the control of high-speed machines, *Endeavour, New Series*, 15:165 (1991).
17. Z.P. Shul'man, R.G. Gorodkin, E.V. Korobko, and V.Y. Gleb, The electrorheological effect and its possible uses, *J. Non-Newtonian Fluid Mech.*, 8:29 (1981).
18. W.M. Winslow, Induced fibration of suspensions, *J. Appl. Phys.*, 20:1137 (1949).
19. C.J. Gow, and C.F. Zukoski, The electrorheological properties of polyaniline suspensions, *J. Coll. Int. Sci.*, 136:175 (1990).
20. H. Block, J.P. Kelly, A. Qin, and T. Watson, Materials and mechanisms in electrorheology, *Langmuir*, 6:6 (1990).
21. W.C. Yu, R.C. Kanu, and M.T. Shaw, Design of anhydrous electrorheological [ER] suspensions and mechanism study, *in*: "Proceedings of the ADPA/AIAA/ASME/SPIE Conference on Active Materials and Adaptive Structures," G.J. Knowles ed., Institute of Physics Publishing, London (1991).
22. D.L. Klass, and T.W. Martinek, Electroviscous fluids. I. Rheological properties, *J. Appl. Phys.*, 38:67 (1967).
23. D.L. Klass, and T.W. Martinek, Electroviscous fluids. II. Electrical properties, *J. Appl. Phys.*, 38:75 (1967).
24. H. Uejima, Dielectric mechanism and rheological properties of electro-fluids, *Jap. J. Appl. Phys.*, 11:319 (1972).
25. H. Conrad, A.F. Sprecher, Y. Choi, and Y. Chen, The temperature dependence of the electrical properties and strength of electrorheological fluids, *J. Rheol.*, 35:1393 (1991).
26. F.E. Filisko, and H-R. Wang, The dynamic character of ER materials, *Polym. Prepr., Am. Chem. Soc. Div. Polym. Chem.*, 35:337 (1994).
27. W. Wong, and M.T. Shaw, The role of water in electrorheological fluids, *in*: "Proceedings of the Second International Conference on Electrorheological Fluids," J.D. Carlson et al., ed., Technomic Publishing Company, Inc., Pennsylvania (1989).
28. Y. Otsubo, M. Sekine, and S. Katayama, Effect of surface modification of colloidal silica on the electrorheology of suspensions, *J. Coll. Int. Sci.*, 146:395 (1991).
29. C.E. Chaffey and S.G. Mason, Particle behavior in shear and electric fields. III. Rigid spheriods and discs, *J. Coll. Sci.*, 19:525 (1964).
30. C.E. Chaffey and S.G. Mason, Particle behavior in shear and electric fields. V. Effects on suspension viscosity, *J. Coll. Int. Sci.*, 27:115 (1968).
31. A. Okagawa, R.G. Cox, and S.G. Mason, Particle behavior in shear and electric fields. VI. The microrheology of rigid spheroids, *J. Coll. Int. Sci.*, 47:536 (1974).
32. A. Okagawa, and S.G. Mason, Particle behavior in shear and electric fields. VII. Orientation distributions of cylinders, *J. Coll. Int. Sci.*, 47:568 (1974).
33. A.P. Gast and C.F. Zukoski, Electrorheological fluids as colloidal suspensions, *Adv. Coll. Int. Sci.*, 30:153 (1989).
34. H. See, H. Tamura, and M. Doi, The role of water capillary forces in electrorheological fluids, *J. Appl. Phys.*, 74:938 (1993).

35. U.S. Department of Energy, Electrorheological Fluids: A Research Needs Assessment, 4-3 (1993).
36. H. Uejima, Dielectric mechanism and rheological properties of electro-fluids, *Jap. J. Appl. Phys.*, 11:319 (1972).
37. N. Sugimoto, Winslow effect in ionic exchange resin dispersion, *Bull. JSME.*, 20:1476 (1977).
38. V.I. Bezruk, A.N. Lazarev, V.A. Malov, and O.G. Us'yarov, Frequency effect of an external electric field on the interaction between dispersed particles in suspensions, *Colloid J.*, 34:276 (1972).
39. P.V. Katsikopoulos and C.F. Zukoski, The electrorhelogical response: The effect of relaxation processes, *in*: "Theoretical and Applied Rheology," edited by P. Moldenaers and R. Keunings, Proc. XIth Int. Congr. on Rheol., Brussels (1992).
40. D.V. Miller, C.A. Randall, A.S. Bhalia, R.F. Newnham, and J.H. Adir, Electrorheological properties of $BaTiO_3$ suspensions, *Ferroelec. Lett.*, 15:141 (1993).
41. T.Y. Chen, and P.E. Luckham, Effect of frequency on a water-activated electrorheological fluid in an ac field, *Coll. Surf. A: Physicochemical & Eng. Asp.*, 78:167 (1993).
42. A. Inoue, Study of new electrorheological fluid, *in*: "Proceedings of the Second International Conference on Electrorheological Fluids," J.D. Carlson et al., Technomic Publishing Company, Inc., Pennysylvania (1989)
43. J.M. Ginder and L.C. Davis, Viscoelesticity of electrorheological fluids: Role of electrostatic interactions, *in*: "Electrorheological Fluids: Mechanism, Properties, Technology, and Applications," R. Tao, ed., World Scientific, Singapore (1994).
44. Y. Kamukura, T. Watanabe, K.Minagawa, and K. Koyama, ER effects in iron particle suspensions, *Polym. Prepr., Am. Chem. Soc. Div. Polym. Chem.*, 35:362 (1994).
45. D. Bhaumik, H.H. Jaffe, and J.E. Mark, Polarizabilities of some benzobisoxazole and benzobisthiazole rigid-rod polymers, *Macromolecules*, 14:1125 (1981).
46. N.J. Bridge and A.D. Buckingham, The polarization of laser light scattered by gases, *Proc. Roy. Soc. Lond., Ser. A*, 295:334 (1966).
47. R.E. Barker, Jr. and D.Y. Chen, Electrical conductivity of anisotropic poly(paraphenylene benzobisthiazole), *CEIDP*, 351 (1981).
48. D. Bhaumik and J.E. Mark, Calculation of electronic band structure for some rigid benzobisoxazole and benzobisthiazole polymers, *J. Polym. Sci:Polym. Phys. Ed.*, 21:1111 (1983).
49. J. Mort, Conductive polymers, *Sci.*, 208:819 (1980).
50. R.J. Farris and L.A. Pottick, Property modifications in poly(p-phenylene benzobisthiazole) fibers with high-tension and high-temperature drying versus postprocessing heat treatment, *Polym. Eng. & Sci.*, 31:1441 (1991).
51. D. Bhaumik, W.J. Welsh, H.H. Jaffe, and J.E. Mark, Interchain interactions in some benzobisozole and benzobisthiazole rigid-rod polymers, *Macromolecules*, 14:951 (1981).
52. A.A. Collyer, Lyotropic liquid crystal polymers for engineering applications, *Mat. Sci. & Tech.*, 6:981 (1990).
53. J.R. Minter, K. Shimamura, and E.L. Thomas, *J. Mat. Sci.*, 16:3303 (1981).
54. T.C. Jordan, M.T. Shaw, T.C.B. McLeish, Viscoelastic response of electrorheological fluids. II. Field strength and strain dependence, *J. Rheol.*, 36:441(1991).

55. L.K.H. van Beek, Dielectric behavior of heterogeneous systems, *in:*Progress in Dielectric, J.B. Birks, ed., CRC Press, Cleveland (1967).
56. D.J. Klingenberg and C.F. Zukoski, Studies on the steady shear behavior of ER suspensions, *Langmuir*, 6:15 (1990).
57. T.B. Jones, R.D. Miller, K.S. Robinson, and W.Y. Fowlkes, Multipolar interactions of dielectric spheres, *J. Electrostat.* 22:231 (1989).
58. C.F. Zukoski, Mechanisms of ER effects, *in*: "Electrorheological Fluids: A Research Needs Assessment," U.S. Department of Energy, (1993).

# STRUCTURE FORMATION IN ELECTRORHEOLOGICAL FLUIDS

R. Tao and Qi Jiang

Department of Physics
Southern Illinois University at Carbondale
Carbondale, IL 62901-4401

## Introduction

The structure is fundamental in understanding the physical mechanism and properties of ER fluids. Many experiments indicate that upon application of electric field, dielectric particles in ER fluids first rapidly form chains between the two electrodes which then aggregate into thick columns.[1-4] A theoretical prediction[5] of a body-centered tetragonal (bct) lattice as the ground state of the thick columns has recently been verified by experiments.[1]

The issue of dynamical structure formation in ER fluids is under extensive investigation. Since the strength of ER fluids depends very strongly on the induced solid structure, information about the structure formation in ER fluids is very useful in search of strong ER fluids. For example, the single chain structure in ER fluids has a limited yield stress while the bct lattice structure has a much higher yield stress. If the knowledge of structure formation may lead us to find a method to accelerate the lattice formation in ER fluids, it will be very useful. The interest in this issue has further been enhanced since Kamien and Nelson recently pointed out that the phase transition in ER fluids is related to the physics of directed polymer melts and quantum mechanics of bosons in 2+1 dimensions.[6-8]

Two-dimensional (2-D) computer simulations were first employed to investigate this issue. If thermal fluctuations are ignored in the simulations, 2-D simulations find separated single chains when the short-range repulsive force is approximated by $\sim 1/r^{13}$ or mostly double-width strands when the repulsive force is exponential (see Fig. 1).[9] Computer simulations on the same model found much thick columns of close-packed triangular lattice when thermal fluctuations were included. (see Fig.2).[10] The extrapolation of results in two dimensions to three-dimensional (3-D) systems, however, is hindered not only by the obvious difficulty, but also by the singular nature of the Coulomb potential.

*Progress in Electrorheology*, Edited by K.O. Havelka
and F.E. Filisko, Plenum Press, New York, 1995

There have been a couple of 3-D simulations of ER fluids reported. Whittle[11] and Molrone[12] only examined the initial aggregation process in their 3-D simulations peripherally. The simulations by Bonnecaze and Brady[13] include the full hydrodynamics and electrostatics, but are performed at zero temperature, thus do not give information on the structure formed. Recently, Hass reported that a regular lattice was not formed in his 3-D simulation.[14] However, his simulation ignores the thermal fluctuations and the effect of two electrodes which have been both proved to be important in formation of thick columns.

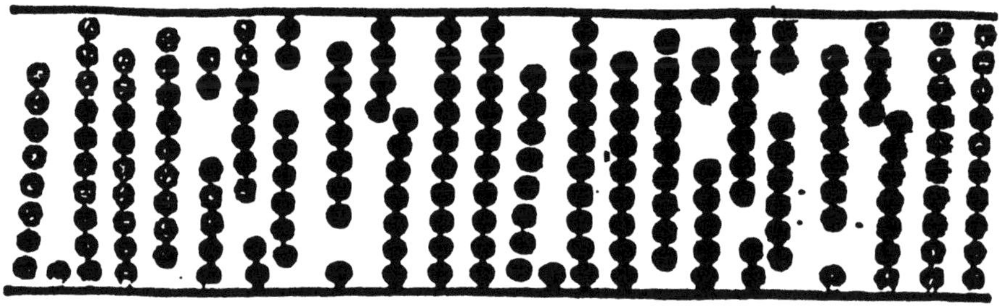

Fig.1 The single-chain structure in 2-D simulations if the thermal fluctuations are not included.

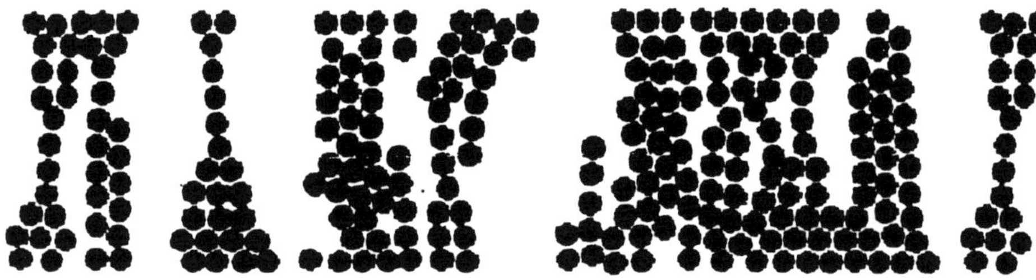

Fig.2 The close-packed triangular lattice is the ER structure found in 2-D simulations if the thermal fluctuations are included.

To clarify the issue, we consider a monodisperse suspension of spherical dielectric particles in a nonconducting liquid.[15] The particles have diameter $\sigma$ and dielectric constant $\epsilon_p$. The liquid has dielectric constant $\epsilon_f$ and viscosity $\eta$. The system is confined between two parallel electrodes which are denoted as planes $z = 0$ and $z = L$. When there is no voltage applied, the particles are randomly distributed throughout the fluid. In an electric field, each particle obtains an induced dipole moment, $\mathbf{p} = \alpha\epsilon_f(\sigma/2)^3\mathbf{E}_{loc}$ where $\alpha = (\epsilon_p - \epsilon_f)/(\epsilon_p + 2\epsilon_f)$ and $\mathbf{E}_{loc}$ is the local field. The motion of the $i$'th particle is described by a Langevin equation

$$md^2\vec{r}_i/dt^2 = \vec{F}_i - 3\pi\sigma\eta d\vec{r}_i/dt + \vec{R}_i(t) \quad (1)$$

where $\vec{F}_i$ is the electric force acting on the particle and $-3\pi\sigma\eta\vec{v}_i$ is the Stokes' drag force. In Eq.(1), we include a random Brownian force $\vec{R}_i(t)$, representing the net effect of collisions of solvent molecules on the particle. As did in other simulations[9–14], we take the dipolar approximation. The role of higher order multipoles will be discussed at the end of the paper. The ER solid structure is determined by the dipolar interactions, viscous drag forces, and Brownian motions.

We have found that the thermal motion plays a very important role in the structure formation of ER fluids. Though the ratio of the dipolar energy to the thermal energy is very high for a typical ER fluid, $(p^2/\epsilon_f\sigma^3)/k_BT \sim 10^6$, the thermal energy cannot be ignored. Because of the Peierls-Landau instability of one-dimensional solid, a ratio of the thermal energy to the lowest phonon excitation energy of a single chain is much than a unity.[4] Therefore, the thermal energy is more than sufficient in creating thermal vibrations of the single chains which accelerate the aggregation process. We will introduce a parameter, $B$, characterizing the ratio of the dipolar force to the Brownian force. Our simulation finds that for a wide range of $B$, with the help of the Brownian force, the ER fluid evolves into thick columns which has the bct lattice structure. When $B$ is very small, the ER system is unable to get out from some trapped local energy minimum state to approach the ground state.[14] The structure at very small $B$ may be related to directed polymers.[6,7] When $B$ is too big, thermal vibrations prevent the system from formation of stable solid structure.

## Computer Simulations

The dipolar force acting on the particle at $\vec{r}_i$ by a particle at $\vec{r}_j$ is given by

$$\vec{f}_{ij} = 3p^2[\vec{e}_r(1-3\cos^2\theta_{ij}) - \vec{e}_\theta \sin(2\theta_{ij})]/(\epsilon_f r_{ij}^4) \tag{2}$$

where $\vec{r}_{ij} = \vec{r}_i - \vec{r}_j$ and $0 \le \theta_{ij} \le \frac{\pi}{2}$ is the angle between the z direction and the joint line of the two dipoles. We use $\vec{e}_r$ as a unit vector parallel to $\vec{r}_{ij}$ and $\vec{e}_\theta$ as a unit vector parallel to $\vec{e}_r \times (\vec{e}_r \times \vec{E}_0)$.

A dipole $\vec{p}$ inside the capacitor at $\vec{r}_i = (x_i, y_i, z_i)$ produces an infinite number of images at $(x_i, y_i, -z_i)$ and $(x_i, y_i, 2Lk \pm z_i)$ for $k = \pm1, \pm2, \cdots$. The force between a dipole and an image has the same form as Eq.(2). The $j$th particle and its infinite images produce an electric force on the $i$th particle,

$$\begin{aligned}
f_{ij,x} &= \frac{p^2}{\epsilon_f L^4}\sum_{s=1}^{\infty}\frac{4s^3\pi^3(x_i - x_j)}{\rho_{ij}}K_1\left(\frac{s\pi\rho_{ij}}{L}\right)\cos\left(\frac{s\pi z_i}{L}\right)\cos\left(\frac{s\pi z_j}{L}\right)\\
f_{ij,y} &= \frac{p^2}{\epsilon_f L^4}\sum_{s=1}^{\infty}\frac{4s^3\pi^3(y_i - y_j)}{\rho_{ij}}K_1\left(\frac{s\pi\rho_{ij}}{L}\right)\cos\left(\frac{s\pi z_i}{L}\right)\cos\left(\frac{s\pi z_j}{L}\right)\\
f_{ij,z} &= \frac{p^2}{\epsilon_f L^4}\sum_{s=1}^{\infty}4s^3\pi^3 K_0\left(\frac{s\pi\rho_{ij}}{L}\right)\sin\left(\frac{s\pi z_i}{L}\right)\cos\left(\frac{s\pi z_j}{L}\right)
\end{aligned} \tag{3}$$

where $\vec{\rho}_{ij} = (x_i - x_j)\vec{e}_x + (y_i - y_j)\vec{e}_y$ and $K_0$ and $K_1$ are modified Bessel functions. The force on the $i$th particle by its own images is in the z direction and denoted as

$\vec{f}_i^{self}$,

$$f_{i,z}^{self} = \frac{3p^2}{8\epsilon_f}\left(-\frac{1}{z_i^4} + \sum_{s=1}^{\infty}[\frac{1}{(z_i - sL)^4} - \frac{1}{(z_i + sL)^4}]\right) \quad (4)$$

To simulate the hard spheres and hard walls, we introduce a short-range repulsive force between two particles

$$\vec{f}_{ij}^{rep} = (3p^2\vec{e}_r/\epsilon_f\sigma^4)\exp[-100(r_{ij}/\sigma - 1)] \quad (5)$$

and a short-range repulsion between a particle and the electrodes

$$\vec{f}_i^{wall} = \frac{3p^2\vec{e}_z}{\epsilon_f L^4}\{\exp[-100(z_i/\sigma - 0.5)] - \exp[-100((L - z_i)/\sigma - 0.5)]\}. \quad (6)$$

Now $\vec{F}_i$ in Eq.(1) is given by

$$\vec{F}_i = \sum_{j\neq i}[\vec{f}_{ij} + \vec{f}_{ij}^{rep}] + \vec{f}_i^{self} + +\vec{f}_i^{wall}. \quad (7)$$

The random force $\vec{R}_i(t)$ has a white-noise distribution,

$$\langle R_{i,\alpha}(0)R_{i,\beta}(t)\rangle = 6\pi k_B T\sigma\eta\delta_{\alpha\beta}\delta(t), \qquad \langle R_{i,\alpha}\rangle = 0, \quad (8)$$

where $k_B$ is Boltzmann's constant and T is the temperature. We introduce a subinterval $\tau$ which is shorter than the time steps used in integration of Eq.(1) but much longer than the molecular collision time.[16] The average of $\vec{R}_i(t)$ over $\tau$, $R_{i,\alpha}(t,\tau) = \frac{1}{\tau}\int_t^{t+\tau} R_{i,\alpha}(t')dt'$, has a Gaussian distribution

$$W(R_{i,\alpha}(t,\tau)) = (2\pi)^{-1/2}\Omega^{-1}\exp[-(R_{i,\alpha}(t,\tau))^2/(2\Omega^2)] \quad (9)$$

where $\Omega = \sqrt{6\pi k_B T\sigma\eta/\tau}$. For a time step $\delta t > \tau$, we can divide it into many subintervals of duration $\tau$, in which all quantities except $\vec{R}_i(t)$ can be treated as constants. Then, for any smooth function $\psi$, $X_\alpha = \int_t^{t+\delta t}\psi(\xi)R_{i,\alpha}(\xi)d\xi$ has a probability distribution

$$W(X_\alpha) = (\pi q)^{-1/2}\exp(-X_\alpha^2/q) \quad (10)$$

where $q = 12\pi k_B T\sigma\eta\int_t^{t+\delta t}\psi^2(\xi)d\xi$ is independent on $\tau$. This implies that our result is independent of a specific choice of $\tau$, though the value of $\tau$ is not very uniquely defined.

The intrinsic time scale in Eq.(1) is $t_0 = m/(3\pi\sigma\eta)$. We take $t = t_0 t^*$, $\vec{F}_i = F_0\vec{F}_i^*$ where $F_0 = 3p^2/(\epsilon_f\sigma^4)$, $\vec{R}_i = \Omega\vec{R}_i^*(t)$, and $\vec{r}_i = \sigma\vec{r}_i^*$ in Eq.(1). The scaling transformation produces a new equation

$$\ddot{\vec{r}}_i^* + \dot{\vec{r}}_i^* = A(\vec{F}_i^* + B\vec{R}_i^*) \quad (11)$$

where $A = F_0/[3\pi\sigma\eta(\sigma/t_0)]$ and $B = \Omega/F_0$. It is clear that $1/A$ is related to the Mason number, a ratio of the viscous force to the dipolar force. $B$ is a ratio of the Brownian force to the dipolar force.

Equation (11) indicates that the final structure of our system is related to the two constants $A$ and $B$. For a real ER system, such as alumina particles in petroleum oil, $\epsilon_f \sim 2$, $\epsilon_p \sim 8$, $\eta \sim 0.2$ poise, $\sigma \sim 10\mu$m, and the mass density of the particle $\rho \sim 3g/cm^3$. At $E_0$ = 3KV/mm and T=300 K, the ER fluid will be solidified in experiments. We estimate $t_0 \sim 8.33\times10^{-7}$ s and $A \sim 10^{-4}$ under these conditions and choose $\tau = 0.4t_0$ to meet the aforesaid condition. Then, $B \sim 10^{-1}$. Our simulation just takes the above values for $A$ and $B$.

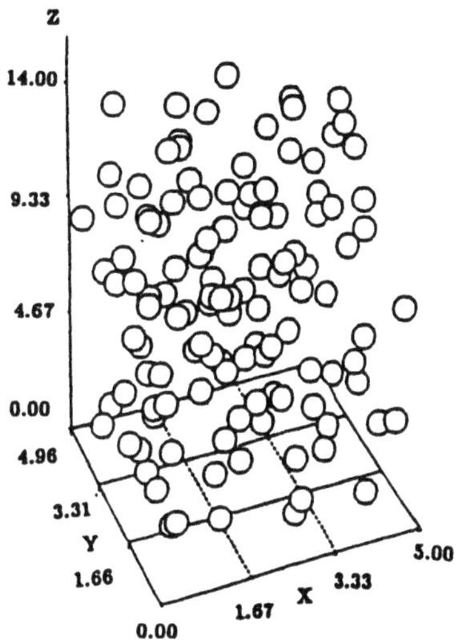

Fig.3 In the initial state, dielectric particles are randomly distributed.

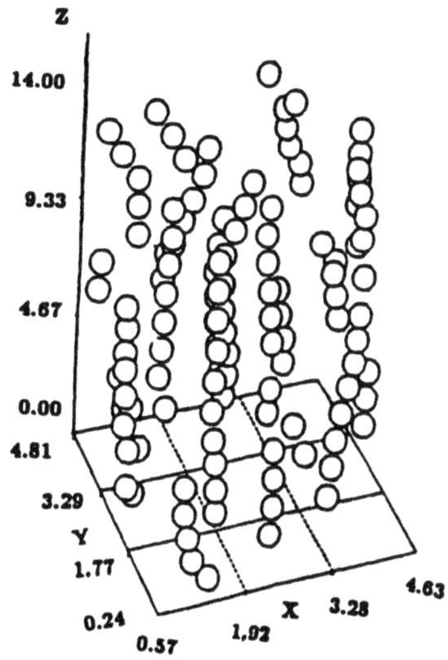

Fig.4 The configuration after first 5000 time steps.

We have applied the Runge-Kutta method to integrate Eq.(11). Our simulation has N=122 particles in a box with $L_x = L_y = 5\sigma$ and $L_z = 14\sigma$. Theoretical calculation has already shown that if $L_z \leq 6\sigma$, the single-chain structure has a lower energy than that of thick columns.[5] Therefore, we take a relatively big $L_z$. A periodic boundary condition is imposed in the x and y directions. This corresponds to a volume fraction $\phi = 0.183$. At every step, we apply the following three order parameters to characterize the structure,[17]

$$\rho_j = \mid N^{-1}\sum_{i=1}^{N} \exp(i\vec{b}_j \cdot \vec{r}_i) \mid \tag{12}$$

where $\vec{b}_1 = (2\pi/\sigma)(2\vec{e}'_x/\sqrt{6} - \vec{e}_z)$, $\vec{b}_2 = (2\pi/\sigma)(2\vec{e}'_y/\sqrt{6} - \vec{e}_z)$, and $\vec{b}_3 = 4\pi\vec{e}_z/\sigma$ are three reciprocal lattice vectors of the bct lattice. The unit vector $\vec{e}_z$ is (001) lattice axis along the field direction, but $\vec{e}'_x$ and $\vec{e}'_y$ are the bct lattice axes (100) and (010). In the structure formation, the ER system may rotate around the z axis, a phenomenon already observed in experiments.[1] Therefore, when measuring $\rho_1$ and $\rho_2$, we always rotate $\vec{e}'_x$ and $\vec{e}'_y$ about the z axis to find a position which maximizes $\rho_1\rho_2$. The order parameter $\rho_3$ characterizes the formation of chains in the $z$ direction. The other two parameters $\rho_1$ and $\rho_2$ characterize the structure in the x-y plane. All these three order parameters are unity if the ER system is the ideal bct lattice.

## Results and Discussion

The initial state is shown in Fig.3 where the dielectric particles are randomly distributed and the three order parameters are vanishing. At $t = 0$, a strong electric field is turned on and the particles begin to move. It can be noticed from Fig.2 that after the first 5000 steps (about 10 milliseconds), single chains are in shape but the lateral ordering is very weak. The structure has $\rho_3 = 0.617$, while $\rho_1 = 0.32$ and $\rho_2 = 0.16$.

At the end of next 15000 steps, we have $\rho_3 = 0.92$, but $\rho_1$ and $\rho_2$ remain around $0.4 \sim 0.5$. The structure is in Fig.4. Clearly, the ordering in the field direction is almost perfect and the lateral ordering is building. The structure obtained at 90000 steps in Fig.5 clearly shows excellent orderings both in the field direction and in the x-y plane. Single chains have been aggregated into a bct lattice structure. Three order parameters are $\rho_3 = 0.991$, $\rho_1 = 0.915$, and $\rho_2 = 0.850$. Further time evolution shows little improvement and the process becomes extremely slow.

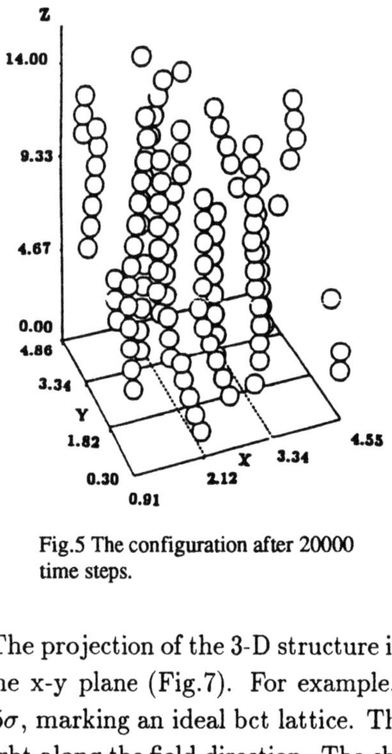


Fig.5 The configuration after 20000 time steps.

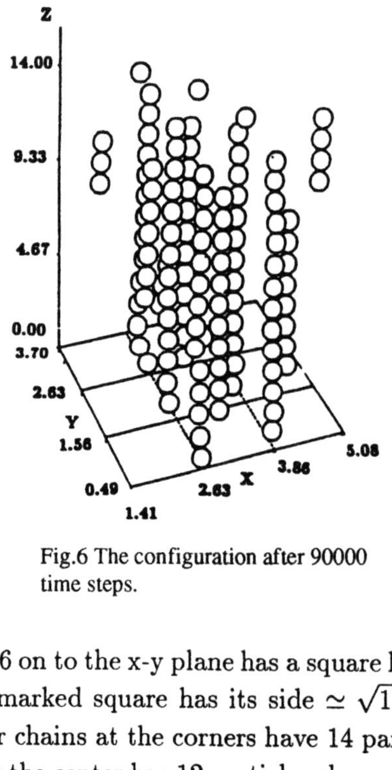


Fig.6 The configuration after 90000 time steps.

The projection of the 3-D structure in Fig.6 on to the x-y plane has a square lattice on the x-y plane (Fig.7). For example, the marked square has its side $\simeq \sqrt{1.5}\sigma = 1.225\sigma$, marking an ideal bct lattice. The four chains at the corners have 14 particles straight along the field direction. The chain at the center has 12 particles close-packed with the four neighbor chains. The structure is an ideal bct lattice if the chain at the center does not have one particle missing. The thick dots in Fig.6 also indicate that the chains are straight in the field direction.

Fig. 7 shows the development of $\rho_i$ $(i = 1, 2, 3)$ with time. The formation of single chains is rapid. In about several milliseconds, $\rho_3$ reaches 0.6. After 0.4s, $\rho_3$ is above 0.9. The building of lateral ordering is relatively slower than the formation of chain. It takes about several seconds to have $\rho_1$ and $\rho_2$ reached 0.8. The strong fluctuations of $\rho_1$ and $\rho_2$ after the formation of chains are consistent with experiment, signifying the aggregation of separated chains into columns.[3]

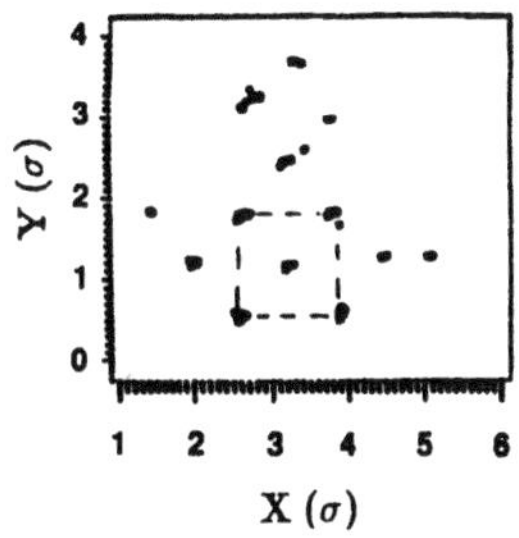


Fig.7 Projection of the configuration in Fig. 6 to the x-y plane.

We have done some tests to verify the importance of Brownian force in development of ER solid structure. If we ignore the Brownian motion completely, or let $B$ be extremely small, such as $B \sim 10^{-6}$, the ER fluid has a rapid chain formation followed by a kinetic trapping into a complicated structure.[14] After a long time, the structure has $\rho_3$ stayed about 0.7 but $\rho_1$ and $\rho_2$ remain 0.3 or below. The structure in such a case needs a further study because it may be related to directed polymers.[6,7] When $B \gg 1$, the ER system has too much vibrations, preventing from formation of a stable structure. However, for a quite a wide range of $B$, the ER fluid has a rapid chain formation, followed by an aggregation of chains into the bct lattice structure with three order parameters close or above 0.9.

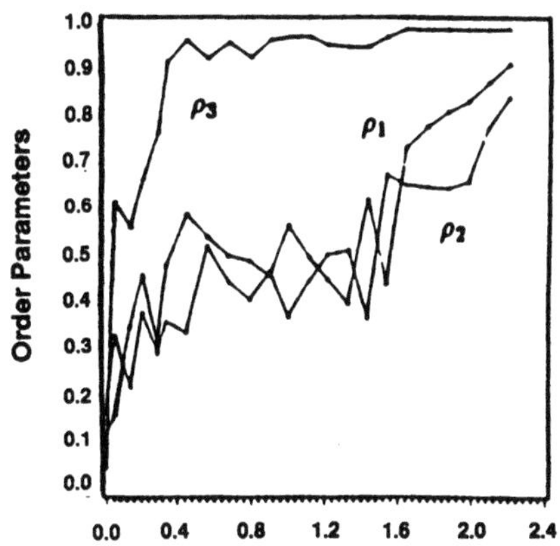


Fig. 8 The order parameters change with time after the electric field is applied. The time unit is second.

When the volume fraction $\phi \geq 20\%$, we have also found the bct lattice. If $\phi$ further increases, the system may be easy to develop into thick columns of a polycrystalline structure which have several bct lattice grains with the same (001) axis but different (100) and (010) axes.[1]

Finally, we compare our simulation with experiments. The rapid chain formation in our simulation takes about 10 milliseconds or longer. This result is consistent with other simulations.[9] However, the experiments on ER fluids, such as alumina particles in petroleum oils, find this chain formation time is only about several milliseconds, shorter than our result.[18] The reason for this difference may lie in the fact that we ignore the contributions from higher multipoles, charges, and currents in ER fluids which begin to draw attention recently.[19]

It is conceivable from the above results that the lattice structure formation may be accelerated by some external vibration force. Our simulation show that under thermal vibration the ER fluid has formed ordered lattice structure around 10 seconds. This finding is consistent with recent experiments.[3] If the lattice formation process can be shortened to several millisecond by some external vibration force, the result will be very significant in ER applications.

**Acknowledgements**

This research is supported by the Office of Naval Research grant N00014-93-1-0582 and a grant from Illinois Consortium for Advanced Radiation Sources.

**References**

1. T. J. Chen, R. N. Zitter, and R. Tao, Phys. Rev. Lett. **68**, 2555 (1992).
2. J. E. Martin, J. Odinek, and T. C. Halsey, Phys. Rev. Lett. **69**, 1524 (1992).
3. J. M. Ginder and L. D. Elie in *Electrorheological Fluids*, edited by R. Tao (World Scientific Publishing Comp., Singapore, 1992), 23.
4. T. C. Halsey and W. Toor, Phys. Rev. Lett. **65**, 2820(1990).
5. R. Tao and J. M. Sun, Phys. Rev. Lett. **67**, 398(1991); Phys. Rev. A, **44**, R6181 (1991).
6. R. D. Kamien and D. R. Nelson, J. of Stat. Phys. **71**, 23 (1993).
7. R. D. Kamien and D. R. Nelson, Phys. Rev. A **45**, 8727 (1992).
8. R. P. Feynman, *Statistical Mechanics* (Benjamin/Cummins, Reading, Mass., 1972).
9. D. J. Klingenberg, F. van Swol, and C. F. Zukoski, J. Chem. Phys. **91**, 7888 (1989).
10. W. R. Toor, J. of Colloid and Interface Science, **156**, 335 (1993); N. K. Jaggi, J. Stat. Phys. **64**, 1093 (1991).
11. M. Whittle, J. Non Newtonian Fluid Mechanics, **37**, 233 (1990).
12. D. M. Heyes and J. R. Molrose, Mol. Simul. **5**, 293 (1990).
13. R. T. Bonnecaze and J. F. Brady, J. of Chem. Phys. **96**, 2183 (1992).
14. K. C. Hass, Phys. Rev. E, **47**, 3362 (1993).

15. R. Tao and Q. Jiang, Phys. Rev. Lett. **73**, 205 (1994).
16. For example, see, S. Chandrasekhar, Re. Mod. Phys. **15**, 1 (1943); F. Reif, *Fundamentals of Statistical and Thermal Physics* (McGraw-Hill, New York, 1965) 560-62; R. K. Pathria, *Statistical Mechanics*, (Pergamon Press, Oxford, 1972) 451-52.
17. R. Tao, Phys. Rev. E **47**, 423 (1993).
18. K. D. Weiss and J. D. Carlson in in *Electrorheological Fluids*, edited by R. Tao (World Scientific Publishing Comp., Singapore, 1992), 264.
19. L. C. Davis, J. Appl. Phys. **72**, 1334 (1992); N. J. Félici, J. N. Foulc, and P. Atten in the *Electrorheological Fluids*, edited by R. Tao and G. D. Roy (World Scientific Publishing Comp. Singapore, 1994) 139-152.

# RELATIONS OF HIGH-FIELD NON-LINEAR DIELECTRIC PROPERTIES AND ELECTRORHEOLOGICAL EFFECTS

L.W. Zhou[1], Z.Y. Qiu[1], Y. Shen[1], Y. Tang[2], Z. Gao[2],
M. Sun[3], J.W. Qiu[1], Z.Z. Yu[1], and L. Hu[1]

[1]T.D. Lee Physics Laboratory and Department of Physics
[2]Department of Chemistry
[3]Department of Polymer Sciences
Fudan University, Shanghai 200433, China

## INTRODUCTION

The non-linear dielectric property of the electrorheological (ER) fluids under high AC electric field still lacks detailed investigation compared with other studies on ER fluids. In its physical principle, ER effect is the result of interaction of high electric field with chemical configurations and liquid environment of solid particles. In the investigation of applications, the response time, power dissipation, strength and stability of ER fluids are closely related to the dielectric properties of the fluids. The study of non-linear dielectric properties of the ER fluids can be a key to reveal their peculiar rheological properties, and would be hopeful to provide information for the research of ER materials.

The investigation of the dielectric properties of ER fluids has been a main trend in this area. Klass and Martinek[1] studied in detail the electric properties of ER fluids. Uejima[2] measured the dielectric spectrum in audio frequencies and the dispersion relation of the shear stress, and explained his experimental results by using Schwarz's theory[3], giving the similarity of dielectric spectrum and shear stress dispersion relation. Conrad[4] studied experimentally the effect of temperature and field strength on the dielectric properties of ER fluids at the frequency of 500 Hz. Filisko[5] discussed the relation of rheological responses of zeolite ER fluids with the cation mobility and the leaking current. Deniga's[6] and Block's[7,8] groups discussed the dielectric properties of flowing ER fluids. These works and many others have laid foundation for understanding the relation between the polarization and rheological effects of ER fluids. However, many of previous measurements on dielectric spectrum were done under electric field less than 1000 V/mm, while the discussions were based on a weak field situation. In the present work, the complex dielectric spectra under high AC electric fields have been measured, which requires the construction of a high AC voltage source and a special bridge, and involves analysis of the non-linear responses of the ER fluids with field.

From the results of previous experiments, it has been suggested that ER effect is due to the polarization of solid particles. However, the upper limit of the yield stress based on simple polarization theory is lower than the measured yield stresses of some best ER fluids.[9] The motivations of this work are, first, to test whether the previous dielectric experiments reflect the real polarization state of ER fluids or not, and furthermore, to measure the dependences of the polarization on field and frequency under high AC electric field. It was hoped that the results of such study would be useful for improving the consistency between experiment and theory, and for getting insight into the mechanism responsible for the ER effect.

The previous dielectric experiments can be classified into two types. In one type, the complex dielectric spectra of ER fluids were measured by using audio frequency bridge with the exciting signal of a few volts, then interaction forces and losses are deduced from the spectra.[2] In the other type, what was measured was the response of ER fluids to a weak AC signal (a few tens of a volt) superimposed on a strong DC electric field (on the order of kV/mm). In this method, the ER fluid is, in fact, set to a DC polarized state, while AC signal is used only to detect the response of the ER fluid under DC polarized state. Since ER fluids are usually multi-phased materials, the dielectric responses detected under both of the above circumstances would be much different from its response to high AC electric field. Charge carrier distributions on the liquid-solid interface under AC and DC field would also be different from each other. The relaxation can not be observed under DC field. The measurements of dielectric properties of ER fluid under high AC field would better reflect the polarized state of an ER system, so one has no way of skirting around the difficulties in the measurements under a high AC electric field.

By using the bridge method under a low AC field superimposed on a high DC field, the dielectric increment, $\Delta\varepsilon'$, was found to be proportional to the square of the field over the range examined[7], namely $\Delta\varepsilon' = \varepsilon_r - \varepsilon_{r0} = \varepsilon_{r2} E^2$.

From this result, one may conclude that, if an ER fluid is under a high AC field, the displacement current $I_c$ and applied AC voltage $U_{AC}$ would satisfy the equation, $I_c = \omega C U_{AC} = \omega \varepsilon_r C_0 U_{AC}$. If $U_{AC} = U_{AC0}\sin\omega t$, and $E = U_{AC}/d$, with d being the gap between two electrodes, then the current can be expressed as

$$I_c = I_{c1}\sin\omega t + I_{c3}\sin 3\omega t, \qquad (1)$$

namely, the first and third harmonic generations caused by non-linear dielectric constant should be observed, while the second harmonic generation is eliminated. If $\varepsilon_r$ can be expressed as

$$\varepsilon_r = \varepsilon_{r0} + \varepsilon_{r1}E + \varepsilon_{r2}E^2 + \varepsilon_{r3}E^3, \qquad (2)$$

the first, second, third and fourth harmonic generations may be observed.

## SAMPLE PREPARATIONS

Various polymer and zeolite ER fluids were used in these experiments, including polyacene quinone (PAQ), sodium alginate polymer-electrolyte, ZSM-5, NaY and NaM. In preparing the materials, we found that the interface polarization played an important role in ER effect. The shear stress is quite weak in those materials where the polarization is mainly due to dipole re-orientation, and such polarization mechanism would not give very strong ER effect. Considering the mechanism of interface polarization, we prepared over a hundred kinds of polymer materials and many zeolite ER fluids.

PAQ has high effective dielectric constant ($\varepsilon'$=78 measured at 1 MHz), and it is non toxic nor corrosive. In its molecular structure, there is a big planar conjugated big $\pi$-bond, and the resistance is very small along the non-local π-orbit with molecular boundary being the only limitation of electron movement. The material is easy to be polarized and quick in response to the external electric field. We synthesized PAQ with the following structure according to the reference.[10]

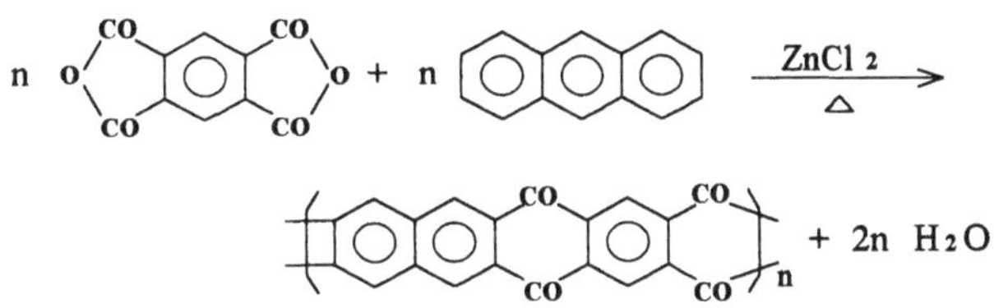


(3)

The starting materials were pyromellitic dianhydride (PMA) and anthracene. In the synthesis process, the equal molar of PMA and anthracene were thoroughly mixed with excessive zinc chloride as catalyst, powdered and then moved into a 250 ml reaction bottle. The mixture was heated slowly to 150°C±10°C under nitrogen gas atmosphere for 2-3 hours, then to 180°C-200°C for 2-3 hours, and then to 260°C-280°C for 20 hours. After it was cooled down, 200ml of 5% hydrochloride acid was added, and the mixture was heated to boiling, and filtered. A black solid was left after drying. The solid was powdered again, washed twice with 5% hydrochloride acid, many times with de-ionized water till it was neutral, then once more with 95% alcohol, soaked and washed three times with DHF, twice with mythylbenzene, and finally washed once with water-free alcohol. The final product was dried for 2 hours at about 100°C, powdered, and dried for another 1-2 hour. The yield is around 60%. The sample was stored in dry box before mixing with liquid paraffin.

The samples were characterized with an electrorheometer, which was remodeled from a digital rheometer. In the electrorheometer both the outer and inner cylinders of the sample cell are insulated from other parts of the rheometer, especially the electronic parts that are very sensitive to high audio frequency electric field. To suit the tremendous change of shear stress in ER fluids with and without electric field, an inner cylinder of diameter of 3 mm was specially designed and built The electrorheometer has 15 speeds of revolution, and the shear rate is as high as 996 $s^{-1}$.

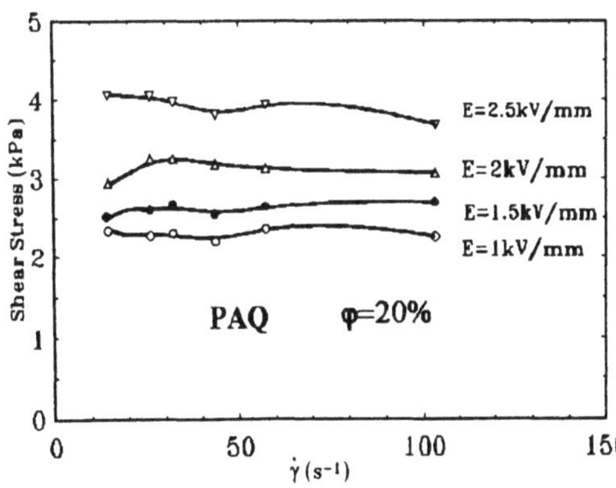


**Figure 1.** Dynamic shear stress versus shearing rate and voltage for PAQ ER fluid with solid particle concentration ϕ=20%.

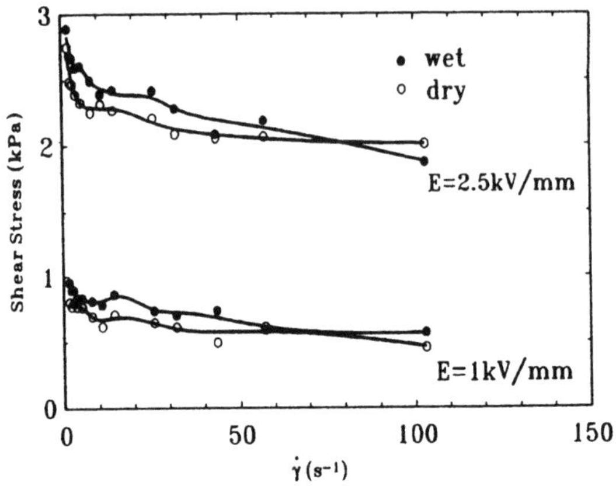


**Figure 2.** Dynamic shear stress of PAQ ER fluid versus shearing rate at different DC electric fields for different water contents (solid circles for the sample with 6% water and open circles for dry sample).

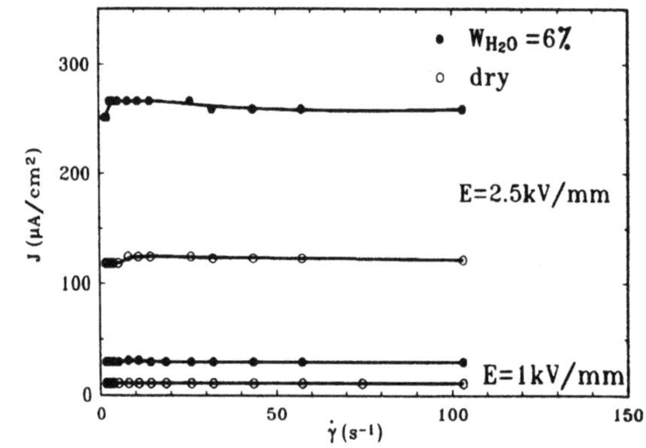


**Figure 3.** Current density of PAQ ER fluid versus shearing rate at different DC electric fields for different water contents (full circles for the sample with 6% water, and open circles for dry sample).

Figure 1 shows the rheological properties of ER fluid of PAQ in liquid paraffin at different DC fields. The dynamic shear stress of the sample does not change much with shear rate when the latter is larger than about 15 $s^{-1}$. Figures 2 and 3 show clearly the effect of water concentration in PAQ ER fluid on the rheological properties and current density. Water content in PAQ ER fluid hardly has the effect on sample's dynamic shear stress, but affects its current density a lot.

## EXPERIMENTS

The setup for measuring dielectric constant consists of an AC high voltage source, a high voltage bridge, a sample cell and a balance indicator. The voltage source contains a low frequency signal generator with a frequency range from 10 to 20000 Hz, a 250 W audio-frequency power amplifier and two transformers designed to work in the frequency ranges from 40 to 5000 Hz and from 1000 to 20000 Hz, respectively. The voltage source can supply up to

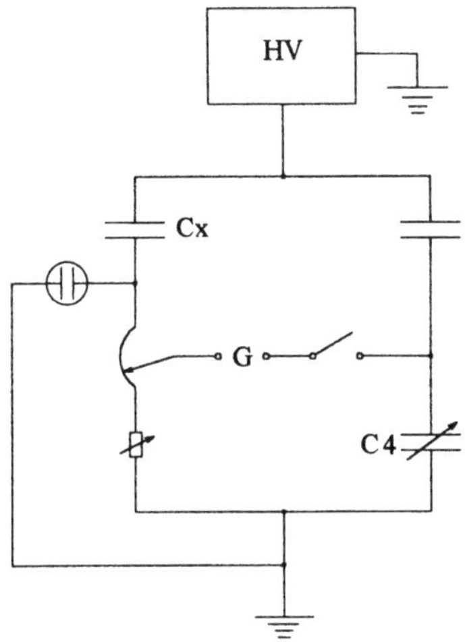


**Figure 4.** Diagram of the setup for measuring dielectric properties with high voltage bridge using high AC electric field. HV represents an audio frequency power supply, which includes a low frequency signal generator, an audio frequency power amplifier and a low frequency transformer.

8000 V in the frequency ranges mentioned above. High voltage bridge QS3 (Huguang Instrument Co., Shanghai) used in these measurements is designed for AC voltage of 0 - 10 kV, but at a single frequency of 50 Hz. The bridge, as shown in figure 4, is improved to be able to work in the frequency range from 50 to 2000 Hz and to eliminate distributive capacitance effect on the circuits. In this case a phase shifter (not shown in figure 4) is removed from the original bridge. This would introduce some error when the experimental data are taken. The bridge system has fairly good frequency response in the frequency range tested. Some improvements were also made on the balance indicating system of the Schering bridge. The bridge system has a standard capacitor of 48.075 pF. The sample cell consists of two coaxial stainless steel cylinders with the gap of 0.5 mm. The capacitance of the empty sample cell is 499.8 pF with standard deviation of about 0.8 pF, and the cell is also used to calibrate the system. The output of the power source and the current through a sample are monitored by using two AC digital voltmeters with frequency response up to 50 kHz, or alternatively, two oscilloscopes. This system can measure the dielectric properties of ER fluids directly and give more reliable results of the polarized states.

To test non-linear polarization of ER fluids, we have measured the second and third harmonics of the current through the ER fluid with exciting signals of base frequencies from 300 to 1500 Hz. The experiments were carried out under high AC field (figure 5) and low AC field superimposed on high DC field (figure 6), respectively. In the first case, the high AC voltage was controlled, tested and then applied to the sample. The current through the sample was measured as the voltage over sampling resister, and the voltage was detected by an analog-digital converter, which was protected by a pair of two positive-to-positive connected diodes. The data collected were then analyzed with fast Fourier transform (FFT) to check if there was any signal of the second, third or even higher harmonic generations. We also measured the non-linear responses of the ER fluids to the low AC field superimposed on the high DC field in order to compare the results of two different situations.

## RESULTS AND DISCUSSIONS

### Dielectric Properties

Figures 7 and 8 show dielectric constants and AC losses as functions of frequency for the PAQ (figure 7) and NaY zeolite (figure 8) ER fluids. When frequency is increased, the dielectric responses to different fields merge, as shown in figure 8. This reflects the time characteristics of polarization of the ER fluid under high field, namely, the low natural frequency of interface polarization, which can also be seen from the dispersion relation of shear stress.

For linear polarization, $\mathbf{P}=\chi_e\varepsilon_0\mathbf{E}$, electric susceptibility, $\chi_e$, and successively, dielectric constant, ε', are not a function of electric field E. However, both figures 8 and 9 show that the dielectric constants ε' and losses tanδ under different electric fields increase with field monotonically, which clearly demonstrates that a non-linear polarization exists.

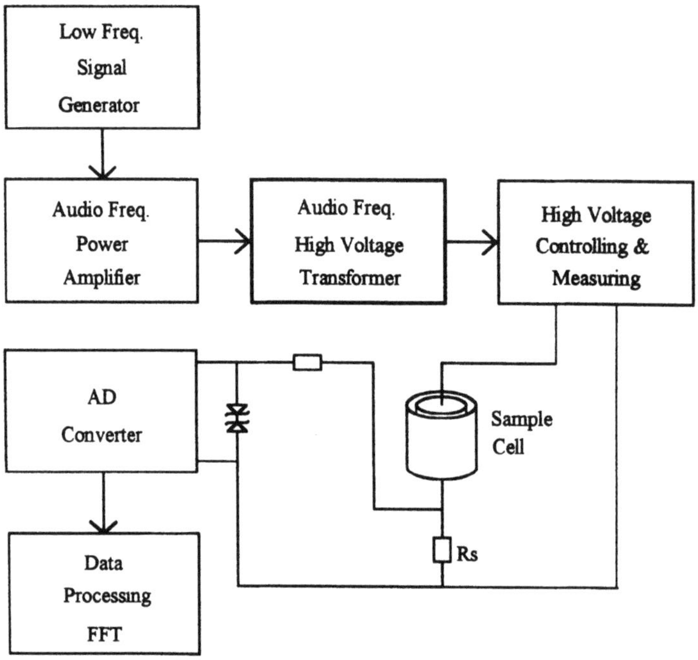


**Figure 5.** Block diagram of setup for FFT measurements under high AC fields.

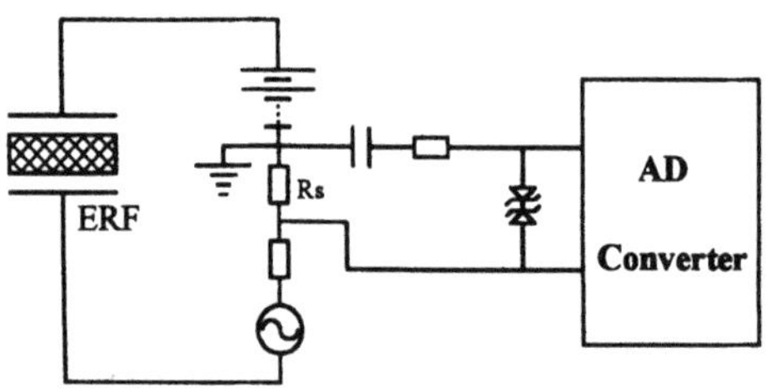


**Figure 6.** Block diagram of setup for FFT measurements under high DC fields superimposed on low AC field.

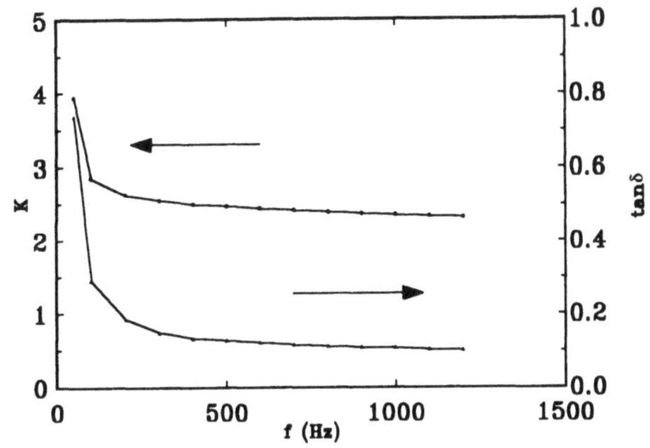


**Figure 7.** K, a constant proportional to dielectric constant, and tanδ, AC loss, versus frequency for PAQ ER fluid.

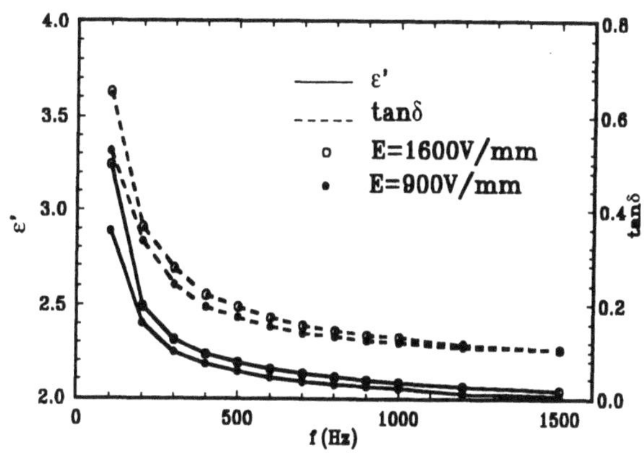


**Figure 8.** Relative dielectric constant, $\varepsilon_r$, and AC loss, tanδ, of NaY zeolite ER fluid versus frequency at different electric AC fields

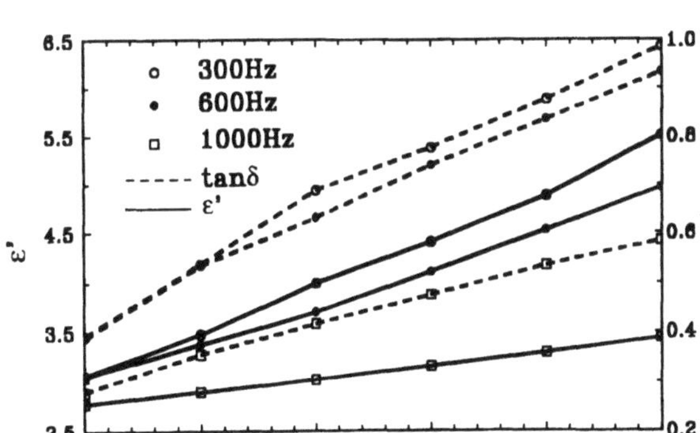


**Figure 9.** Relative dielectric constant, $\varepsilon_r$, and AC loss, tanδ, of NaY zeolite ER fluid with different particle concentration versus field for different frequencies.

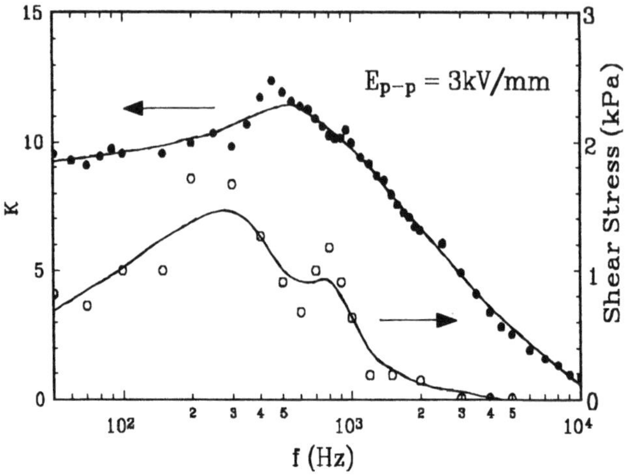


**Figure 10.** K, a value proportional to dielectric constant, and shear stress versus frequency of external field for NaY zeolite ER fluid when Ep-p=3kV/mm.

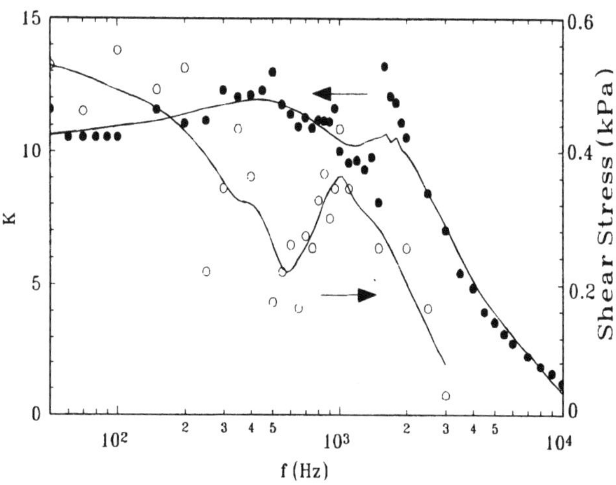


**Figure 11.** K, a value proportional to dielectric constant, and shear stress versus frequency of external field for sodium alginate, when Ep-p, peak-to-peak electric field, for shear stress measurement is 2.4 kV/mm and that for K measurement, 3 kV/mm.

Figures 10 and 11 show that the non-linear dielectric spectra of the ER fluid measured under high electric fields are similar to the dispersion curves of their shear stresses. It can be suggested that the shear stress is caused by the polarization of the particles. When the frequency of the exciting electric field is above a certain value, the fluid does not have enough time to be polarized, so the interaction between the particles is weakened, and the shear stress is lowered. Thus there is a lower limit of the response time of an ER fluid. The limit can be defined as the inverse of the frequency at which the shear stress is decreased to $1/\sqrt{2}$ of that at the lowest frequency. From figures 10 and 11, one may find the lower limit of response time to be about 1 ms and 0.7 ms, respectively.

In the zeolite crystals, there are mobile metallic cations, which may transport in a certain direction under external electric field and electric field inside the zeolite framework. Meanwhile, the cations may also diffuse due to the influences of density gradient and temperature. We introduce an equation to describe the motion and distribution of the metallic cations under the influences of temperature, external electric field and framework electric field. The polarizability and dielectric response can be calculated by using the equation. The motion of the cations inside zeolite channels is assumed to be one dimensional for simplification, i.e., moving only along the axis of the channel. Suppose that in the ER fluid the concentration of the solid phase is low, the external electric field changes slowly inside the crystals. In this case, the cation current density is the sum of the diffusion current density,

$$-D\nabla n = -D\frac{\partial n}{\partial x}, \tag{4}$$

and the transport current density,

$$\mu n E = -\mu n\frac{\partial V}{\partial x}. \tag{5}$$

From electric charge conservation law,

$$\frac{\partial n}{\partial t} = -\frac{\partial j}{\partial x}, \tag{6}$$

Under constant temperature the cation current density satisfies the following equation:

$$\frac{\partial n}{\partial t} = \mu\frac{\partial}{\partial x}(n\frac{\partial V}{\partial x}) + D\frac{\partial^2 n}{\partial x^2}, \tag{7}$$

where $n$ is cation number density, $j$, cation current density, $\mu$, cation mobility, $D$, diffusion coefficient and $V = V_1(x)+V_2(x,t)$ with $V_1(x)$ as framework potential and $V_2(x,t)$, potential of external field. Our results of dielectric measurements are quantitatively consistent with the numerical solution of Eq. 7, which will be discussed in another paper.

**Non-Linear Polarization**

The non-linear responses of various ER fluids to the high AC field have been measured. Figure 12 shows a typical FFT result of the non-linear response of the current through NaY zeolite ER fluid. The liquid phase is liquid paraffin and the particle concentration of the sample is about 25% by weight. The applied AC electric field is 1400 V/mm with frequency of 315 Hz. In the upper part of the figure 12 is the amplitude-frequency spectrum, and in the lower part is the frequency spectrum of the initial phase angle. As shown in the figure, the third harmonic generation (THG) is very large, with the ratio of its amplitude to that of the base frequency signal being about 0.035. However, the second harmonic generation (SHG) is much smaller, with the ratio of its amplitude to that of the base frequency signal being about 0.007. Same experiments have also been done under the field of 1000 V/mm for several other base frequencies. The results are shown in figure 13, in which the ratios of the amplitudes of THG or SHG to those of the base frequency signals are presented. It is noted that THG disappears under the lower field of 1000 V/mm.

It is interesting to compare the non-linear dielectric properties with the rheological properties. The dynamic shear stresses of NaY ER fluid were measured under the field of 1130 V/mm and the shear fields of various shear rates. Figure 14 shows part of the results. It seems that the frequency dependencies of the dynamic shear stress at some shear rates are similar to dependence of the THG amplitude ratio on frequency shown in figure 13.

The measurements for the non-linear responses of ER fluids to the low AC superimposed on the high DC electric field were done. Neither SHG nor THG was observed. However, an initial phase change of about 180° appeared in the phase-frequency curve. In this case, we think that the high DC field caused the main polarization, while the low AC field did a supplemental pulsating polarization.

To rule out the possible sources of the high order harmonic generations in the signal generator, power amplifier or transformers, the related tests were also performed. Care was taken to avoid any distortion of the signal from either the low frequency signal generator or the audio

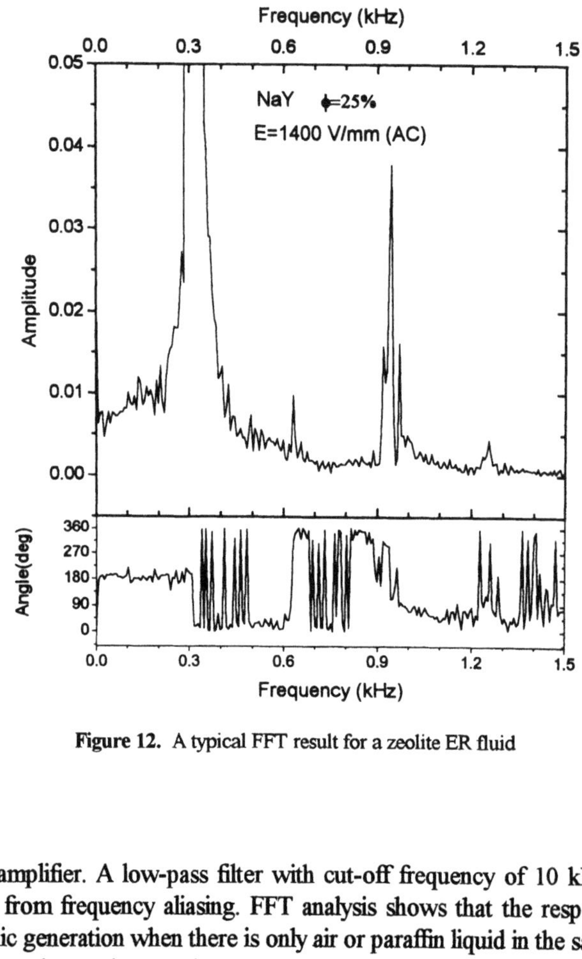


**Figure 12.** A typical FFT result for a zeolite ER fluid

frequency power amplifier. A low-pass filter with cut-off frequency of 10 kHz is also used to prevent the signal from frequency aliasing. FFT analysis shows that the response signal has no high order harmonic generation when there is only air or paraffin liquid in the sample cell.

It is also interesting to know whether the observed non-linear dielectric responses come from non-linear electric conductance or from non-linear polarization. Thus NaY powder was calcined in the temperature range from 500°C to 580°C for 26 hours, and then mixed with paraffin oil and characterized. The ER activity of the sample disappeared, the leaking current was very weak, and SHG or THG were not observed. The zeolite fluid sample was left in open air with relative humidity of 75% for one day. The ER activity of the sample remained unobservable, but its leaking current became larger than a day ago. The curves of its electric conductance and

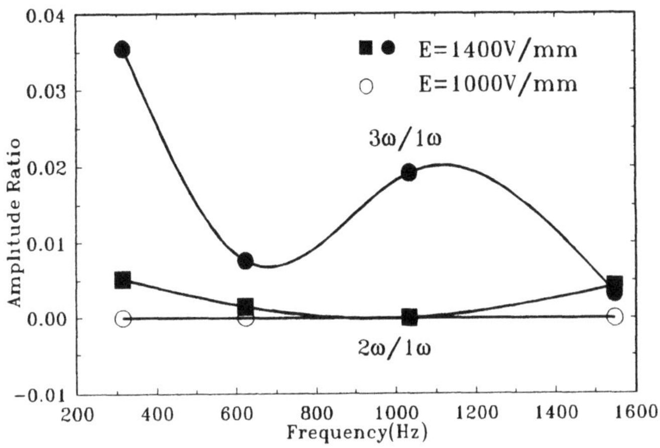


**Figure 13.** The ratios of amplitudes of THG and SHG to those of base frequency versus frequency for NaY ER fluid.

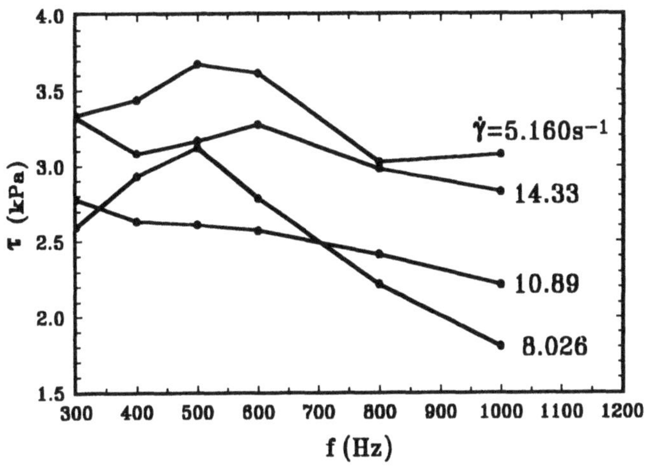


**Figure 14.** Dynamic shear stress of NaY ER fluid versus frequency at lower shearing rates.

its relative dielectric constant versus applied field were fitted as the following polynomial expressions

$$\sigma = \sigma_0 + \sigma_1 E + \sigma_2 E^2 + \sigma_3 E^3 = 0.1403445 + 0.000739E + 0.0000008E^2 \quad (8)$$

$$\varepsilon_r = \varepsilon_{r0} + \varepsilon_{r1} E + \varepsilon_{r2} E^2 + \varepsilon_{r3} E^3 = 2.4234748 + 0.0004625E + 0.0000011E^2 \quad (9)$$

with the unit of $\sigma$ being μA/V, and that of E, V/mm. $\sigma_3$ and $\varepsilon_{r3}$ were almost equal to zero. Namely, the SHG should be observed, and the amplitude ratio of SHG to base frequency signal should be apparently larger than that of THG to the base frequency signal. This was verified by the FFT analysis of non-linear dielectric responses, as shown in figure 15. This result strongly suggests that the observed non-linear dielectric response is probably mainly from the non-linear electrical conductance since the observed non-linear component of the relative dielectric constant is very weak. ER activity is still not detected in this sample.

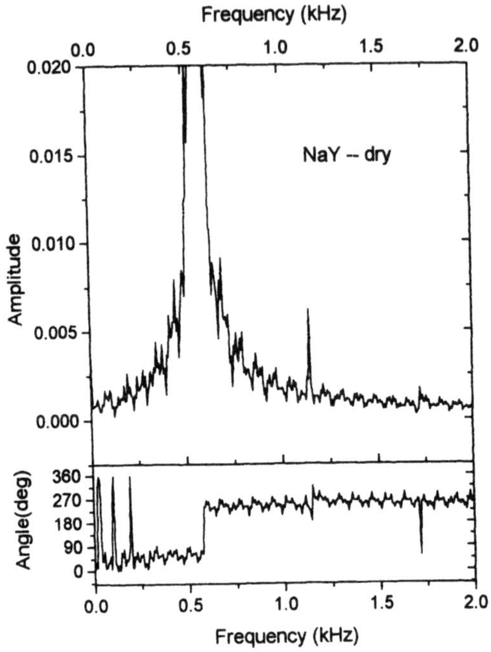


**Figure 15.** FFT result for a zeolite fluid, calcined, but left in open air for overnight to increase its water content.

In order to find out the possible source of non-linearity we treat the sample fluid as a non-linear resister parallel connected with a capacitor, and its conductance and dielectric constant can be expressed as equations (8) and (9), respectively. The current through the non-linear resister, $I_r$, and the displacement current through the capacitor, $I_c$, can be written as

$$I_r = \sigma U \approx \sigma_0 dE + \sigma_1 dE^2 + \sigma_2 dE^3 + \sigma_3 dE^4 \tag{10}$$

and

$$I_c = i\omega C_0 \varepsilon_r U \approx i\omega C_0 d(\varepsilon_{r0} E + \varepsilon_{r1} E^2 + \varepsilon_{r2} E^3 + \varepsilon_{r3} E^4), \tag{11}$$

respectively. If $E=E_0\sin\omega t$, the non-linear current is expressed as

$$I_r + I_c \approx I_0 + I_1 \sin\omega t + I_2 \cos 2\omega t + I_3 \sin 3\omega t \tag{12}$$

where $I_0$, $I_1$, $I_2$ and $I_3$ are the complex constants, and

$$|I_1| = [(\sigma_0 E_0 + \frac{3}{4}\sigma_2 E_0^3)^2 + (\omega C_0 d\varepsilon_{r0} E_0 + \frac{3}{4}\omega C_0 d\varepsilon_{r2} E_0^3)^2]^{\frac{1}{2}}, \tag{13}$$

$$|I_2| = \frac{1}{2}E_0^2(\sigma_1^2 d^2 + \omega^2 C_0^2 d^2 \varepsilon_{r1}^2)^{\frac{1}{2}} \tag{14}$$

and

$$|I_3| = \frac{1}{4}E_0^3(\sigma_2^2 d^2 + \omega^2 C_0^2 d^2 \varepsilon_{r2}^2)^{\frac{1}{2}}. \tag{15}$$

From equations (13), (14) and (15), the ratios of $I_2/I_1$, $I_3/I_1$ and $I_3/I_2$ can be obtained, which may be compared with the experimental results.

$$\frac{I_3}{I_2} = \frac{E_0}{2}\frac{(\sigma_2^2 + \omega^2 C_0^2 \varepsilon_{r2}^2)^{\frac{1}{2}}}{(\sigma_1^2 + \omega^2 C_0^2 \varepsilon_{r1}^2)^{\frac{1}{2}}} \approx \frac{E_0}{2}\cdot\frac{\sigma_2}{\sigma_1}. \tag{16}$$

The approximation is valid because the second terms in the both parentheses are much smaller than the first ones. Equation (16) gives that $I_3/I_2$ equals about 0.3, which is consistent with the experimental result of 0.2 from figure 12.

However, the source of high amplitude ratio of THG to the base frequency signal shown in figure 12 has not been identified yet.

## CONCLUSIONS

Non-linear dielectric properties of various polymer and zeolite ER fluids under high electric field of audio-frequency have been studied. Their dynamic shear stress, static complex dielectric spectrum and non-linear dielectric responses to strong electric field of audio frequency were measured. The lower limit of the rheological response time of the ER fluids is given theoretically according to the similarity of the relation of shear stress versus frequency to the complex dielectric spectra. A model is developed to understand non-linear dielectric behaviors. It is found that the possible source of non-linear dielectric properties is non-linear conductance. Good ER fluid shows obvious THG under a high AC field. The further research includes: (1), to determine the types and properties of charge carriers and their migrating region; (2), to study the effect of temperature on ER effect; (3), to understand the energy transferring process in ER effect. (4), to study the origin of leaking current and its contribution to ER effect; (5), to determine the contribution of non-linear conductance and non-linear dielectric properties to SHG and THG and the relations of high order harmonic generation to ER effect.

## ACKNOWLEDGMENTS

Authors would like to sincerely thank Prof. Z.F. Zhao of Tongji University for his momentous contribution to this work.

The subject was supported by National Natural Science Foundation of China, No. 19374018.

## REFERENCES


1. D.L. Klass and T.W. Martinek, Electroviscous fluids I rheological properties, *J. Appl. Phys.* 38:67 (1967); D.L. Klass and T.W. Martinek, Electroviscous fluids II electrical properties, *J. Appl. Phys.* 38:75 (1967).
2. H. Uejima, Dielectric mechanism and rheological properties of electrofluids, *Jap. J Appl. Phys.* 11:319 (1972).
3. G. Schwarz, A theory of low frequency dielectric dispersion of colloidal particles in electrolytes solution, *J. Phys. Chem.* 66:2636 (1963).
4. H. Conrad, A.F. Sprecher, Y. Choi, and Y. Chen, The temperature dependence of the electrical properties and strength of electrorheological fluid, *J. Rheol.* 35:1393 (1991).
5. F.E. Filisko and L.H. Radzilowski, An intrinsic mechanism for the activity of alumino-silicate based electro-rheological materials, *J. Rheol.* 34:539, (1990).
6. Y.F. Deinega and G.V. Vinogradov, Electric fields in the rheology of disperse systems, *Rheol. Acta* 23:636, (1984).
7. H. Block, P. Rattray and T. Watson, Semi-conducting polymers as ER fluid substrates, *in* "Electrorheological Fluids," R. Tao, ed., World Scientific, Singapore (1991).
8. H. Block and J. P. Kelly, Electrorheology, *J. Phys. D: Appl. Phys.*21:1661, (1988).
9. P.A. Adriani and A.P. Gast, A microscopic model of electrorheology, *Phys. Fluids* 31:2757 (1988).
10. R. Rosen and H.A. Pohl, Some polymers of high dielectric constant, *J. Polymer Scie. Part A-1* 4:1135 (1966).
11. R. Coelho, "Physics of Dielectrics for the Engineer," Elsevier Scientific Publishing Company, Amsterdam (1979).

# ASYMMETRICAL POLARIZATION OF SPHERICAL COLLOIDAL PARTICLE DOUBLE LAYER IN ELECTRIC FIELDS

Yue Hu[1], Vidya Kumar[1], and Seth Fraden[2]

[1] Physics Department, Wellesley College
Wellesley, MA 02181
[2] Physics Department, Brandeis University
Waltham, MA 02254

## INTRODUCTION

It has long been accepted that the drastic increase in the viscosity of electrorheological fluids subjected to electric fields is due to interactions between the induced dipole moments of the colloidal particles in the suspensions. Computer simulations of dipole-dipole interactions have produced chain and column formations of particles in the direction of the applied field that agree well with experimental observations.[1]

In an aqueous suspension of colloidal particles, however, the polarization of particles is very sensitive to the frequency of the applied electric field, and the dynamics of the particles can differ drastically from the dynamics that would be expected from a simple induced dipole model. In certain frequency regimes, instead of forming chains and columns in the direction of a linear applied field, the particles are dynamically unstable and circulate in bands that tilt at a significant angle relative to the direction of the field.[2] This indicates that the interactions between the particles are no longer symmetrical with respect to the applied electric field.

We have proposed a phenomenological model to explain the mechanism of this symmetry breaking. We have carried out a computer simulation of colloidal particle dynamics resulting from our model,[2] and the results of the simulation agree well with experimental observations. The main features of the model are summarized in the following:

A charged colloidal particle in an aqueous solution attracts a cloud of ions of opposite charge (double layer). When a linear alternating electric field is applied to the suspension, the double layer becomes polarized. This polarization, however, does not follow the applied field instantaneously, because the motion of the counterions in the double layer is governed by diffusion. In other words, the induced polarization lags behind the applied field by a phase shift $\theta$. The characteristic time constant of the diffusion process, $\tau_o$, is on the order of the diffusion time for an ion to move the length of a particle.

At frequencies close to $1/\tau_o$, the phase lag between the induced dipole and the applied electric field becomes important. It has been shown[3] that this phase lag causes neighboring

*Progress in Electrorheology*, Edited by K.O. Havelka
and F.E. Filisko, Plenum Press, New York, 1995

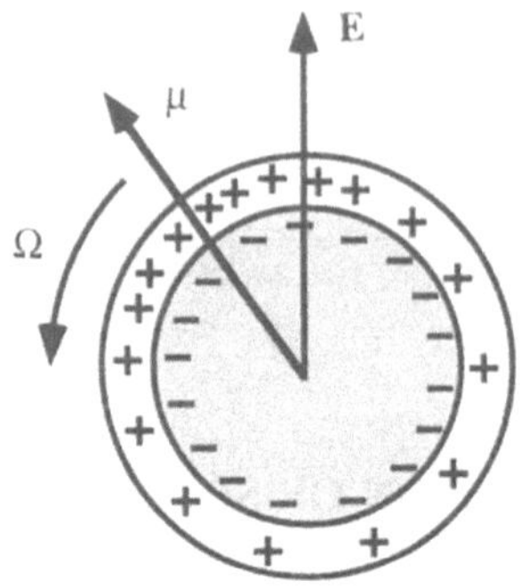


Fig. 1 Misalignment of the induced dipole moment of the double layer relative to the direction of the electric field due to the asymmetric diffusion of the counterions around a spinning sphere.

particles to exert torques on each other when the mutual polarization is taken into account, and this torque causes the particles, as well as the double layers, to spin.

When a particle is spinning, the diffusion of the ions in the double layer becomes asymmetrical with respect to the electric field, with the ions diffusing faster in the direction of spinning than against it. This causes the induced dipole moment of the particle to be misaligned with respect to the applied field (see Fig. 1).

Our earlier simulation results show that this misalignment between the polarization of the particle and the applied electric field plays a crucial role in the formation of a band of particles tilted at an angle relative to the direction of the applied electric field.

We report here the results of simulation of the asymmetric polarization of the double layer using a biased random walk method.

## DYNAMICS OF COUNTERIONS IN APPLIED ELECTRIC FIELDS

The disassociated surface ion groups of a colloidal particle in an electrolyte attract a layer of counterions (double layer), with the thickness of the double layer determined by the ionic strength of the electrolyte. For a 1 mM 1:1 electrolyte, the double layer thickness is on the order of 100 Å,[4] which is very small compared to the diameter of a micron-size particle. When an electric field is applied, the counterions diffuse in the direction of the local field. Because the counterions are bound to the surface of the colloidal particle by the Coulomb attractive force between the counterions and the surface ion groups of the particle, we can assume that the motions of the counterions are tangential along the surface of the particle and therefore ignore motion in the radial direction.

To study the motion of counterions in an applied electric field, we need to calculate the forces exerted on each counterion by the polarized colloidal particle and by the other counterions. We consider the case where the particle is a sphere of radius $R$ and dielectric constant $\varepsilon_p$. The dielectric constant of the solution is $\varepsilon_s$. A uniform applied field of strength $E$ is in the z-direction, and a counterion of charge $q$ is located on the surface of the sphere at $\theta = 0$ in a spherical coordinate system (see Fig. 2). We now consider the potential at an arbitrary point on the surface of the sphere. This is a standard boundary-value problem of electrostatics, and the exact solution for the potential is:

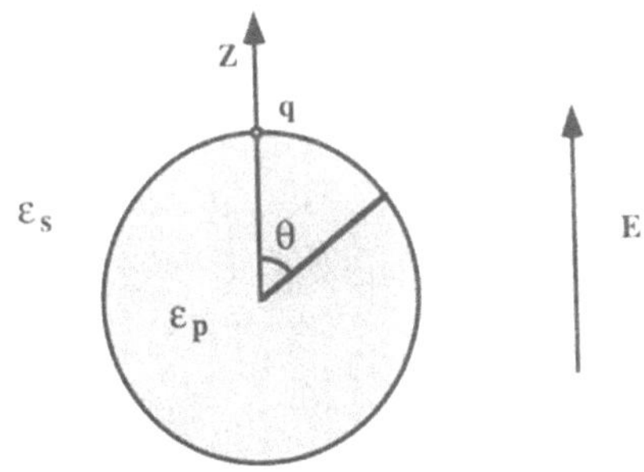


Fig. 2 A charge q on the surface of a dielectric sphere in a dielectric medium in a uniform applied electric field.

$$\varphi = -\frac{3\varepsilon_s}{\varepsilon_p + 2\varepsilon_s} ER\cos\theta + \frac{q}{4\pi\varepsilon_o\varepsilon_s R}\left(\frac{1}{2\sin(\theta/2)} + \sum_{n=0}^{\infty} \frac{n(\varepsilon_s - \varepsilon_p)}{(n+1)\varepsilon_s + n\varepsilon_p} P_n(\cos\theta)\right), \quad (1)$$

where $P_n(\cos\theta)$ is a Legendre polynomial of order n. The first term of equation (1) comes from the applied field and the polarization of the dielectric sphere due to this field, the second term is the potential of the point charge, and the third term is the result of the polarization of the media due to the point charge. Using the Legendre polynomial generating function, the sum can be expressed as

$$\sum_{n=0}^{\infty} \frac{n(\varepsilon_s - \varepsilon_p)}{(n+1)\varepsilon_s + n\varepsilon_p} P_n(\cos\theta) = \frac{\varepsilon_s - \varepsilon_p}{\varepsilon_s + \varepsilon_p} \frac{1}{2\sin(\theta/2)}$$

$$+ \sum_{n=0}^{\infty} \left( \frac{n(\varepsilon_s - \varepsilon_p)}{(n+1)\varepsilon_s + n\varepsilon_p} - \frac{\varepsilon_s - \varepsilon_p}{\varepsilon_s + \varepsilon_p} \right) P_n(\cos\theta). \quad (2)$$

The first term on the right-hand side of equation (2) is proportional to the potential of the image charge of the point charge if the interface between the two media is a flat surface,[5] and the second term (the sum) is the correction for the spherical interface. In Fig. 3 we plot the contribution of the point charge and its image charge with and without the correction of the sum of Legendre polynomials in equation (2), and we can see that the difference is small enough that we can keep just the first term on the right-hand side of equation (2) to make the calculation needed for the simulation a lot less time-consuming. Equation (1) can thus be simplified to

$$\varphi \cong -\frac{3\varepsilon_s}{\varepsilon_p + 2\varepsilon_s} ER\cos\theta + \frac{1}{4\pi\varepsilon_o(\varepsilon_p + \varepsilon_s)} \frac{q}{R\sin(\theta/2)}. \quad (3)$$

For the potential at the position of the *i*th counterion due to the external field, the other counterions, and the polarization of the media, we therefore obtain:

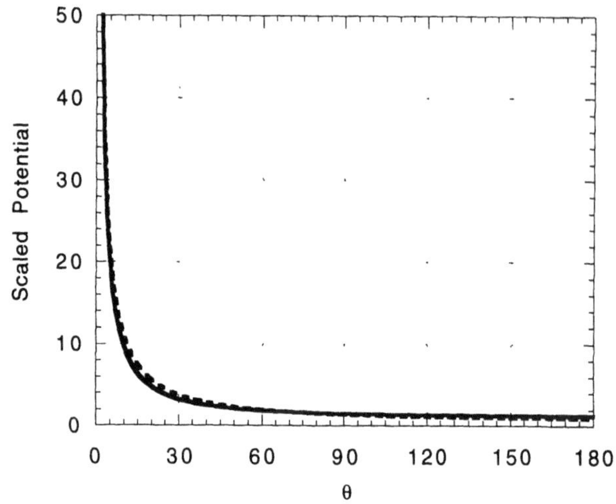


Fig. 3 The scaled potential of the point charge and its image charge. The solid curve is the more precise solution, which includes the sum of Legendre polynomials in the right-hand side of equation (2) of orders from 1 to 30. The dashed curve is the approximation which omits these terms.

$$\varphi_i = -\frac{3\varepsilon_s}{\varepsilon_p + 2\varepsilon_s} ER\cos\theta_i + \frac{q}{4\pi\varepsilon_o(\varepsilon_p + \varepsilon_s)R} \sum_{j \neq i} \frac{1}{\sin(\gamma_{ij}/2)}, \tag{4}$$

where $\gamma_{ij}$ is the angle between the position vectors of the *i*th and the *j*th counterions:

$$\cos\gamma_{ij} = \cos\theta_i \cos\theta_j + \sin\theta_i \sin\theta_j \cos(\phi_i - \phi_j). \tag{5}$$

The velocity of the *i*th counterion is

$$v_{i,ele} = -\mu \nabla \varphi_i, \tag{6}$$

where $\mu$ is the mobility of the counterions (velocity per unit field).

To facilitate the simulation, we scale length, electric field, and time by

$$r_o = R, \tag{7}$$

$$E_o = \frac{q}{4\pi\varepsilon_o\varepsilon_s R^2}, \tag{8}$$

$$t_o = \frac{R}{\mu E_o} = \frac{4\pi\varepsilon_o\varepsilon_s R^3}{q\mu}, \tag{9}$$

and use asterisks to denote scaled quantities. The scaled velocity of the $i$th counterion is

$$v_{i,ele}^{*} = -\nabla\left( -\frac{3\varepsilon_s E^*}{\varepsilon_p + 2\varepsilon_s}\cos\theta_i + \frac{\varepsilon_s}{\varepsilon_p + \varepsilon_s}\sum_{j\neq i}\frac{1}{\sin(\gamma_{ij}/2)} \right). \tag{10}$$

Of course we keep in mind that we are only interested in the tangential components of equation (10).

## BIASED RANDOM WALK ON A SPHERE

It has been well established that a diffusion process can be simulated with a random walk.[6] In a one-dimensional case, if a step of size $\Delta x$ is made in time $\Delta t$, and if the probability of moving in the positive direction is $p_+$ and in the negative direction is $p_-$, the average drift velocity is determined by

$$v_d = (p_+ - p_-)\frac{\Delta x}{\Delta t}, \tag{11}$$

where $p_+ + p_- = 1$, and the diffusion coefficient is

$$D = \frac{1}{2}\frac{(\Delta x)^2}{\Delta t}. \tag{12}$$

In a two-dimensional case, a step can be taken in either the x- or y-directions if we adopt a Cartesian coordinate system. The probabilities and drift velocities satisfy the following relations:

$$v_x = (p_{x+} - p_{x-})\frac{\Delta x}{\Delta t}, \qquad v_y = (p_{y+} - p_{y-})\frac{\Delta y}{\Delta t}, \tag{13}$$

$$p_{x+} + p_{x-} = p_{y+} + p_{y-} = \frac{1}{2}. \tag{14}$$

Equation (12) for the diffusion coefficient still holds.

Performing a random walk on the surface of a sphere involves more than simply substituting $\theta$ and $\phi$ of the spherical coordinates for x and y in the equations above. To avoid biases towards the north and south poles, steps have to be taken *along two orthogonal great circles* at every starting point — one chosen to be a constant-longitude circle (tangential to the $\mathbf{e}_\theta$ direction), and the other chosen perpendicular to this (tangential to the $\mathbf{e}_\phi$ direction). Let $(\theta_o, \phi_o)$ be the initial position before a step, $(\theta, \phi)$ be the position after a step, and $\Delta\delta$ the angular step size. If a step is taken along the great circle in the $\pm\mathbf{e}_\theta$ direction then,

$$\theta = \theta_o \pm \Delta\delta, \qquad \phi = \phi_o. \tag{15}$$

If a step is taken along the other great circle in the $\pm$ direction, however,

$$\cos\theta = \cos(\Delta\delta)\cos\theta_o, \qquad \tan(\phi - \phi_o) = \pm\frac{\tan(\Delta\delta)}{\sin\theta_o}. \tag{16}$$

We have verified that by using this method in the absence of any drift velocity (i.e., in the unbiased case), a uniformly distributed set of walkers on a sphere will remain uniform after thousands of random walk steps.

## SIMULATION RESULTS

We start with a fixed number of counterions randomly distributed on a sphere of unit radius. The velocity of each counterion due to the applied field and the neighboring counterions is determined by equation (10). To see the effect of spinning of the sphere on the polarization of counterions, we add an additional velocity

$$\boldsymbol{v}^*_{i,spin} = \boldsymbol{\Omega}^* \times \boldsymbol{r}^*_i, \tag{17}$$

where $\Omega^*$ is the normalized angular velocity due to spinning. The probabilities are calculated according to

$$v^*_\theta = (p_{\theta+} - p_{\theta-})\frac{\Delta\delta}{\Delta t}, \qquad v^*_\phi = (p_{\phi+} - p_{\phi-})\frac{\Delta\delta}{\Delta t}, \tag{18}$$

$$p_{\theta+} + p_{\theta-} = p_{\phi+} + p_{\phi-} = \frac{1}{2}, \tag{19}$$

where components of velocities are determined by the sum of equations (10) and (17).

The diffusion coefficient in equation (12) can also be written as[7]

$$D = \frac{kT}{f}, \tag{20}$$

where k is the Boltzmann constant, T the temperature, and f the friction factor. For a counterion in an electric field E, when the electric force is balanced by the frictional force,

$$qE = fv, \tag{21}$$

where v is the velocity of the counterion. Using equations (6) and (12), we have, in normalized units,

$$\Delta t^* = \frac{qE_oR}{2kT}(\Delta\delta)^2. \tag{22}$$

For counterions of unit electron charge and spheres of one micron diameter, the factor in equation (22) is $7\times10^{-4}$.

For the simulation runs, we choose a step size of 0.5° and a total of about 50000 random walk steps to be performed. To reduce the computation time, we also introduce a cut-off angle of 45° between neighboring particles, beyond which we no longer consider the pair-interactions. We have verified that this does not introduce errors beyond ordinary random walk statistical errors. The components of the polarization per counterion of the double layer are calculated at regular time intervals according to:

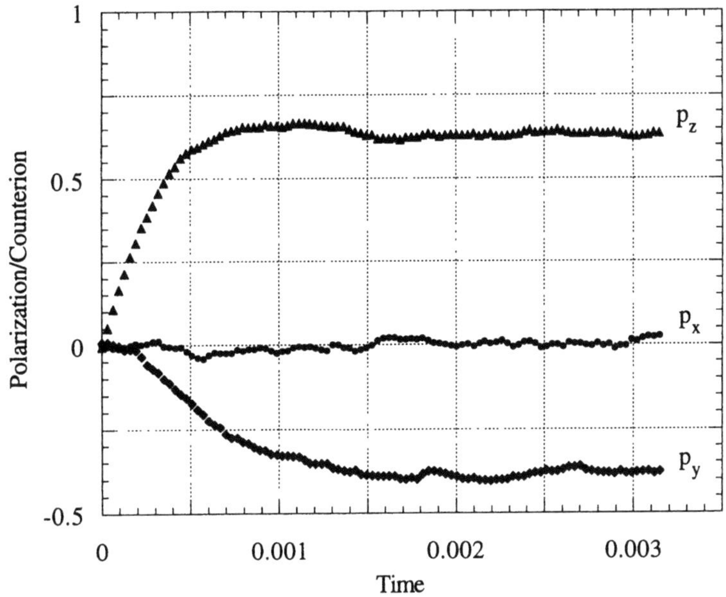


Fig. 4 Time evolution of components of polarization per counterion of the double layer of a spinning colloidal sphere in a uniform applied electric field. There are a total of 500 counterions, and the normalized field strength is 2000. The spinning angular velocity is 0.8 times field strength.

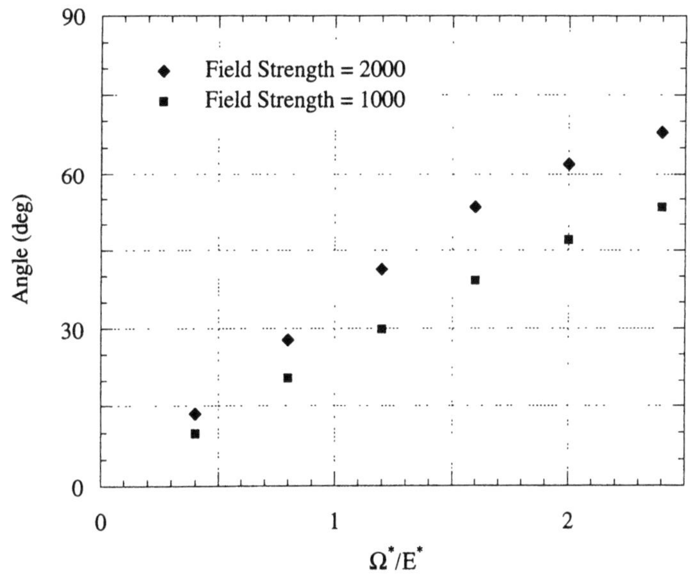


Fig. 5 The angle between the directions of the polarization of the double layer and the applied field at two field strengths for 500 counterions.

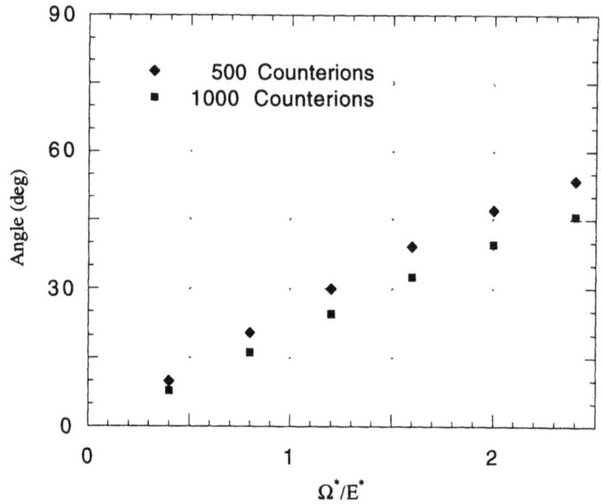


Fig. 6 The angle between the directions of the polarization of the double layer and the applied field at the same field strength but for different numbers of counterions.

$$p_x = \frac{1}{N}\sum_{i=1}^{N} x_i, \quad p_y = \frac{1}{N}\sum_{i=1}^{N} y_i, \quad p_z = \frac{1}{N}\sum_{i=1}^{N} z_i, \tag{23}$$

where $(x_i, y_i, z_i)$ is the position of the $i$th counterion and N is the total number of counterions used in the simulation. The final equilibrium values of each component are computed by taking an average of the component at different times after equilibrium has been established.

Fig. 4 shows the time evolution of each component when a static field of normalized field strength 2000 is applied in the z-direction. The sphere spins about the x-direction at an angular velocity of $0.8E^*$, where $E^*$ is the normalized field strength. By scaling the angular velocity of the spinning to $E^*$, we ensure that the drift velocity of the counterion due to spinning is comparable to that due to the applied field. The z-component of the polarization, $p_z$, gradually reaches an equilibrium value under the influence of the applied field. The y-component, $p_y$, also reaches a non-zero value due to the spinning of the sphere. The x-component remains close to zero, as expected, because neither the applied field nor the spinning of the sphere will induce a polarization in this direction.

If we examine Fig. 4 carefully, we can see that the change in $p_y$ does not follow the change in $p_z$ instantaneously — $p_z$ starts to rise immediately after the field is applied, but $p_y$ only starts to deviate from zero after a short time interval. The decrease in the magnitude of $p_z$ is also for the same reason — $p_z$ reaches a saturation value first and is unable to grow further, but $p_y$, which lags behind $p_z$, is still changing, and this causes $p_z$ to decrease.

Fig. 5 shows the angle between the directions of the polarization of the double layer and the applied field at two different field strengths for 500 counterions. Even though the spinning speed is scaled to the field strength, we can see that the larger field produces a larger angle for a given ratio of $\Omega^*/E^*$. This is because the polarization is larger for a larger field.

Fig. 6 shows the angle between the directions of the polarization of the double layer and the applied field at the same field strengths but for different total numbers of counterions. Because of the Coulomb repulsive force between the counterions, polarization per counterion is less for more counterions, and this also causes the angle between the polarization and the applied field to be smaller for more counterions.

## CONCLUSIONS

We have demonstrated here that the polarization of the double layer of a spherical colloidal particle can be simulated by a biased random walk process and that a misalignment between the directions of the polarization and the applied electric field indeed develops if the sphere spins. The degree of the misalignment, however, is not a simple function of the ratio of the drift velocity of the counterions caused by the spinning to that caused by the applied field, but depends on the magnitude of polarization per counterion. There is also a phase lag between the polarization in the y-direction (due to the spinning) and the polarization in the z-direction (due to the applied electric field). The degree of the misalignment between the directions of the polarization of the double layer and the applied electric field is significant, in support of our earlier phenomenological model.

## ACKNOWLEDGMENTS

This work has been supported by grants from the Henry Luce Foundation, the Research Corporation, and the National Science Foundation (grant no. DMR-92-12090).

## REFERENCES

1. D. J. Klingenberg, Frank van Swol, and C. F. Zukoski, *J. Chem. Phys.* **91**, 7888 (1989).
2. Y. Hu, J. L. Glass, A. E. Griffith, and S. Fraden, *J. Chem. Phys.* **100**, 4674 (1994).
3. C. Holzapfel, J. Vienken, U. Zimmermann, *J. Membrane Biol.* **67**, 13 (1982).
4. R. J. Hunter, *Foundations of Colloidal Science*, Sect. 6.3, Oxford University Press (1991).
5. J. D. Jackson, *Classical Electrodynamics*, 2nd Ed., Sect. 4.4, John Wiley & Sons (1962).
6. C. V. Heer, *Statistical Mechanics, Kinetic Theory, and Stochastic Processes*, Sect. 3.8, Academic Press (1972).
7. R. J. Hunter, *Foundations of Colloidal Science*, Sect. 2.1, Oxford University Press (1991).

# THE POTENTIAL ENERGY OF A PAIR OF POLYSTYRENE SPHERES IN ALTERNATING ELECTRIC FIELDS

San Lin,[1] Seth Fraden,[1] and Yue Hu[2]

[1]Department of Physics
Brandeis University
Waltham, MA 02254-9110

[2]Department of Physics
Wellesley College
Wellesley, MA 02181

## INTRODUCTION

The drastic change in the viscosity of electrorheological (ER) fluids when an external electric field is applied has intrigued scientists from many different fields including engineering, chemistry, and physics for over half a century. It has been generally understood that the microscopic attractive forces between induced dipole moments of the suspended particles have important effects on macroscopic fluid properties. Great effort has been given recently to maximizing the strength of the attractive forces between the particles in ER fluids to achieve practical fluids, and important progress has been made. Many questions, however, need to be investigated further. For example, it has been known that the addition of a small amount of water, that is, the addition of mobile charge carriers to an ER fluid, increases the ER effect, although the exact role that the mobile charges play in strengthening the induced dipole moments of the suspended particles is still not very well understood.[1] It has also been shown in recent work on the stress response of ER fluids to transient electric fields that the relaxation process of an ER fluid is more complicated than just a simple fluid-particle dielectric mismatch model, and the surface conductivity of impurities adsorbed on particle surfaces seems to be a contributing factor.[2]

In the preliminary work presented here, we focus on trying to understand the strength of the microscopic interactions between colloidal particles in alternating electric fields. The system used is a model system of aqueous suspensions of monodisperse polystyrene spheres. By using digital microscopy techniques[3,4] to measure the inter-particle separations between two spheres at thousands of different times and at several different field strengths, we have been able to determine the probability density function $P(r)$ for the separation distance $r$ between the two particle centers.

*Progress in Electrorheology*, Edited by K.O. Havelka
and F.E. Filisko, Plenum Press, New York, 1995

Under equilibrium statistical mechanics, the probability density function is related to the pair potential $U(r)$ by

$$P(r) = Ce^{-U(r)/kT}, \tag{1}$$

where $C$ is a normalization constant. From this, we can quantitatively deduce the mutual interaction potential energy of the particles and have found qualitative agreement between our observations and theoretical predictions. This agreement seems to validate the experimental approach here as being of use for studying inter-particle interactions in colloidal suspensions.

Even though an aqueous suspension is not a conventional ER fluid (where one normally uses a non-conducting suspension), there are several advantages. Firstly, because of the monodispersivity of the polystyrene particles, the strengths of the interactions between different pairs of particles are more uniform compared to a polydisperse system, and this makes interpretation of our experimental results more straightforward. Secondly, since we are interested in general in studying the role of mobile charges in a suspension, having an aqueous suspension is not a disadvantage. However, as will be explained below, in this preliminary experiment, we operate at a frequency of the applied field where the effect of the mobile charges is minimized and the polarization of the particles is dominated by the dielectric mismatch between the particles and the suspending fluid. We expect that under these conditions the polystyrene/water system is equivalent to a non-conducting suspension. Therefore, the significance of the experimental results presented here, as well as the usefulness of our experimental technique, are not limited to just this model system, but may also apply to the typical ER suspension of non-conducting particles and solvent.

We begin this paper with a theoretical discussion in which the contributions of electrostatic repulsion, electric-field-induced dipole interaction, and van der Waals attraction to the pair potential are first considered. A probability density function for the distance between a pair of particles is then obtained from the pair potential using equilibrium statistical mechanics. The theoretical discussion is followed by a detailed description of the experimental apparatus — one that allows us to isolate pairs of polystyrene spheres in an essentially two-dimensional geometry by restricting the particles to a narrow gap between parallel glass plates. Finally, we examine the results of our experiments and compare our findings with theoretical predictions.

## THEORY

### Double Layer Repulsion

To calculate the total mutual potential energy of a pair of charged spheres in our experiment, we first consider the electrostatic interaction in the absence of an external electric field, known as double layer repulsion. The charges originate from the polystyrene sulfanate particles losing positive surface ions into the water, leaving the particles with a negative surface charge. Positive free ions are then attracted to the spheres and form a positively charged layer around each sphere, which screens the repulsive force. The layer of free positive ions around a colloidal particle and the negative surface charge on the colloidal particle together comprise the electric double layer. The thickness of the free ions screening the repulsion between colloidal particles is the Debye screening length, $\kappa^{-1}$, and for water at room temperature,[5,6]

$$\kappa^{-1} = \left( \frac{\varepsilon_0 \varepsilon'_{rw} kT}{\sum_i Z_i^2 e^2 n_{i0}} \right)^{\frac{1}{2}} \approx \frac{3 Å}{\sqrt{I}}, \tag{2}$$

where $\varepsilon'_{rw}$ is the real part of the relative dielectric constant of water and $I$ is the molar ionic strength of the solution. The sum here is over different types of ions, each with its own valence $Z_i$ and its own numerical density $n_{i0}$.

Proceeding from the assumption that the average concentration of free ions at a given point can be calculated from the average value of the electric potential at the same point, one can use the Poisson-Boltzmann equation for the electric potential in the vicinity of a colloidal particle. One can get approximate analytical solutions of the Poisson-Boltzmann equation only in certain limiting situations.[5] One of these situations — of relevance to our experiment — occurs when the diameter of the spheres is both much greater than the Debye screening length and also much greater than the separation $r-d$ between the surfaces of the spheres. In this limit, the double layer repulsive energy between a pair of particles is

$$U_d(r) = \pi\varepsilon_0\varepsilon'_{rw} d V_s^2 \exp(-\kappa(r-d)), \tag{3}$$

where $V_s$ is the surface potential of the sphere. For our experiments, with $d$ = 2.16 µm, $V_s \sim 30\mathrm{mV}$,[7] $I$ = 0.1 mM, $\kappa d$ = 42 (»1), and $0 < (r-d)/d < 0.3$, the repulsive energy reaches a peak of $\sim 1000kT$ when two particles touch each other.

## Electric Dipole Attraction

There are three contributing factors to the induced dipole moment of a particle. One is due to the mobile charges in the double layer, which we will call double layer polarization. The second and third are due to the dielectric and conductivity mismatch between the particle and the suspending fluid, which are called dielectric and Maxwell-Wagner polarization, respectively. The field-induced attraction between particles arises from this field-induced polarization of the colloids. A *single* dielectric sphere immersed in a conducting solvent distorts a uniformly applied electric field in such a way that the field outside the sphere can be described as the superposition of the undistorted field and the field due to a point dipole located at the center of the sphere with moment $\mu = \alpha K \boldsymbol{E}$, where $\boldsymbol{E}$ is the applied electric field, $\alpha \equiv 4\pi\varepsilon_0\varepsilon'_{rw}$, and $K \equiv [(\varepsilon_p - \varepsilon_w)/(\varepsilon_p + 2\varepsilon_w)](d/2)^3$,[8] where $\varepsilon_p$ and $\varepsilon_w$ are the complex dielectric constants of the particle and water, respectively.

Double layer polarization arises from the fact that the applied electric field affects the distribution of the positive mobile charges in the double layer of a polystyrene sphere. At low frequencies, the particle can be thought of as having a thin conducting layer on its surface, and in this case, the effective particle dielectric constant $\varepsilon_p$ is greater than that of the water. The applied electric field polarizes the double layer and creates a dipole moment in the direction of the applied electric field, since $K > 0$. However, at high frequencies, when the field alternates faster than the double layer can follow, there will be no induced dipole moment in the double layer, and only the dielectric polarization of the polystyrene sphere remains. In this limit, one thinks of the particle as being a neutral plastic sphere immersed in water, because the charges in the double layer no longer respond to the rapidly alternating field. Since polystyrene has a dielectric constant much less than that of water, the particle behaves as a dielectric hole, and $K < 0$. The cutoff frequency for the crossover from double layer to dielectric polarization is approximately the inverse of the time for the ions to diffuse a distance equal to the radius of the sphere: $\omega_c = D/a^2$, with $D$ the ion diffusion constant. For 1 µm spheres, this occurs at about 10 kHz. We have observed this crossover between the two forms of polarization by constructing electrodes which produce a large gradient in the electric field. In this case, there is a net force on the field-induced dipoles, and the spheres move in the field. Dielectric holes ($K < 0$) are repelled from the high-field region, but dielectric inclusions ($K > 0$) are attracted to the high field. Through observation of the motion of spheres in the field gradient, we experimentally observe $\omega_c$ to be ~ 30 kHz for the spheres in our study.

A single sphere in a uniform electric field is exactly equivalent to a single dipole, with no other multipoles of higher order. However, a second sphere in the vicinity of the first experiences a non-uniform field, which is the sum of the external field and the dipolar field of the first sphere. Thus the induced dipole moment is different from that of a single sphere in a uniform electric field. One could argue that higher order multipole moments should also be considered, but computer simulations show that as long as $K < 0.5$, which is the case for the dielectric constants of polystyrene and water at high frequencies, the error due to ignoring the higher order moments is less than a few percent.[9] Because of this, we ignore higher order moments and only consider dielectric dipole interactions between polystyrene spheres here.

The frequency of the applied electric field used in this experiment is 360 kHz — high enough for us to neglect the double layer polarization and consider only the dielectric dipole interaction between two polystyrene spheres. We consider two particles positioned such that the relative position vector connecting their centers has magnitude $r$ and is oriented at an angle $\theta$ relative to the direction of the applied electric field. To calculate the strength of the induced dipole moment of each sphere, both the strength of the applied electric field and the field created by the dipole moment of the neighboring particle are considered. Because of the symmetry of the dipolar potential, the potential energy of two induced dipoles can be solved exactly, accounting for the mutual polarization of the two spheres:

$$U_p(r,\theta)=\frac{\alpha}{r^3}\left|\frac{K|E_0|}{1+\frac{K}{r^3}}\right|^2\left\{1-\left[2\left|\frac{\frac{3K}{r^3}}{1-\frac{2K}{r^3}}\right|^2+4\,\mathrm{Re}\left(\frac{\frac{3K}{r^3}}{1-\frac{2K}{r^3}}\right)+3\right]\cos^2\theta\right\}. \tag{4}$$

## Van der Waals Attraction

Van der Waals forces also play an important role in the interactions between colloidal particles.[5] The energy of the non-retarded van der Waals attraction between two spheres is[5,6]

$$U_V(r)=-\frac{A}{12}\left\{\frac{1}{\left(\frac{r}{d}\right)^2-1}+\frac{1}{\left(\frac{r}{d}\right)^2}+2\ln\left[\frac{\left(\frac{r}{d}\right)^2-1}{\left(\frac{r}{d}\right)^2}\right]\right\}, \tag{5}$$

where $A$ is the Hamaker constant, determined by the composition of the colloidal particles. For polystyrene, the Hamaker constant is about $1.0\times10^{-21}$J.[10] The energy of the van der Waals attraction diverges as two spheres approach each other and $r$ approaches $d$. In our experiment, however, the van der Waals attraction is a very small effect, because double layer repulsion prevents the particles from approaching too close to each other, and this results in the induced dipole-dipole attraction being dominant over the range of distances where our measurements are done.

## Total Potential

In our model of polystyrene spheres interacting in a uniform electric field, the total potential energy is a combination of double layer repulsive energy, dipole-dipole attractive energy, and the van der Waals attractive energy:

$$U_t(r,\theta)=U_d(r)+U_p(r,\theta)+U_v(r). \tag{6}$$

The double layer repulsion creates an energy barrier of about $1000kT$ for $(r-d)/d\approx0.01$. If two particles bump into each other and break through this energy barrier, the two particles will be permanently bound by the van der Waals attraction, and thermal energy usually cannot break them apart. Outside this barrier is the induced dipole-dipole attractive potential well created by the applied electric field. In our experiment, this well can be

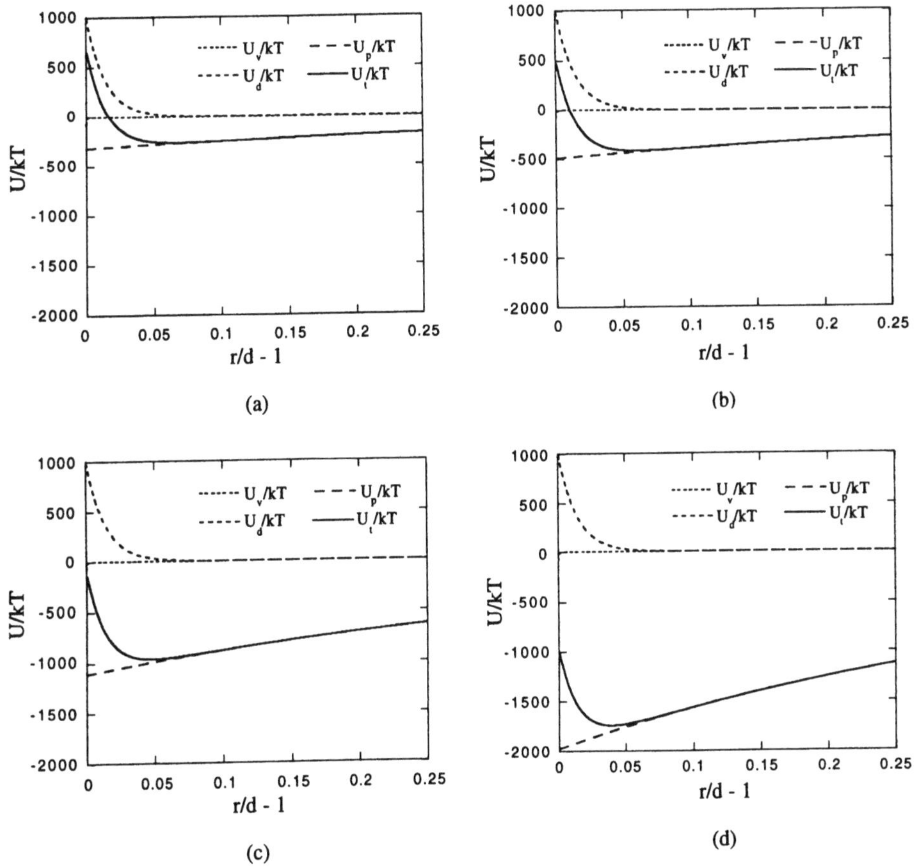


**Figure 1**. The pair potential energy in units of $kT$ due to double layer repulsion ($U_d(r)$), van der Waals attraction ($U_v(r)$), and the dipole-dipole electric field-induced attraction, evaluated at $\theta = 0$ ($U_p(r,0)$), as well as the total potential ($U(r,0)$) for a pair of 2.16 μm diameter polystyrene spheres, are plotted as a function of the fractional separation of the surfaces (*r/d* - 1) of the two spheres. The high-frequency electric field has RMS amplitudes of (a) 51 V/mm, (b) 63 V/mm, (c) 95 V/mm, and (d) 126 V/mm, respectively.

as deep as $1180kT$. (See Figure 1.) It should be mentioned here that glass is also negatively charged in water, so that there is double layer repulsion between the polystyrene spheres and the surface of the glass, as well. This helps to confine the polystyrene spheres to the mid-plane between the two glass plates. Once it has broken through the double layer repulsive energy barrier, a polystyrene sphere can become permanently bound to the glass surface.

In Figure 2, $P(r)$ is plotted against $r$. $P(r)$ is obtained by numerically evaluating the following expression:

$$P(r) = \frac{\int_0^{2\pi} e^{-U_t/kT} d\theta}{\int_0^{2\pi} \int_{1.01d}^{R} e^{-U_t/kT} r' dr' d\theta}. \tag{7}$$

Here $R$ is the upper limit of r in the experiment, and this distribution is normalized so that[12]

$$\int_{1.01d}^{R} P(r) r dr = 1. \tag{8}$$

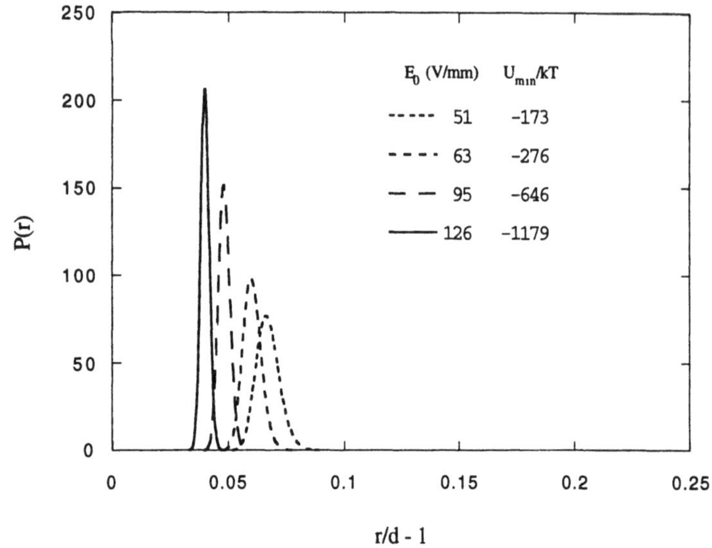


**Figure 2.** Normalized probability density function $P(r)$ of the separations of a pair of charged 2.16 μm diameter polystyrene spheres as a function of the fractional separation of the sphere surfaces, $r/d$ - 1. The probability density functions are calculated from the theoretical pair-potential energy shown in Fig. 1. The RMS value of the applied electric field and the minimum of the pair potential for each curve are also indicated.

## EXPERIMENTAL DETAILS

### The Cell

The cell used in this experiment is shown in Figure 3. It consists of parallel glass plates clamped between two aluminum holders. An O-ring serves both as a seal for the sample cell and also as a compliant spacer for adjusting the gap between the glass plates. The bottom glass plate has a coating of indium-tin-oxide (ITO), a transparent conductor. Two parallel electrodes have been etched onto the ITO surface, providing a uniform electric field between the electrodes when connected to a power supply. The distance between the electrodes is 0.560 mm in this experiment. The samples used in this experiment are comprised of 2.16 μm-diameter polystyrene spheres[11] suspended in a 0.1 mM NaCl aqueous solution, and are injected into the cell through one of the syringes. The other syringe keeps the pressure inside the sealed space the same as outside.

### Data Acquisition

The cell is placed under an Olympus microscope with a 50x long working distance objective lens. The images are first acquired using a CCD camera and recorded by a SVHS Sony tape recorder. The even-fields of the recorded images are then analyzed by a Macintosh computer that has an 8-bit, 480×640 digitizing board.

We adjust the microscope to have the centers of the spheres appear brighter than the background. By setting a threshold above the background and below the light intensity around the centers of the spheres, the spheres can be identified as bright clusters of pixels with intensities above the threshold. The computer finds the centers of the bright clusters and

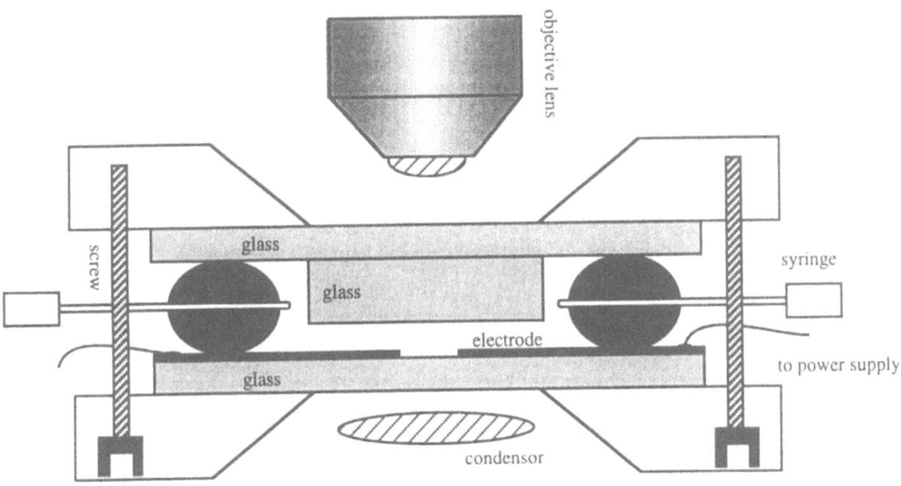


**Figure 3**. Illustration of a side view of the cell placed under an optical microscope.

records them as the coordinates of the spheres. The distance calibration is done with a stage micrometer with 10-μm scales. Both x and y directions are calibrated before the experiment.

We analyze 8,000 images for each of four different RMS field strengths (51V/mm, 63V/mm, 95V/mm, and 126V/mm) at 360 kHz. When the pair breaks up, *i.e.*, when the distance between the two spheres is so great that the binding energy is less than $kT$, the two particles can diffuse away from each other and do not recombine, and the data taking has to stop. This break-up prevents us from taking data below about 50 V/mm.

## RESULTS

The measured pair distance as a function of image frame number is shown for each field strength in Figure 4. We can see that the fluctuation of the pair distance decreases as the electric field becomes stronger.

In Figure 5, we show $P(r)$ vs. $r/d-1$ for our experiment. Here $P(r)$ is defined as:

$$P(r)\Delta r = \frac{1}{r}\frac{\text{counts in }[r,\ r+\Delta r]}{\text{counts in}[1.01d,\ 1.25d]}. \tag{9}$$

Here $r$ is restricted to lie between $1.01d$ and $1.25d$, and the distribution is normalized so that[12]

$$\int_{1.01d}^{1.25d} P(r)r\,dr = 1. \tag{10}$$

The finite resolution of the optics system and electronic noise in the imaging system in the experiment lead to some uncertainty in the measurement of the distance between two spheres. When a pair of spheres has surmounted the double layer repulsive energy barrier and has

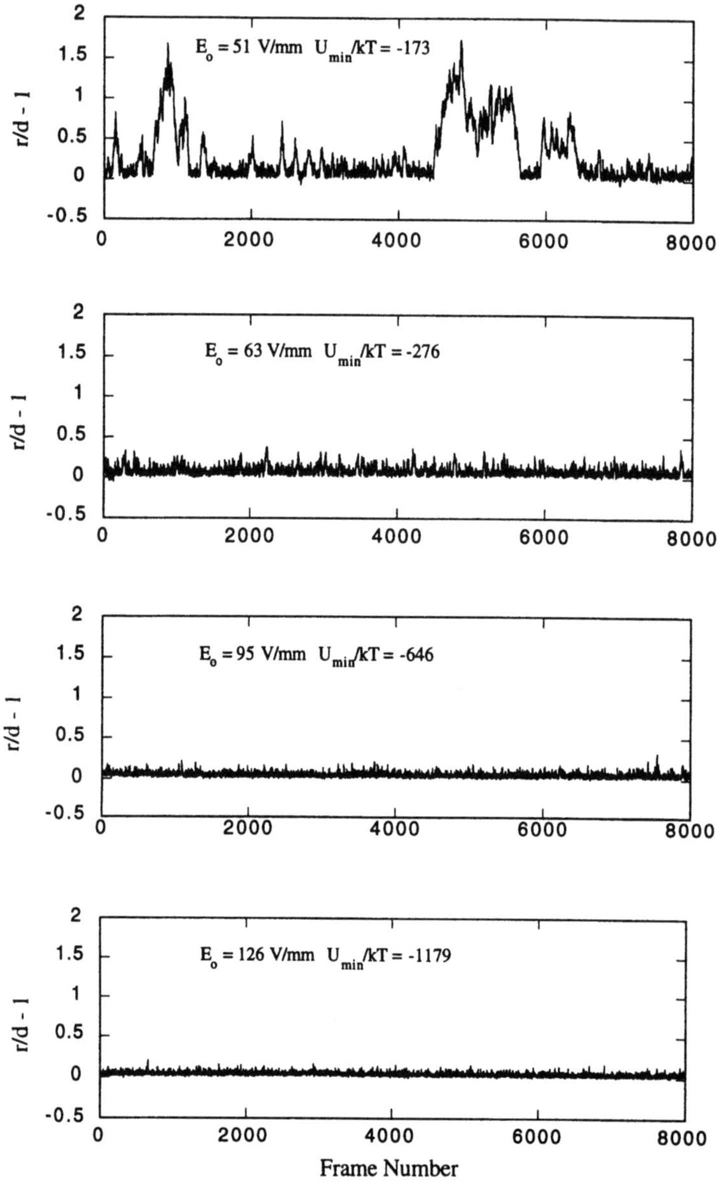


**Figure 4**. Experimental measurements of the fractional separation, *r/d* - 1, of the surfaces of a pair of charged 2.16 μm diameter polystyrene spheres. 1000 measurements were taken at a rate of 10/sec. The measured RMS value of the applied electric field and the calculated minimum of the pair potential for each curve are also indicated.

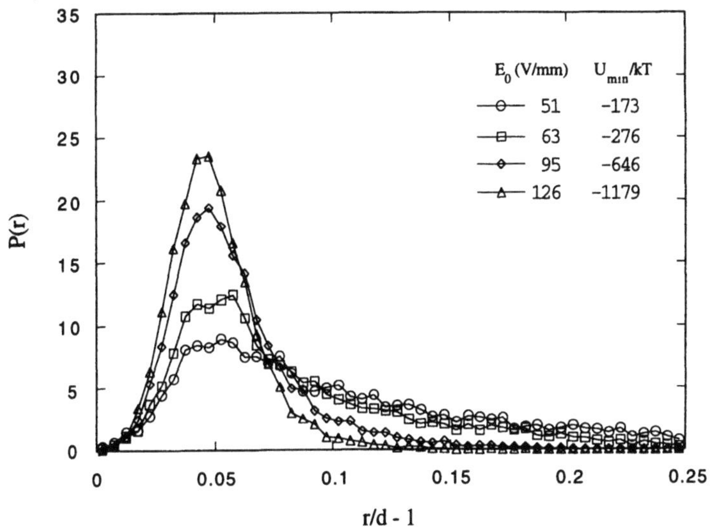


**Figure 5**. Measured probability density function of the separation of a pair of polystyrene spheres as a function of the fractional separation of the sphere surfaces. The probability density functions are obtained by integration of the displacement data shown in Fig. 4. The measured field amplitudes and the calculated minima in the corresponding pair-potential are indicated in the figure.

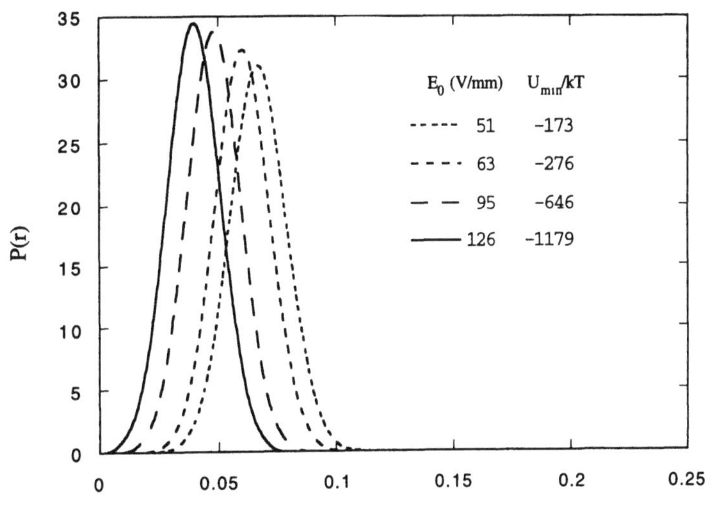


**Figure 6**. The theoretical probability function of the fractional separation between the surfaces of a pair of spheres, shown in Fig. 2, is convolved with the experimental point-spread function for the four experimental values of the applied field. Reasonable qualitative agreement between the theory (Fig. 6) and experiment (Fig. 5) is evident.

become permanently bound by a van der Waals attraction to the surface of a glass plate, the distance between the spheres does not change with time, and the probability density function $P(r)$ should be a delta function. Instead of this, however, the measured $P(r)$ has a Gaussian distribution with a finite width (point-spread function). Therefore, to compare experimental results with predictions from our model, we have to take the effects of this uncertainty into account by performing a convolution of the theoretical results with the experimental point-spread function, which is shown in Figure 6.

As we can see from Figures 5 and 6, our experimental results agree qualitatively with the predictions from the simple model proposed in this paper, but it is obvious that a more sophisticated model is needed to explain the long tails of the experimental probability distribution function.

The results of this preliminary experiment also show that the digital image microscopy technique is a useful tool for quantitatively measuring the pair potential between a pair of colloidal particles under the influence of a uniform alternating electric field. We should emphasize that an aqueous suspension of polystyrene particles is only one of the systems that can be studied by this technique. Based on the positive results of this work, we plan to investigate many other systems to study the interactions between colloidal particles in electric fields and thereby advance our understanding of ER on a microscopic level.

## REFERENCES


1. A. P. Gast and C. F. Zukoski, "Electrorheological Fluids As Colloidal Suspensions", *Adv. Colloidal Interface Sci.* **30**, 153 (1989).
2. J. M. Ginder, L. C. Davis, and S. L. Ceccio, "Transient Stresses in Electrorheological Fluids: Role of Particle Polarization Dynamics", Article to appear in this volume.
3. G. M. Kepler and S. Fraden, "Attractive Potential between Confined Colloids at Low Ionic Strength", *Phys. Rev. Lett.* **73**, 356 (1994).
4. G. M. Kepler and S. Fraden, "Video Microscopy Study of the Potential Energy of a Colloidal Particle Confined between Two Plates", *Langmuir* **10**, 2501 (1994).
5. E. J. W. Verwey and J. Th. G. Overbeek, "Theory of the Stability of Lyophobic Colloids", Elsevier Publishing Company, Inc., 1948.
6. W. B. Russel, D. A. Saville, and W. R. Schowalter, "Colloidal Dispersions", Cambridge University Press, 1989.
7. K. H. Lim and E. I. Franses, "Electrical Properties of Aqueous Dispersions of Polymer Microspheres", *J. Colloid Interface Sci.* **110**, 201 (1986).
8. J. D. Jackson, "Classical Electrodynamics, Second Edition", John Wiley & Sons, 1975.
9. T. B. Jones and R. D. Miller, "Multipolar Interactions of Dielectric Spheres", *Journal of Electrostatics* **22**, 231-244 (1989).
10. R. Evans and D. H. Napper, "On the Calculation of the van der Waals Attraction Between Latex Particles", J. Colloid Interface Sci. **45**, 138 (1973).
11. Sample purchased from Interfacial Dynamics Corporation, 17300 SW Upper Boones Ferry Road, Suite 120, Portland, Oregon 97224.
12. The reason that the lower limits of the integrals in equations (7) and (8) are $1.01d$ instead of $d$ is the following: At $r = d$, the van der Waals attractive force creates a potential well of infinite depth, and a pair of particles form a permanent bond there. However, because of the double layer repulsion, two particles normally cannot get close enough for this to take effect. Because we exclude the pairs that are permanently bound together in our experimental measurements, we can also eliminate the contributions of van der Waals attractions at $r = d$ to the probability distribution function in our theoretical calculations.

# INDEX

MIX
Papier aus verantwortungsvollen Quellen
Paper from responsible sources
FSC® C105338

If you have any concerns about our products,
you can contact us on
**ProductSafety@springernature.com**

In case Publisher is established outside the EU,
the EU authorized representative is:
**Springer Nature Customer Service Center GmbH**
**Europaplatz 3, 69115 Heidelberg, Germany**

Printed by Libri Plureos GmbH
in Hamburg, Germany